Design of Hydroelectric Power Plants – Step by Step

Design of Hydroelectric Power Plants – Step by Step

Geraldo Magela Pereira

CRC Press is an imprint of the
Taylor & Francis Group, an **informa** business

Cover image: Tucuruí Hydroelectric Power Plant, Tocantins river, Brazil

CRC Press/Balkema is an imprint of the Taylor & Francis Group, an informa business

© 2022 Taylor & Francis Group, London, UK

All rights reserved. No part of this publication or the information contained herein may be reproduced, stored in a retrieval system, or transmitted in any form or by any means, electronic, mechanical, by photocopying, recording or otherwise, without written prior permission from the publisher.

Although all care is taken to ensure integrity and the quality of this publication and the information herein, no responsibility is assumed by the publishers nor the author for any damage to the property or persons as a result of operation or use of this publication and/ or the information contained herein.

Library of Congress Cataloging-in-Publication Data
Names: Pereira, Geraldo Magela, author.
Title: Hydroelectric power plants : step by step / Geraldo Magela Pereira.
Other titles: Projeto de usinas hidrelétricas. English
Description: Boca Raton : CRC Press, [2022] | Translation of: Projeto de usinas hidrelétricas. | Includes bibliographical references and index.
Subjects: LCSH: Hydroelectric power plants--Design and construction.
Classification: LCC TK1081 .P397 2022 (print) | LCC TK1081 (ebook) | DDC 621.31/2134—dc23
LC record available at https://lccn.loc.gov/2021010794
LC ebook record available at https://lccn.loc.gov/2021010795

Published by: CRC Press/Balkema
           Schipholweg 107C, 2316 XC Leiden, The Netherlands
           e-mail: Pub.NL@taylorandfrancis.com
           www.crcpress.com – www.taylorandfrancis.com

ISBN: 978-0-367-75172-2 (Hbk)
ISBN: 978-0-367-75173-9 (Pbk)
ISBN: 978-1-003-16132-5 (eBook)

DOI: 10.1201/9781003161325

Typeset in Times New Roman
Typeset by codeMantra

Contents

About the author	xiii
Preface	xv
Acknowledgments	xix
Acronyms	xxi
Symbols	xxvii
Greek symbols	xxxi

1 Hydroelectric powerplants   1

1.1	*Introduction*	*1*
1.2	*The history*	*2*
1.3	*Hydroelectric plants – outstanding events*	*6*
1.4	*Hydroelectric powerplants in Brazil*	*8*
1.5	*Energy transformation*	*19*
1.6	*Component structures of a hydroelectric*	*20*
1.7	*Largest hydroelectrics in the world*	*21*

2 Planning hydropower generation   25

2.1	*Catchment areas and multiple uses of water*	*25*
2.2	*Generation expansion planning*	*28*
2.3	*Phases of studies*	*29*
	2.3.1 Inventory hydroelectric studies	32
	2.3.2 Integrated environmental assessment	34
	2.3.3 Basic project of mini plants	34
	2.3.4 Basic project of small plants	34
	2.3.5 Feasibility studies	35
	2.3.6 Environmental impact studies	35
	2.3.7 Consolidated basic engineering project	35
	2.3.8 Environmental basic project	36
	2.3.9 Detailed project	36

vi Contents

2.4   *Budget and evaluation of plant's attractiveness*   *36*
   2.4.1   Standard budget   37
   2.4.2   Budgets after privatization   37
   2.4.3   Assessment of plant's attractiveness   38

3 Types of power plants and layouts   39

3.1   *Introduction*   *39*
3.2   *Types of power plants*   *39*
   3.2.1   Function of the type of operation   39
   3.2.2   Function of type of use   40
   3.2.3   Function of the head   41
3.3   *Types of layouts*   *41*
   3.3.1   Dam layouts   42
   3.3.2   Canal drop layouts   42
3.4   *Notes on the spillway positon in the layout*   *47*

4 Hydrological studies   53

4.1   *Introduction*   *53*
4.2   *Hydrological studies*   *54*
   4.2.1   Basin characterization   54
      4.2.1.1   Drainage area   54
      4.2.1.2   Shape of the basin   54
      4.2.1.3   Mean bed slope   55
      4.2.1.4   Time of concentration   55
   4.2.2   Hydrometeorology   56
      4.2.2.1   Temperature   57
      4.2.2.2   Relative humidity   57
      4.2.2.3   Precipitation   57
      4.2.2.4   Climate classification   57
   4.2.3   Fluviometric measurements   57
   4.2.4   Tailwater elevation curve   59
   4.2.5   Flow-duration curves   60
   4.2.6   Extreme flows   64
      4.2.6.1   Powerhouse design flow   67
      4.2.6.2   Diversion flows   67
      4.2.6.3   Risk analysis   67
   4.2.7   Minimum flows   67
   4.2.8   Regularization of discharges   67
   4.2.9   Determination of sanitary flow   68
4.3   *Curves quota×area×volume*   *68*
4.4   *Reservoir flood routing*   *69*
4.5   *Backwater studies*   *69*
4.6   *Free board*   *70*
4.7   *Reservoir filling studies*   *74*
4.8   *Reservoir useful life studies*   *75*

Contents   vii

5 Power output 81
5.1 *Available head* *81*
5.2 *Power output* *81*
5.3 *Turbine type selection* *83*
5.4 *Energy simulation* *83*
5.5 *Energy-economic dimensioning* *85*
5.6 *Number of generating units* *85*
5.7 *Determination of physical guarantee* *87*

6 Geological and geotechnical studies 91
6.1 *Introduction* *91*
6.2 *Investigations/study phases* *92*
6.3 *Material parameters* *109*
6.4 *Foundation treatment methods* *112*
6.5 *Drainage systems* *117*
    6.5.1 Drainage system of earth and rockfill dams 117
    6.5.2 Drainage system of the concrete dams 122
6.6 *Instrumentation of foundations* *122*
6.7 *Construction materials* *123*

7 Dams 125
7.1 *Types of dams* *125*
7.2 *Earth dams* *125*
    7.2.1 Design criteria and section type 127
        7.2.1.1 Principle of flow control 127
        7.2.1.2 Principle of stability 127
        7.2.1.3 Principle of compatibility of deformations of the various materials 127
    7.2.2 Percolation analysis 130
        7.2.2.1 Internal drainage system 132
        7.2.2.2 Transitions 132
        7.2.2.3 Foundation waterproofing 133
    7.2.3 Stability analyses 133
    7.2.4 Tension and strain analysis 134
        7.2.4.1 Deformability and displacements 135
    7.2.5 Slopes protection 138
7.3 *Rockfill dams* *138*
    7.3.1 Rockfill dam with clay core 140
    7.3.2 Concrete face rockfill dams 143
    7.3.3 Asphalt concrete face rockfill dams 146
    7.3.4 Asphalt core rockfill dams 149
7.4 *Concrete gravity dam* *150*
    7.4.1 Gravity dam – conventional concrete 150
    7.4.2 Gravity dam – roller compacted concrete (RCC) 155
7.5 *Concrete arch dam* *160*

viii Contents

8 Spillways — 171

8.1 *Types of spillways and selection criteria* — 171
8.2 *Hydraulic design* — 175
  8.2.1 Design of the tucuruí spillway — 182
  8.2.2 Physical model studies — 185
8.3 *Energy dissipation* — 185
  8.3.1 Ski jump dissipators — 187
  8.3.2 Hydraulic jump energy dissipators – stilling basins — 197
  8.3.3 Efforts downstream of dissipators — 208
  8.3.4 Erosion pit dimensions assessment — 208
8.4 *Cavitation* — 211
  8.4.1 Conceptualization and characteristic parameters — 211
  8.4.2 Cavitation caused by irregularities — 212
  8.4.3 Protective measures specifications — 213
  8.4.4 Cavitation cases — 218
8.5 *Aeration* — 223
8.6 *Operating aspects in spillway monitoring* — 234

9 Hydraulic conveyance design — 239

9.1 *Introduction* — 239
9.2 *Power canal* — 239
9.3 *Intake* — 241
  9.3.1 Geometry — 241
  9.3.2 Minimum submergence — 243
  9.3.3 Ventilation duct — 243
  9.3.4 Vibration in the trashracks — 244
  9.3.5 Head losses — 244
9.4 *Penstocks* — 246
  9.4.1 Head losses — 246
  9.4.2 Economic diameter — 248
    9.4.2.1 Annex support and anchor blocks — 250
  9.4.3 Waterhammer — 258
    9.4.3.1 Overpressure calculation due to instant closing — 261
    9.4.3.2 Calculation of overpressure (h) due to gradual closure without surge tank — 264
9.5 *Tunnel* — 269
  9.5.1 General design criteria — 269
    9.5.1.1 Tunnel alignment — 269
    9.5.1.2 Covering criteria — 271
  9.5.2 Criteria for hydraulic tunnel dimensioning — 274
  9.5.3 Design application — 278
  9.5.4 Assumptions for tunnel lining dimensioning — 281
9.6 *Surge Tanks* — 282
  9.6.1 Types of surge tanks — 282
  9.6.2 Criteria used in inventory studies (Canambra) — 283
  9.6.3 Canambra criteria — 284

Contents   ix

	9.6.4	Rotating masses inertia	284
	9.6.5	Interconnected system operation	286
	9.6.6	Surge tank need – summary	288
	9.6.7	Minimum dimensions of the surge tank	288
9.7	*Powerhouse*		*289*
	9.7.1	Outdoor powerhouses	292
		9.7.1.1 Powerhouse at the foot of the dam	292
		9.7.1.2 Powerhouse as part of the dam	297
		9.7.1.3 Powerhouse downstream of the dam	297
	9.7.2	Underground powerhouses – examples	299
9.8	*Tailrace*		*302*

10 Mechanical equipment   **309**

10.1	*Gates and valves*		*309*
	10.1.1	Preliminary considerations	309
	10.1.2	Gates	310
		10.1.2.1 Types of gates	311
		10.1.2.2 Gate classification	312
		10.1.2.3 Selection of the type of gates	312
		10.1.2.4 Usage limits	313
		10.1.2.5 Outlet discharge coefficients	314
		10.1.2.6 Discharge coefficients – spillways segment gates	318
	10.1.3	Valves	322
10.2	*Turbines*		*330*
	10.2.1	Generalities	330
		10.2.1.1 Action turbines	331
		10.2.1.2 Reaction turbines	331
	10.2.2	Design conditions and data	331
	10.2.3	Turbine efficiency and plant efficiency	335
	10.2.4	Turbine equation	336
	10.2.5	Hydraulic similarity and speed number	338
	10.2.6	Specific numbers	339
	10.2.7	Operation out of design head	339
	10.2.8	Runaway speed	340
	10.2.9	Hydraulic thrust	341
	10.2.10	Suction height and cavitation	341
	10.2.11	Cavitation limits	343
10.3	*Pelton Turbines*		*344*
	10.3.1	Application range	344
	10.3.2	Basic principle	345
	10.3.3	Dimensions	345
	10.3.4	Performance data	345
10.4	*Francis turbines*		*346*
	10.4.1	Application range	346
	10.4.2	Basic principle	346
	10.4.3	Dimensions	346
	10.4.4	Performance data	348

x Contents

10.5		Kaplan turbines	349
	10.5.1	Application range	349
	10.5.2	Basic principle	350
	10.5.3	Dimensions	350
	10.5.4	Performance data	351
10.6		Bulb turbines	352
	10.6.1	Application range	352
	10.6.2	Basic principle	353
	10.6.3	Dimensions	353
	10.6.4	Performance data	353
10.7		Tubular turbines	354
10.8		Straflo turbines	354
10.9		Open flume turbine	356
10.10		Turbine performance tests	357
	10.10.1	Performance guarantees	357
	10.10.2	Field test	358
	10.10.3	Model tests	358
10.11		Turbine control	359
10.12		Mechanical auxiliary equipment	360

11 Electrical equipment: operation and maintenance — 361

11.1		Synchronous generator	361
	11.1.1	Synchronous machines	361
	11.1.2	The energy conversion	363
	11.1.3	Generator main elements	365
	11.1.4	Generator rated capacity	366
	11.1.5	Dimensioning factors	367
	11.1.6	Design principles	370
		11.1.6.1 The stator core	372
		11.1.6.2 The stator winding	373
		11.1.6.3 The poles and pole windings	373
		11.1.6.4 The bearings	373
		11.1.6.5 The cooling system	375
	11.1.7	Monitoring and instrumentation	377
	11.1.8	Transport of turbine-generator and assembly	377
	11.1.9	Tests	379
11.2		Layout of the generating unit	379
11.3		Main transformers	388
11.4		Auxiliary electrical systems	389
	11.4.1	Alternating current system (AC)	390
	11.4.2	Direct current system (DC)	390
11.5		Protection systems	391
	11.5.1	Protective relays	391
	11.5.2	Current protection criteria	391
	11.5.3	Protection of generating nits	392
		11.5.3.1 Electrical faults	392
		11.5.3.2 Mechanical faults	393

	11.5.4	Protection of elevator transformers	393
	11.5.5	Transmission line protection	394
	11.5.6	Breaker failure protection	395
	11.5.7	Substation bar protection	395

11.6 *Substation interconnection of the plant to the system* — 396
    11.6.1 Switchyard, or substation, equipment — 396
    11.6.2 Other components and installations — 397
    11.6.3 Switchyard types — 397
    11.6.4 Equipment arrangements — 397
    11.6.5 Maneuvering schemes — 397
        11.6.5.1 Simple bar — 398
        11.6.5.2 Main transfer bar, single breaker — 398
        11.6.5.3 Double bar, single breaker — 398
        11.6.5.4 Double bar, single circuit breaker with bypass disconnecting switches — 400
        11.6.5.5 Double bar and transfer bar — 401
        11.6.5.6 Double bar, one breaker and a half — 401
        11.6.5.7 Double bar, double breaker — 402
    11.6.6 Maneuvering scheme selection criteria — 403
    11.6.7 Powerplant connection to electrical system — 404
        11.6.7.1 Receiving substation — 405
        11.6.7.2 Transmission line — 405
11.7 *Operation and maintenance* — 405

12 Construction planning — 407

12.1 *Construction phases* — 407
    12.1.1 First phase diversion — 407
    12.1.2 Second phase diversion — 407
12.2 *River diversion design* — 409
    12.2.1 Discharges and risks — 409
    12.2.2 Phases of river diversion — 415
    12.2.3 River diversion dimensioning — 416
    12.2.4 River diversion – execution — 417
    12.2.5 Hydraulic models — 418
12.3 *Construction planning* — 418
12.4 *Assembly or erection planning* — 422
12.5 *Accesses to the construction site* — 423
12.6 *Contracting procedures* — 423
    12.6.1 Classical modality — 423
    12.6.2 Turn-Key — 425
    12.6.3 Alliance — 428
    12.6.4 Guaranteed maximum price — 428
    12.6.5 Final considerations — 429

13 Risks and management of patrimony — 431

13.1 *Introduction* — 431
13.2 *Dam breaks causes statistics* — 431

xii Contents

13.3 *Main accidents in the world* *432*
    13.3.1 Malpasset dam (Southeast France) 433
    13.3.2 Vajont dam (Italy) 437
    13.3.3 Teton dam (USA) 438
        13.3.3.1 US dam safety 439
    13.3.4 El Guapo dam (Venezuela) 440
    13.3.5 Lower San Fernando dam (USA) 440
    13.3.6 Sayano-Shushensk accident (Russia) 443
    13.3.7 Bieudron plant – breakdown of the penstock (Switzerland) 445

13.4 *Risks associated with hydroelectric plants* *448*
    13.4.1 Risks of dam breaks – submersion waves 448
    13.4.2 Dam breaks risk prevention – regulatory and legal aspects 455
    13.4.3 Flood risks 456
    13.4.4 Geological and geotechnical risks 456
    13.4.5 Risks related to the constructive aspects 457
    13.4.6 Risks related to penstocks 457
    13.4.7 Risks related to turbine start-up 457
    13.4.8 Risks during operation and maintenance 457

13.5 *Management of hydroelectric patrimony* *457*
    13.5.1 Context evolution 457
    13.5.2 The three issues of asset management in
        hydraulic production 458
    13.5.3 Risk management: key issues 458
        13.5.3.1 The technical questions 458
        13.5.3.2 The coordination of actions 458
        13.5.3.3 Decision support for measurement of issues posed 459
        13.5.3.4 principles governing the development of decision
            approaches 459
    13.5.4 Risk hierarchy 459
        13.5.4.1 Operations prioritization process 460
        13.5.4.2 Define unwanted events 460
        13.5.4.3 Evaluate occurrences 461
        13.5.4.4 The impacts per question 462
    13.5.5 A multicriteria decision support 464

13.6 *Conclusion* *465*

References 467
Glossary 473
Appendix: Chapter 3 – Additional examples of layouts 497
Index 585

About the author

Geraldo Magela Pereira is a Civil Engineer, graduated from the Brasília University (July 1974), with 45 years of experience in hydroelectric powerplants projects (HPPs). He has received his Master's degree in Civil Engineering and Civil Defense Protection at the Fluminense Federal University, Niterói – Rio de Janeiro State (2017).

He has worked in the geotechnical and hydraulic areas, including studies on hydraulic models, arrangements and plan-coordination and direction of projects, in its various phases: Inventory Studies, and Feasibility Studies, along with Basic Projects and Executive Projects. He has also worked in the commercial area between 1998 and 2012, developing business for the implementation of projects in EPC Contracts. He published one paper in the XVI ICOLD of San Francisco (USA) in 1988, about the "Historic Food During the 2nd Phase of Tocantins River Diversion for the Construction of Tucuruí HPP", Q. 63, R. 2, and several papers in Brazilian seminars of large dams. He also published three books in Portuguese language: *Hydroelectric Power Plants Design Step by Step* (2015), *Spillways Design Step by Step* (2016), and *Accidents and Ruptures of Dams* (in pdf form) (2018). He also published again the book *Spillways Design Step by Step* in 2020 in English language.

The main companies/projects he has been part of are as follows:

Federal Railway Network, Rio de Janeiro (1974/1975): third line in the eastern region of São Paulo city;

Hidroesb, Rio de Janeiro (1975/1976): Physical Hydraulic Models,

Enge-Rio, Rio de Janeiro (1977/1979): Balbina HPP (250 MW), Eletronorte;

Milder Kaiser Engineering, São Paulo (1980/1981): Rosana HPP (320 MW), Cesp;

Engevix, Rio de Janeiro (1981–2012): Tucuruí HPP (first phase -4,000 MW), Eletronorte, including 2 years training the team of operation of the spillway gates of the spillway; Santa Isabel HPP (2,200 MW), Eletronorte; Salto da Divisa HPP and Itapebi HPP Studies, Furnas; Canoas I HPP (72 MW) and Canoas II HPP (82.5 MW), Cesp; Capim Branco I HPP and Capim Branco II HPP – EPC; Baguari HPP – EPC.

Magela Engenharia, Rio de Janeiro since October 1994:

- Design of several small hydroelectrics (PCHs) for several clients, totaling 1,000 MW, approx.
- Inventory studies of the rivers Sucuriú, Verde, Iguatemi and Paraíso (MS): Araguaia/Tocantins – revision (TO); Teles Pires (MT); Fetal and Prata (MG).
- Consultant of Engevix Eng.: the implementation of the Monte Serrat (25 MW), Bonfante (19 MW) and Santa Rosa (30 MW) SHPs in the Rio de Janeiro;
- Consultant of Leme Eng, Belo Horizonte. (1997) for projects in Chile (Laja I/Laja II) and Panama (Teribe and Changuinola);
- Coordinated for Eletrobras the revision of the Manual of Small Hydroelectric Plants (PCHs-1997/1998);
- Consultant of SGH – ANEEL, for analysis of SHP projects (2000);
- Consultant of CPFL Energia to develop business with SHPs (2008–2010).

Preface

This book presents the "complete" methodology for the Step-by-Step Hydroelectric Power Plants Design. The author is originating from Brazil, a developing country with a wide hydrographic network. The first edition was published in Spanish language in São Paulo in 2015.

It is important to highlight that the author "tried" to observe all the recommendations contained in ICOLD's bulletins that refer to the theme.

It should be made clear that, as a principle, the basic concepts of all disciplines involved in these projects must be understood by the technicians of the project teams, each one in their respective area, as well as by the professionals of the construction companies of these plants.

Since ancient times, humanity has been using reservoirs to accumulate water for its survival. In addition to this primary purpose, reservoirs are also indispensable for better management of risks associated with extreme hydrological events.

Certainly, at first, the reservoirs affect the ecological balance of the region, but the correct mitigation measures will be able to adapt the lives of affected population to the new environment with a better quality of life. From the point of view of energy generation, hydroelectricity is certainly less impacting than most of the applicable alternative sources, in addition to being a renewable source with the ability to be stored in large quantities for use in periods of low availability.

The design of a hydroelectric plant, that is an installation that transforms the potential energy of water into electrical energy, is an activity that is not standardized. Each new project is an interesting engineering challenge and teams need to work with different topographic and geological conditions of each site in an integrated way, to design a functional, economical, and environmentally sustainable project.

The development of the project, here understood as the dam and the reservoir, the spillway, the plant and enclosed areas, the switchyard and the associated transmission line, is a multidisciplinary activity that encompasses Civil Engineering, Geology, Mechanical Engineering and Electrical, Environmental, Economic Engineering, Construction and Assembly, as well as the Operation and Maintenance of Civil Works and of the Electromechanical Equipment.

In Civil Engineering, the disciplines of Topography and Hydrology stand out because they characterize the gross head of the plant (H) and the "average" flow rates available (Q) – parameters that are key to determining the power of the plant.

The hydrological studies will also determine the flow rates to be applied in the river diversion, as well as in "extreme floods" for the sizing of the spillway.

Special attention needs to be paid to the sizing of the reservoir and to the sedimentation studies, since the transport and deposition of solid materials will reduce its volume over time, which can eventually compromise the generation capacity and the lifespan of the plant.

Still with regard to the reservoir, it is important to note that the author took into account the papers *"Water Usage for Sustainable Development and Poverty Eradication"* (Gomide, 2012) and *"World Water Resources, Usage and the Role of Man-Made Reservoirs"* (White, 2010), both available on the internet.

Geology and geological studies, of course, also stand out as they enable the knowledge of the sub-soil which is of fundamental importance for the definition of the foundations of the structures where a large part of the safety of dams is deposited.

Regarding the safety of dams, the following works of Ralph Peck from 1973 and 1980 should be highlighted:

- "The influence of non-technical factors on the quality of dams" (1973), and
- "Where has all the judgment gone?" (1980),

which remain current for the technical environment involved in dam design and construction.

As will be seen in Chapter 13, it should be noted that only three causes are responsible for 94% of all the dam breaks, which coincidentally are associated with the main and most complex disciplines involved in the project:

- 59% of accidents are caused by geological/geotechnical causes, including varying foundation problems, settlement, high neutral pressures, slope landslides, and poor materials;
- 23% of the disasters are caused by overtopping of the structures caused by floods that surpassed the design flood (underestimated flood); and
- 12% to various construction problems, either in foundations treatment, in compacted landfills, or in the concreting of the various structures' components.

With this in mind and observing the lack of bibliography that generally encompasses the different disciplines involved in the design of a hydroelectric plant, the author decided to write this book, with the objective of facilitating the professional life of the new generations of engineers that ingress the electric sector or other sectors that demand the knowledge of hydraulic works.

It is noteworthy that there are still many hydroelectric plants to be built in the world, mainly in South America, Africa, and Asia.

It must be said that Brazilian hydroelectric estimated potential is of approximately 250 GW (EPE, 2018), being 56% in the Amazon region (Figure 1.15). Of this total, 100 GW have already been installed. So, there are still 150 GW left to install, which is a very large number anywhere in the world.

There is much to be done in Brazil that depends on the market growth and on the decision of the federal government opting to make HPPs, which produce clean, renewable energy that is generally cheaper than the energy from other sources.

It must also be said that the reservoirs of all existing Brazilian hydroelectric plants flood an area of 37,000 km^2 in a country that has an estimated surface of 8.6 million km^2. That is, they flood a relatively very small area.

The author, a member of the BCOLD, Brazilian Committee on Dams, based on his experience of almost 50 years of working with hydroelectric projects in the country and abroad, would like to conclude by recording his opinion that the Brazilian plants are, in fact, environmentally sustainable and that he always defends this point of view in any international forum.

Acknowledgments

The author thanks his wife Cristine, daughter Marina, and the daughter Ligia, mother of my granddaughter Luiza for existing. They are the reason of my life. And, in memoriam, thanks to my parents Geraldo and Luiza.

Special thanks to my best friend and colleague since 1970, from the Engineering School of the University of Brasília (Brazil), Murilo Lustosa Lopes. Murilo is a Civil Engineer and effectively contributed by making, with patience, several suggestions, comments, and criticisms in both Brazilian and international editions of the book.

Special thanks also to my friend Alberto Sayão, Professor of Geotechnical Engineering at the Pontifical Catholic University of Rio de Janeiro, for his objective comments.

Thanks to the colleagues from Brazil I have listed who, in their respective specialties, effectively contributed by commenting on the Brazilian version of 2015: Roneí Vieira de Carvalho, my advisor in Hydraulic Engineering and Layouts; Sérgio Correia Pimenta, Civil Engineer; Roberto Corrêa, Guido Guidicini and Paulo Fernando Guimarães, Engineering Geologists; Ewaldo Schlosser, Electrical Engineer; Helmar Alves Pimentel, Mechanical Engineer; Paulo C. F. Erbisti, Consulting Engineer; Biologist and Environmental Consultant Carlos Roberto Bizerril; Vantuil Ribeiro, Polytechnic Engineer, for drafting the text on physical guarantee of the plant's energy; and Raul Odemar Pithan, Civil Engineer, for the general revision of the text.

Thanks to the designers Keren Feitosa and Welliton dos Santos Rodrigues, from Florianópolis, Santa Catarina, Brazil.

The author thanks Angela Medina, Senior Affairs Specialist, United States Bureau of Reclamation, who responded in August 12, 2020 to my request for a writing permission to use data from USBR. She reported that USBR publications are not copyright.

The author also thanks Andreas Noteng, Officer of Student Affairs of the Department of Civil and Environmental Engineering of the NTNU – Norwegian University of Science and Technology, that authorized in writing in August 19, 2020 to quote and paraphrase the data in any way as long as the sources were referenced.

Finally, the author would like to register his thanks to the BCOLD, Brazilian Committee on Dams, and to ELETROBRAS, holding company of the Electric Sector in Brazil, for the permission in writing to use the data of its publications (August 19, 2020). It should be noted that both were the sponsors of the book edition in 2015.

Geraldo Magela Pereira
Rio de Janeiro, Brazil

Acronyms

AAI	Integrated Environmental Assessment
ABCE	Brazilian Association of Engineering Consultants
ABGE	Brazilian Association of Engineering Geology
ABMS	Brazilian Association of Soil Mechanics
ABNT	Brazilian Association of Technical Standards
ABRH	Brazilian Association of Water Resources
AC	Alternating current
AIS	Air-Insulated Substations
ANA	National Water Agency of Brazil
ANEEL	National Electric Energy Agency, Brazil
ANP	National Oil, Natural Gas and Biofuels Agency, Brazil
ArcGIS	Geographic Information System to work with maps and geographic information
ASCE	American Society of Civil Engineers
ASME	American Society of Mechanical Engineers
ASTM	American Society for Testing Materials
AT	High voltage
BC	Before Christ
BNDES	National Bank for Economic and Social Development, Brazil
BoQ	Bill of Quantity
BOT	Build, Operate, Transfer
BOOT	Build, Own, Operate, Transfer
BOO	Build, Own, Transfer
BOOST	Build, Own, Operate, Sales, Transfer
CANAMBRA	Canambra Engineering Consultants Ltd.
CBDB	Brazilian Committee on Dams
CBT	Brazilian Committee of Tunnels
CCC	Construction Loading Condition
CCE	Exceptional Loading Condition
CCN	Normal Loading Condition
CCL	Load Condition Limit
CCEE	Electric Energy Trading Chamber, Brazil
CCV	Conventional concrete
CCR	Compacted Concrete Roller Dam
CCV	Conventional Gravity Concrete Dam

CE	Cost of energy generated
CEEE	State Electricity Company of Rio Grande do Sul, Brazil
CELG	Goiás State Energy Company, Brazil
CELUSA	Urubupungá Power Plants, Brazil - extinct
CEMAR	Maranhão State Energy Company, Brazil
CEHPAR	Hydraulic Center Prof. Parigot de Souza. Federal University of Paraná, Brazil
CEMIG	Minas Gerais State Energy Company, Brazil
CESP	São Paulo State Energy Company, Brazil
CGH	Micro Power Plant (<5 MW);
CHERP	Pardo River Hydroelectric Company, Brazil – extinct
CHESF	São Francisco River Hydroelectric Company, Brazil
CI	Owner Total Investment, including JDC (interest during construction)
CIRIA	Construction Industry Research and Information Association
CLAIM	Formal request for special compensation for quantity variations
CME	Marginal cost of energy
CMP	Marginal cost of peak
CMS	Marginal cost of secondary energy
CMSE	Committee for the Monitoring of the Electric Sector, Brazil
CNPE	National Council for Energy Policy, Brazil
CNRH	National Water Resources Council, Brazil
CONAMA	National Council for the Environment, Brazil
CPFL	Paulista Company of Force and Light, Brazil
CPRM	Mineral Resources Research Company, Brazil
CSN	National Steel Company, Brazil
CVRD	Vale do Rio Doce Company, Brazil
DC	Direct current
DNAEE	National Department of Water and Electric Energy (extinct), Brazil
DNIT	National Department of Transport Infrastructure, Brazil
DNPM	National Department of Mineral Production, Brazil
DRDH	Declaration of Water Availability Reserve, Brazil
DRM-RJ	Department of Mineral Resources – Rio de Janeiro, Brazil
EF	Firm Energy/Guaranteed Energy during the critical period
EG	Annual Generated Energy
EI	Infiltration Test
EIA	Environmental Impact Study
EIH	Hydrographic Basin Inventory Studies (see IHE)
El	Elevation abbreviation (quota)
ELETROBRAS	Holding of the Electric Sector in Brazil
ELETRONORTE	Holding of the Electric Sector – North of Brazil
ELETRONUCLEAR	Nuclear Power Plants, Brazil
ELETROSUL	Holding of the Electric Sector – South of Brazil
ENERSUL	Mato Grosso do Sul Energy Company, Brazil

EPA	Water Test
EPC	Engineering, Procurement and Construction
EPC-M	Engineering, Procurement, Construction and Management
EPE	Energy Research Company, Brazil
EPRI	Electric Power Research Institute
EPUSP	Polytechnic School of the University of São Paulo, Brazil
ESCELSA	Espírito Santo Power Stations, Brazil
FCTH	Foundation Hydraulic Technological Center, São Paulo University, Brazil
FEED	Front-End Engineering Design (or a Detailed Basic Project)
FIDIC	International Federation of Consulting Engineers
FRC	Capital Recovery Factor
FSF	Floating safety factor
FUNAI	National Indian Foundation, Brazil
FURNAS	Furnas Power Plants Company, Brazil
GCOI	Interconnected Operation Coordinating Group, Brazil
GCPS	System Planning Coordinator Group, Brazil
GIS	Geographic Information System
GPR	Ground Penetration Radar
GPS	Global Positioning System
HDSBED	Hydraulic Design of Stilling Basin and Energy Dissipators
HEC-RAS	Hydrologic Engineering Center-River Analysis System
HEH	Hydropower Engineering Handbook
HIDROESB	Hydrotechnical Laboratory Saturnino de Brito – Rio de Janeiro, Brazil
HPP	Hydroelectric Power Plant (>30 MW); CGH – Micro Power Plant (<5 MW);
SHP	Small Hydroelectric Power Plant (5 MW<PI<30 MW)
IBAMA	Brazilian Institute of Environmental and Renewable Natural Resources
IBGE	Brazilian Institute of Geography and Statistics
ICOLD	International Commission on Large Dams
IEC	International Electrotechnical Commission
IEEE	Institute of Electrical and Electronics Engineers
IEF	State Forest Institute, Brazil
IHE	Inventory Hydroelectric Studies
INEA	State Institute of Environmental – Rio de Janeiro, Brazil
IP	Scheduled Unavailability
IPHAN	Institute of National Historical and Artistic Heritage, Brazil
IPT	Institute of Technological Research, Sao Paulo, Brazil
JDC	Interest Rate During Construction
LI	Installation License, Brazil
LIGHT	Light Electricity Services, Rio de Janeiro, Brazil
LNEC	National Laboratory of Civil Engineering, Lisbon, Portugal
LP	Preview License, Brazil
LO	Operating License, Brazil
MAB	Movement of the Affected by Dams, Brazil

MAE	Wholesale Energy Market, Brazil
MEF	Finite Element Method
MIT	Massachusetts Institute of Technology, USA
MMA	Ministry of the Environmental, Brazil
MME	Ministry of Mines and Energy, Brazil
MRE	Energy Reallocation Mechanism, Brazil
MST	Movement of Landless Workers, Brazil
NATM	New Austrian Tunnelling Method
NIT	Norwegian Institute of Technology
NPSH	Net Positive Suction Head
NUCLEBRÁS	Nuclear Power Plants, Brazil – extinct in 1989
NUST	Norwegian University of Science and Technology
O&M	Annual Cost of Operation and Maintenance
ONG	Non-Governmental Organization, Brazil
ONS	National Electrical System Operator, Brazil
OPE	Standard Eletrobras Budget, Brazil
PBA	Environmental Basic Project
PBC	Consolidated Basic Project
PBE	Basic Engineering Project
PI	Test Pit
PMF	Probable Maximum Flood
PMP	Probable Maximum Precipitation
PNSB	National Security Policy for Dams – Brazil
POC	Gates Operation Plan
PPA	Power Purchase Agreement
PSB	Dam Safety Plan
RAS	Simplified Environmental Report
RCC	Roller compacted concrete
RDH	Water Availability Reserve
RDPA	Environmental Programs Detail Report
RIMA	Environmental Impact Report
RMR	Rock Mass Rating
SCG	Generation Concession Superintendence, ANEEL, Brazil
SEV	Vertical Electrical Sounding
SFG	Superintendent of Inspection of Generation Services, ANEEL, Brazil
SHP	Small Hydroelectric Power Plant (5 MW<PI<30 MW)
SIN	National Interconnected System, Brazil
SNISB	National Safety Information System for Dams, Brazil
SPC	Specific Purpose Company
SPT	Standard Penetration Test
SP/SR/ST:SP=	Boring Sounding; SR = Drilling Sounding; ST = Auger Boring Sounding
SRG	Superintendent of Regulation of Generation Services, ANEEL, Brazil
SSARR	Streamflow Synthesis and Reservoir Regulation
SSM	Standard Step Method for Backwater Computations
SUCS	Unified Soil Classification System

TAC	Term of Conduct Adjustment
TBM	Tunnelling Boring Machine
TEIF	Equivalent Rate of Forced Unavailability
TIR	Internal Rate of Return
TR	Recurrence Time
TRAFO	Transformer
UFRGS	Federal University of Rio Grande do Sul, Brazil
UFRJ	Federal University of Rio de Janeiro, Brazil
UNCITRAL	United Nations Commission on International Trade Law
USA	United States of America
USACE	United States Army Corps of Engineers
USBR	United States Bureau of Reclamation
USCOLD	United States Committee on Large Dams
USP	University of São Paulo, Brazil
UTE	Thermoelectric Plant
VMP	Maximum Probable Flood
WES	Waterways Experiment Station (Corps of Engineers – USA)
WL	Water Level
WMO	World Meteorological Organization
WP&DC	Water Power and Dam Construction

Symbols

A	area (m^2)
A	duct area under the water lamina (m^2)
A_d	drainage area (km^2)
B	chute width, corresponding to each aerator (m)
B	slope angle
c	cohesion (kgf/cm^2)
C	Poisson constant
C	Isbash stability coefficient
C	sediment concentration in suspension (mg/l)
CE	cost of energy generated (R\$/MWh);
C_i	value of total investment, including JDC (interest rate during construction)
C_{int}	intern consumption of energy by the plant (MWmed)
Cc	coefficient of compressibility ($l/kg/cm^2 \times 10^{-6}$)
CLI	cutter life index
Co	flow or discharge coefficient
$\cos \phi$	power factor
D	average depth of the reservoir along the fetch (m)
d	depth (m)
d_c	critical depth (m)
d	characteristic block diameter (m)
D/h	characterizes the effect of the aerator
D	diameter (m)
DRI	drilling rate index
d	erosion depth (m)
D	inner diameter of stator (m)
e	overall efficiency (%); void ratio
E	deformability module (GPa)
E	energy
Ec	thickness of the dam ar crest
ef	thickness of the dam at the foundation
EF	firm energy/guaranteed energy during the critical period (MW med)
EG	annual generated energy (MWh)
f	alternating current frequency (Hz); in Brazil, $f = 60$

xxviii   Symbols

f	free board (m)
F	fetch (km)
Fe	fetch (km)
fp	generator power factor
Fr	Froude number
H	height (m)
H	available or gross head (m)
H	hydraulic head (m)
H_d	design head (m)
H_o	design head (m)
h	height of water (m)
Hs	significant wave height (m)
ht	velocity head (m)
i	hydraulic gradient
IP	scheduled unavailability (%)
Ir	number of Iribarren
k	coefficient of permeability (cm/s)
K_a	coefficient of contraction of the extreme spillway spans
K_c	coefficient of compactness
K_f	conformation index
K_p	coefficient of contraction of the pillars
L	length; effective crest length (m)
L'	net crest length (m)
L	length of stator core (m)
Lo	wave length in deep water (m)
M	momentum (kg.m/s)
n	coefficient of the site of the dam: $n = L/H$
n	slope of the ramp
n	rotational speed of the turbine-generator set (rpm)
N	number of pillars
P	height of the spillway crest (m)
P	perimeter of the basin (km)
P	hydrostatic pressure (m)
p	number of poles (always in even numbers)
P	power output (kW)
p	absolute pressure at a point in the system outside the cavitation zone (m)
p_v	vapor pressure within the bubbles, equal to the vapor tension at water temperature (m)
Q	flow (m^3/s)
q	specific discharge (m^3/s/m)
q_a	specific air flow per unit chute width (m^3/s/m)
Q_a	total air flow (m^3/s)
Q_L	net flow (m^3/s)
Qmáx	peak or maximum flow (m^3/s)
Q_s	suspended load discharge (t/day)

qr	remaining flow of the plants (m³/s)
qu	flow of consumptive uses (m³/s)
r	attenuation factor
Ri	radial distance from the contour line of the reservoir to the dam (m)
Ru	run-up, vertical height above the reservoir water level (m)
S	safety factor
S	wind tide (m)
T	minimum wind duration (min)
TEIF	equivalent rate of forced unavailability (%)
Tp	peak wave period (s)
t	period of time (s)
t	ramp height (m)
t/h	measure of the relative height of the step or ramp
T	shear stress (kgf/cm²)
t_c	time of concentration (min)
TW	tailwater level (m)
U	wind velocity (m/s)
V	velocity (m/s)
V	volume (m³)
Δ*V*	volumetric variation
W	weight (t/m³)
WL	water level
WL	normal water level
WLmax	maximum water level
WLmax max	maximum maximorum water level
WLmin	minimum water level
WLres	reservoir water level
WLtw	tailwater or downstream water level
WLtw-min	downstream minimum water level
WLtw-max	downstream maximum water level

Greek symbols

α_i — angle between the wind direction and the radial direction radial (°)

α — chute inclination angle

β — aerator performance is measured by the ratio of air to water flow

β — form factor of the concrete dam

σ — cavitation characteristic parameter

σ_i — incipient cavitation index

ϕ — friction angle (°)

g — gravity acceleration (9.8 m/s^2)

φ — ramp angle with respect to the bottom of the chute

γ — specific weight of water (N/m^3)

ρ — specific mass of water (1,000 kg/m^3)

σ — normal stress (kgf/cm^2)

σ_c — compression strength (kgf/cm^2)

μ_T — turbine efficiency (0.90)

μ_G — generator efficiency (0.98)

μ — efficiency of the set (0.88)

Chapter 1

Hydroelectric powerplants

1.1 INTRODUCTION

This book presents the methodology for the Hydroelectric Power Plants Design, Step by Step (HPPs). The first edition was published in Brazil in 2015. It was written to fill a gap in this type of bibliography, aiming to reach young engineers at the beginning of their careers.

In this English version, the author sought to observe the ICOLD's recommendations for these projects. The author has 50 years of experience in hydroelectric projects in Brazil, that has a great hydrographic network with several HPPs installed (100 GW). One of the projects in which the author participated was the Tucuruí HPP in the Tocantins river, 8.085 MW, which will be one of those to be explored in the book. In spite of this fact, it will be seen in this book that Brazil still has a large and attractive hydroelectric potential. This yet to be explored potential is environmentally sustainable and was estimated at 162 GW, a considerable value anywhere in the world.

The basic concepts of the several disciplines involved, namely, cartography, hydrology, hydraulics, geology, geotechnics and environmental, are out of the scope of the book. As a principle, it is understood that they must be known by the technical who should use the book, each one in their respective area, as well as by construction engineers.

It must be emphasized that the interaction in harmony between all the specialists involved in these projects, designers, suppliers, builders and assemblers, is fundamental to the successful implementation of the enterprises.

After these considerations, it should be noted that several acronyms, which make up a particular section at the beginning of the book, are used in the text to simplify it. Some of the examples are as follows:

ICOLD, International Commission on Large Dams;
BCOLD, Brazilian Committee on Dams;
HPP, Hydroelectric Power Plant;
WL, Water Level; WLres, Reservoir Water Level; WLmax, Maximum Water Level; TW, Tailwater Level;
HDC, Hydraulic Design Criteria;
EPC, Engineering, Procurement and Construction; etc.

DOI: 10.1201/9781003161325-1

1.2 THE HISTORY

Although the earth and the solar system originated 4.6 billion years ago, it was only during the Holocene period of the Quaternary era, 10,000 years ago, that humans developed agriculture in the hills of Syria and Iraq. Dams have served people for at least 5,000 years, as evidenced by the cradles of civilization, in Babylonia, Egypt, India, Persia and Far East. The remains of these ancient structures exist in both the old and the new worlds, marking the attainments of societies which have long disappeared.

Many of the outstanding waterworks of antiquity declined into disuse because the knowledge of their designers and builders was not preserved by the generations who inherited them. And without water, the civilizations which it had supported faded away (Jansen, 1983).

Table 1.1 presents a chronology of water knowledge. The ancient history of dams is not well known. Many dates before 1,000 BC can only be estimated. This is particularly true in ancient Egypt, whose peculiar chronology casts only a faint light on many dynasties and their engineering achievements.

Ruins of ancient works in India and Sri Lanka offer some evidence of how reservoirs were created by ancient civilizations. The methods involved building barriers across the rivers. Some lakes formed occupied vast areas. The construction materials were transported in baskets or other containers. Compaction was carried out by the trampling of worker's feet, a method still used in some countries where the cost of labor is low.

The ancient builders made free use of soils and gravel. As they only had a little knowledge of the mechanics of materials and of the floods, they had no definite methods and often their works collapsed. In ancient Egypt, the channel construction was the greatest effort of the pharaohs.

Table 1.1 Chronology of Knowledge about Water (Mays, 1999)

Period	Type of use
Prehistoric	Springs
3rd to 2nd millennium BC	Cisterns
3rd millennium BC	Dams and wells
2nd millennium BC	Gravity flow supply pipes of channels and drains, pressure pipes
8th to 6th centuries BC	Long-distance water supply lines with tunnels and bridges, as well as intervention in and harnessing of karst water systems
6th century BC at the latest	Public as well as private bathing facilities, consisting of bathtubs or showers, footbaths, washbasins, latrines or toilets, laundry and dishwashing facilities
6th century BC at the latest	Use of about two or three qualities of water: drinkable, sub-drinkable, and non-drinkable, including irrigation using storm runoff, probably combined with waste waters
6th to 3rd centuries BC	Pressure pipes and siphon systems

The first duties of the governors of the provinces were the excavation of canals, which were used to irrigate large areas of lands in times of floods in the Nile river. The land was cut into small basins through a system of dikes.

One of the earliest records of engineering works is that of the city of Memphis by the Nile river, estimated to be built between 5,700 and 2,700 BC (USBR, Jansen, 1983).

The historian Herodotus attributes this construction to Menes, the first Egyptian pharaoh. According to him, Menes built a masonry dam on the Nile river in Kosheish, 20 km upstream from the site planned for the capital.

Considered a legend by some historians, this version reports that before founding the capital, Menes diverted the Nile river to the east side of the valley and constructed a great dam near the Lybian hills. The purpose was to secure space for the construction of the city on the west side, and to enhance its protection from the attacks of the enemies.

Translation of Herodotus' writings indicates that the masonry dam of cut stones reached 15 m in height and 450 m in length. Modern historians think that the ancient builders did not have the capacity for so much.

However, it is reported that in the Wadi el-Garawi, there is a well-preserved Sadd al-Kafar masonry dam (Figure. 1.0A). It is near Helwna, 32 km south of Cairo. This dam is 11 m high and 107 m long, and was built 2,600 years ago to reserve water for the region's quarry workers.

The base of the Sadd al-Kafar masonry dam was 84 m wide, and the crest was 61 m long. The upstream and downstream faces of the dam, distant 36 m at the base, were 24 m thick walls of gravel masonry. The core was filled with gravel, stones and earth. The dam, of course, did not have a cutoff trench in the foundation. The upstream slope was lined with blocks of limestone, and were stepped, without grout joints. The stones with an average weight of 50 kg were placed in steps of 30 cm height on a slope of 0.75 V:1.00 H.

Figure 1.0A Ruins of Sadd al-Kafar masonry dam (Structurae.net).

The dam did not have a spillway, which was its greatest deficiency. The small reservoir, with a volume of $570,000\,m^3$, was not enough to control the floods. The dam was climbed and collapsed before the completion of its construction, since there were no signs of sedimentation of the reservoir. Although the builders expected the lower crest at the center of the dam to function as a spillway, the core at this point was not protected against erosion by overtopping. This error was repeated, and the dams broke. The rupture probably discouraged the Egyptians from building this type of dams.

It is recommended to consult the chapter "Dams from the beginning" of the book *Dams and Public Safety* by USBR – Jansen (1983). The Sumerians, in southern Mesopotamia, also built cities, temples and canals that are included in the first works of engineering. It is worth remembering that these peoples have been fighting for water rights since the beginning of history. Irrigation was vital for Mesopotamia, between the Tigris and Euphrates rivers, and to Egypt and Greece, as the remnants of these works in the region attest.

In the 8th century BC, the Assyrians copied from the Armenians the system of tunnels used to bring water from the underground fountains of the hills to the city at the foot of these.

Over the centuries, this system has spread throughout the region to northern Africa and is still in use today. The works in the Middle East and the prehistoric works in Mexico and America are discussed in Jansen (1983) and Mays (1999). Jansen (1983) cited the case of the Proserpina dam, 19 m high and 427 m long, built by the Romans 6 km north of Mérida in Spain in the 2nd century AD.

The Proserpina dam has been ranked as a classic among the structures of this type. An upstream section is comprised of a concrete core sandwiched between two masonry walls. The original thickness of the composite wall has been estimated to be about 3.75 m at the top.

According to Jansen (1983), some reports indicate that this may have been as little as 2.1 or 2.4 m in few places. At the foundation, the thickness may have been as much as 5 m. The upstream face is battered steeply at 1–10, while the downstream masonry free is vertical. The wall extends about 6 m into the foundation. An earth fill was placed against the downstream side of the wall, with slopes 5:1.

The records do not reveal much about the nearly 2,000-year lifespan of Proserpina dam. It is assumed to have suffered long periods of neglect. There is information on repairs and modifications accomplished in the years 1617, 1689 and 1791.

This would indicate that the reservoir has been in possibly continuous service for nearly 400 years. It has not suffered diminution by siltation. Major repairs were made in 1942, including rehabilitation of the masonry. Water is still supplied via a Roman aqueduct between the dam and Mérida.

Waterwheels have been used to do work for thousands of years. The direct action of the jet from a fall on a waterwheel produces mechanical energy. The Romans had known these devices since the 1st century BC.

However, waterwheels only became extensively used from the 14th century on grinding, sawing, and feeding textile factories, among other uses.

By the end of the 18th century, there were about 10,000 waterwheels in New England. During the 18th and 19th centuries, with the new technologies such as steam engine, engine, dynamo, lamp and hydraulic turbine, it became possible to convert mechanical energy into electric energy.

Hydroelectric powerplants 5

Figure 1.0B Proserpina dam view.

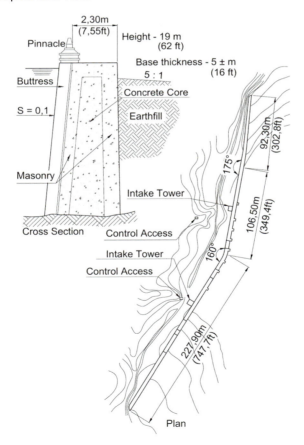

Figure 1.0C Proserpina dam plan (Jansen, 1983).

6 Design of Hydroelectric Power Plants – Step by Step

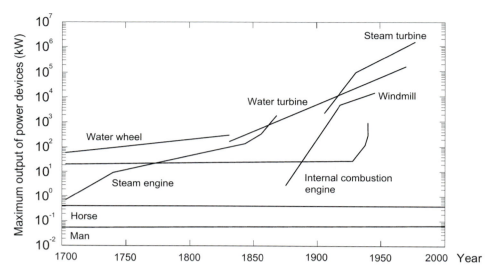

Figure 1.1 Maximum output of selected power devices along 1,700–1,970 (Gulliver and Arndt, 1991).

Mills and factories were built close to the hydroelectric power plants to be able to directly use the available energy.

The power of these first plants, limited to 100 kW (134 CV), is compared with those of the sources of energy in the period between the years 1,700 and 1,970 in Figure 1.1, extracted from Gulliver and Arndt (1991).

1.3 HYDROELECTRIC PLANTS – OUTSTANDING EVENTS

Although a turbine was developed in 1750, the first of these machines is credited to the French engineer Benoit Fourneyron in 1827.

This researcher made a wheel under constant pressure, completely submerged, where the flow entered without shock and with low speed. Installed in a 1.4 m drop under the bridge over the Ognon river, in Haute-Saône, it produced 4.5 kW with a yield of 83%, more than triple the old reed wheels.

This technology was perfected by Fourneyron himself and by Jonval, Fontaine and Girard in France, by Thomson in England, and by Pelton and Francis in the United States of America.

In 1831, Michel Faraday (1791–1867) developed the induction ring, accomplishing its goal of producing electricity from magnetism. This ring, two electrically separate coils, is considered the first transformer.

In 1837, Fourneyron installed turbines fed by pressure conduits in St-Blaise, in the Black Forest, in a head of 108 m.

In 1843, Jean Bernard Léon Foucault, inventor of Foucault's pendulum (1851), occasionally lit the Square of La Concorde, in Paris, with electric arc lamps.

In 1856, James Maxwell (1831–1879) mathematically translated the vision of Faraday in relation to electricity and magnetism.

In 1882, Thomas Edison (1849–1931) manufactured incandescent lamps and all the other components required to install electric light in homes.

Edison and his team set up boilers and dynamos in a New York building. In addition, they also installed transmission lines to distribute electrical energy to the Wall Street region.

In 1882, the operation of the first hydroelectric plant in USA, Vulcan, Vulcan Street Plant, 12.5 kW, in the Fox River, Appleton – Wisconsin (Figures 1.2 and 1.3) started.

Figure 1.2 America's first power plant (1882). Vulcan Street Plant, 12.5 kW. Fox river, Appleton. Wisconsin (wisconsinconfidential.blogpost.com).

Figure 1.3 Vulcan street plant, 12.5 kW (Royalbroil, 2010).

Figure 1.4 (a) Upton plant – Minneapolis brush electric company (1882). (b) Minneapolis City – USA (pvnworks.com).

At the same time, operation of the Upton HPP, of the Minneapolis Brush Electric Company, started in the Mississippi river, in the Upton Island down the St. Anthony Falls, Minneapolis city (Figure 1.4). The plant had five generators connected to a waterwheel and illuminated the Washington Avenue.

In 1888, Nikola Tesla (1856–1943) invented the induction motor, which paved the way for use of AC sources. He also invented the transformer called Tesla coil.

In 1901, the transmission of energy over long distances became economical in America after the installation of an alternating current equipment in Niagara Falls Power Plant, New York, by George Westinghouse. This allowed expansion of the use of hydroelectric powerplants.

In the 1930s, the capacity of the turbines increased significantly, which coincided with the increase in energy demand. Large plants (>15 MW) have become the norm. The power of the steam turbines was also increasing rapidly, and the cost of electricity continued to fall.

In the period 1940–1970, in América, the cost of operation and upkeep of old and small hydroelectric plants became greater than the revenue they produced. Thus, the small plants (<15 MW) were retired and the large hydro plants (>15 MW) were climbed, as depicted in Figure 1.5.

A similar trend occurred in Europe. The growth of the HPPs in the rest of the world only happened in the 1960s.

1.4 HYDROELECTRIC POWERPLANTS IN BRAZIL

The author is a Brazilian civil and hydraulic engineer with 45 years of experience in hydroelectric projects. He considered it important to introduce this item to highlight the Brazilian hydroelectric potential including the remaining that still needs to be studied, developed and implemented. It's recommended to see on the internet "The story of the Dams in Brazil Centuries XIX, XX and XXI", published by CBDB in 2011 in commemoration to the 50 years of its foundation.

The first HPP in Brazil, Ribeirão do Inferno, came into operation in 1883. It was built to take advantage of a 5 m drop in this tributary of river Jequitinhonha, in Diamantina, Minas Gerais State. The plant had two Gramme dynamos of 5 HP each

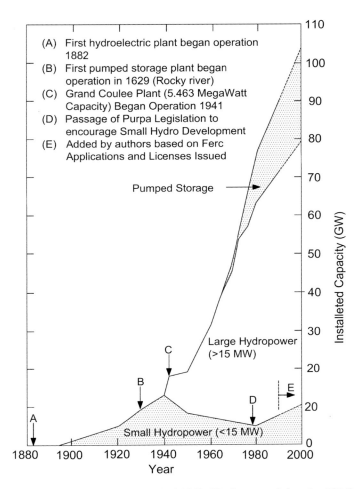

Figure 1.5 Installed capacity in USA (1882–2000) (Gulliver and Arndt, 1991).

which generated energy to move the water pumps to dismantle rock formations of the diamond mines. Later, the plant started to generate energy to supply the city. The transmission line was 2 km long.

In 1887, a 500 HP plant, with a drop of 40 m, was put in operation in the Macacos river, Nova Lima, also in Minas Gerais, to attend to a gold mine and for illumination.

Two years later, in 1889, the Marmelos Zero plant was put in operation in the Paraibuna river, Juiz de Fora, Minas Gerais, of the electricity company, with two generators of 126 kW each, from Max Nothman & Co, to supply the city (Figure 1.6). In 1891 a third group of 125 kW was installed for industrial purposes.

This was the first big plant in South America. The growing demand for energy led to successive expansions of the plant with its current capacity at 4 MW. In 1983, the plant was dropped by the municipal patrimony, and the Marmelos Museum installed at the plant is now administered by the university.

Figure 1.6 Marmelos Zero HPP: (a) upper view, (b) the powerhouse and the spillway in the back, (c) dam spillway and (d) powerhouse (CBDB, 2011).

In 1905, Light Electricity Services was founded in Rio de Janeiro, by the Canadian lawyer Alexandre Mackenzie and the American engineer Frederick Stark Pearson (CBDB, 2011). Figure 1.7 shows the Light's HPPs between São Paulo and Rio de Janeiro.

In 1908, Light inaugurated the Fontes Velha HPP with a capacity of 12 MW, the largest in Latin America and the second largest in the world (Figure 1.8).

The concrete arc dam was completed in 1906, which was 32 m high and 234 m long. The dam included an uncontrolled spillway, 134 m wide (Figure 1.9). In 1909, the expansion to 24 MW was completed (CBDB, 2011).

Hydroelectric powerplants 11

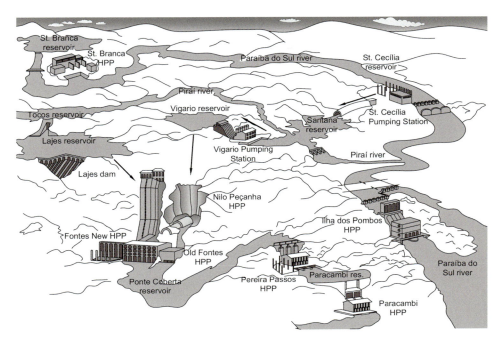

Figure 1.7 Light power plants. (Adapted from the "General scheme of the HPPs of the Paraíba, Piraí and Ribeirão das Lajes rivers", Brazil (Light, 1981).)

Figure 1.8 Fontes Velha HPP. 24 MW (CBDB, 2011).

Figure 1.9 Fontes Velha HPP. Dam and spillway (CBDB, 2011).

In 1914, the 25 m high Tócos Dam (Figure 1.10) was completed in the Piraí river to divert 25 m³/s of flow through an 8.4 km long tunnel to the Lajes reservoir, making it possible to increase the power of the Fontes plant to 54 MW.

Figure 1.11 shows the exit portal of this tunnel at that time with the longest hydraulic tunnel in the world. Light rehabilitated and repowered this plant in the 1990s.

In 1924, the Ilha dos Pombos HPP was completed on the Paraíba do Sul river, 150 km from Rio de Janeiro (Figure 1.12). The plant has a 2.5 km power canal consisting of compacted soil dikes and concrete stretches. The main spillway has three radial gates, 45 m wide × 7.4 m high, and is operated normally (CBDB, 2011). After enlargement in 1937, the plant reached 170 MW, with a gross head of 31 m.

Hydroelectric powerplants 13

Figure 1.10 Tócos dam in 1914 (CBDB, 2011).

Figure 1.11 Tócos tunnel exit portal (CBDB, 2011).

14 Design of Hydroelectric Power Plants – Step by Step

Figure 1.12 (a–b) Ilha dos Pombos HPP. 170 MW (CBDB, 2011).

In 1940, Light was authorized to expand the Lajes buttress dam to increase de energy generation. The height was raised from 26 to 63 m. The raising was completed in 1958. Reservoir volume increased to 1.0 billion cubic meters.

The expansion, with three more units of 39 MW, increased the power to 172 MW. It should be noted that the elevation of the dam led to construction of the Cacaria dyke, on the dam in the Prata river and on Dykes 4 and 5. These projects are not shown in Figure 1.7.

Today, Light has 620 MW of power installed in HPPs (Figure 1.13) on the Paraíba do Sul river, between the two big cities of São Paulo and Rio de Janeiro.

Figure 1.13 (a–b) Lajes dam and Fontes New HPP. 100 MW (ub.edu).

After the founding of the electric sector holding company Eletrobras in 1961, the installed capacity increased significantly (Figure 1.14). In 1995, 34 years later, privatization and restructuring of the electric sector started. In 1996, the National Electric Energy Agency (ANEEL) was created, and in 1998, the National System Operator (ONS) came into existence. The restructuring was carried on till the year 2004 when EPE, Energy Research Company under the Ministry of Mines and Energy (MME) started functioning.

Table 1.2 summarizes important events for the Brazilian Electricity Sector, highlighting the foundation of several energy companies.

16 Design of Hydroelectric Power Plants – Step by Step

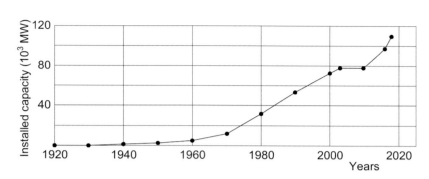

Figure 1.14 Evolution of the installed capacity in Brazil (MW).

Table 1.2 Events of the Brazilian Electricity Sector: 1880–2020

Year	Event
1883	Ribeirão do Inferno, First HPP – 7.46 kW, Diamantina, Minas Gerais
1889	Ribeirão dos Macacos, 372.85 kW, Nova Lima (Minas Gerais)
	Marmelos Zero HPP, 252 kW, Juiz de Fora (Minas Gerais)
1891	Marmelos Zero HPP: + one group of 125 kW; the installed capacity reaches 377 kW
1903	National Congress approves text disciplining the use of electric energy
1905	Founded LIGHT and Power Co. Ltd., Rio de Janeiro
1908	Fontes Velha HPP, 12 MW, Rio de Janeiro; in 1909 was increased to 24 MW
1912	Unification of LIGHT Rio de Janeiro/São Paulo
	Company Paulista of Power and Light (CPFL)
1913	Angiquinho HPP, 1.1 MW, first of Paulo Afonso complex, San Francisco river (Bahia)
1920	Output reached 360 MW
1924	Ilha dos Pombos HPP, 1ª máquina 36.6 MW. Em 1937 output reached 167 MW (Rio de Janeiro)
1930	Output reached 780 MW (541 hydros, 337 thermals and 13 mixed)
1934	Edition of the Water Code
1940	Output reached 1,250 MW
1941	National Steel Company (CSN) – Rio de Janeiro
1942	Vale do Rio Doce Company (CVRD) – Rio de Janeiro
1943	Start year of the creation of several State and Federal Energy Companies
1945	San Francisco Hydroelectric Company (CHESF). Start of Paulo Afonso I HPP
1950	Output reached 1,900 MW
1952	BNDES – National Bank for Economic/Social Development
	CEMIG – Minas Gerais Energy Company
1953	USELPA – Paranapanema River Powerplants
1954	Paulo Afonso I HPP start operation. CELESC – Santa Catarina Powerplants Company
1955	CHERP – Pardo River HPP Company; CELUSA – Urubupungá HPP; CELG – Goiás HPP
1956	ESCELSA – Espírito Santo HPP
1957	FURNAS – Start the projects of P. Colômbia, Marimbondo, Estreito, Volta Grande HPPs in the Grande river
1960	Output reached 4,800 MW. Ministry of Mines and Energy
1962	June, 11, ELETROBRAS – Holding – Brazilian HPP S. A. Company
	CANAMBRA Engineering Company: conducted studies of Brazil hydroelectric potential
1963	FURNAS HPP, 1,216 MW in the 1970s, between São Paulo, Rio de Janeiro and Belo Horizonte

(Continued)

Hydroelectric powerplants   17

Table 1.2 (Continued) Events of the Brazilian Electricity Sector: 1880–2020

Year	Event
1965	DNAEE – National Department of Water and Electric Energy
1966	CESP-São Paulo Energy Company. Start the projects in Paranapanema, Tietê, Grande and Paraná rivers. HPPs: Jupiá (1,550 MW), Ilha Solteira (3,444 MW), Água Vermelha (1,396 MW);
1968	ELETROSUL Company; S. Cruz UTE; ENERAM – Amazon Studies Coordinating Committee
1969	GCOI – Interconnected Operation Coordinator Group
1970	Output reached 11,460 MW
1973	ITAIPU; ELETRONORTE; NUCLEBRÁS – Nuclear Powerplants – extinct in 1989
1978	CEMIG – operation of the 1st machine of São Simão HPP (2,680 MW)
1979	Nationalization of LIGHT. Sobradinho HPP (1,050 MW)
1980	Output reached 31,300 MW
1982	GCPS – System Planning Coordinator Group; CEMIG ends Emborcação HPP (1,192 MW)
1984	Itaipu HPP (14,000 MW) and Tucuruí HPP (8,085 MW)
1985	Angra I Nuclear Power Plant (640 MW)
1990	Output reached 53,000 MW
1995	Privatization Auctions; 1996 was created ANEEL – National Electric Energy Agency
1997	ELETRONUCLEAR-Nuclear Powerplants, replaced NUCLEBRÁS
1998	MAE – Wholesale Energy Market; ONS – National System Operator
2000	Output reaches 72,200 MW. Program of UTEs. ANA – National Water Agency
2001	Crisis, rationing; Lajeado HPP (902 MW)
2002	Canabrava HPP (450 MW) and Machadinho HPP (1,140 MW)
2004	EPE – Energy Research Company
2005	Antas River Energy Complex Company – Ceran (360 MW)
2007	Campos Novos HPP (880 MW) and Barra Grande HPP (708 MW)
2010	Output reached 78,658 MW
2016	Belo Monte HPP – start operation; Installed capacity reaches 96,925 MW
2020	Output reaches ~110,000 MW

Table 1.3 Hydroelectric Potential in the World (EPE, 2016)

China	26.7%
USA	8.2%
Brazil	7.8%
Canada	6.4%
Russia	4.1%
Japan	4.0%
India	3.8%
Norway	2.6%
Turkey	2.1%
France	2.1%
Others	32.2%

Table 1.3 shows that Brazil has the third largest hydroelectric potential in the world. The Brazilian hydroelectric estimated potential is about 250 GW, as it consists of 56% of the Amazon region (Figure 1.15). It is emphasized that there is still more potential to generate energy to the tune of 150 GW. There is much to be done depending on the market growth and on the decision taken by the federal government opting to make

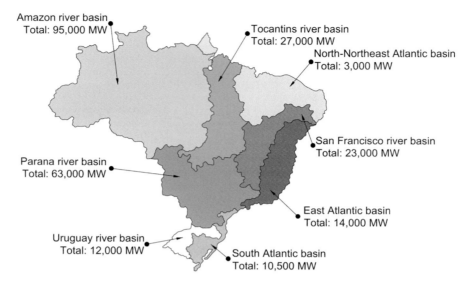

Figure 1.15 Brazilian hydroelectric potential by hydrographic regions (EPE, 2018).

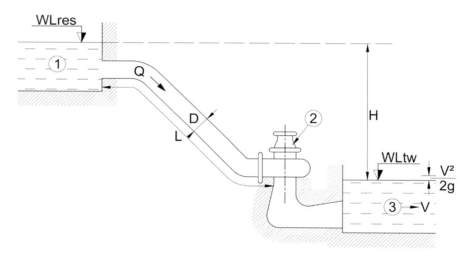

Figure 1.16 HPP typical scheme: (1) reservoir, (2) turbine/generator and (3) tailrace.

hydroelectric power plants, which produce clean, renewable and cheaper energy than energy from other sources.

It must be said that the reservoir of all Brazilian HPPs flood only 37.000 km² of the surface of the country, which has 8.6 million square kilometers. The author is of the opinion that hydroelectric generation is absolutely environmentally sustainable (Figure 1.16).

1.5 ENERGY TRANSFORMATION

The following illustration shows that the typical scheme of installation of a hydroelectric plant is to convert the potential energy into electrical energy. Container 1 represents the reservoir, created by a dam, which feeds the turbine. Installation 2 comprises the power house, housing the turbine generator group. Container 3 represents the tailrace channel, through which the flow is restored to the natural bed of the river.

The gross head H, representing the potential energy per unit weight, is equal to the difference between WLres and the WLtw, neglecting the velocity head $V^2/2g$. The water from the reservoir theoretically has a potential energy compared to the tailwater given by the equation:

$$E = \gamma HV = \rho gVH \tag{1.1}$$

Energy per unit of time is power, P. As V/T is the flow, Q:

$$P = \gamma QH = \rho gQH \tag{1.2}$$

According to the international system MLT (mass, length, time) or the practical system FLT (force, length, time), we must use the water parameters listed in Table 1.4.

With the plant in operation, the hydraulic energy is converted in mechanical energy by the turbine, which in turn is converted into electric energy by the generator.

It is necessary to consider the turbine efficiency, $\mu_T \approx 0.90$, the generator efficiency, $\mu_G \approx 0.98$, and in the set $\mu = 0.88$.

It is necessary to consider also the head losses in the hydraulic circuit, including the tailwater, which can be estimated by Equation (8.6), $H_f = f \dfrac{L}{D} \dfrac{V^2}{2g}$, as discussed in Chapter 8.

These losses are in the order of 2%–5% of the gross head, depending on the characteristics of the plant. So, we have $H_L = H - \Sigma H_f$. (H_f = sum of head losses).

Therefore, the power output is:

$$P = \mu \gamma QH_L = \mu \rho gQH_L \tag{1.3}$$

In international system:

$$P = 1,000 \times 9.8\,\mu \times Q \times H_L \left(\text{kg/m}^3 \cdot \text{m/s}^2 \cdot \text{m}^3/\text{s} \cdot \text{m}\right)$$
$$P = 9,800\,\mu QH_L \left(\text{kg} \cdot \text{m/s}^2 \cdot \text{m/s}\right) \tag{1.4}$$

Table 1.4 International System MLT and Practical System FLT

Parameter	International System	Practical System
χ – Specific weight of water	$9,810\,\text{N/m}^3$	$1,000\,\text{kgf/m}^3$
ρ – Specific mass of water	$1,000\,\text{kg/m}^3$	$102\,\text{kgf s}^2/\text{m}^4$
g – Gravity acceleration	$9.8\,\text{m/s}^2$	$9.8\,\text{m/s}^2$
e – Overall plant efficiency	$\%$	$\%$
V – Reservoir volume	m^3	m^3
H – Gross head	m	m
Q – Nominal flow	m^3/s	m^3/s

20 Design of Hydroelectric Power Plants – Step by Step

In practical system:

$$P = 1,000\,\mu\,QH_L\left(\text{kgf/m}^3 \cdot \text{m}^3/\text{s} \cdot \text{m}\right)$$
$$P = 1,000\,\mu\,QH_L\left(\text{kgf} \cdot \text{m/s}\right)$$

(1.5)

The electrical power is expressed in kilowatts (1 kW = 1,000 W). By definition, 1 W = 1 J/s = 1 Nm/s; 1 kW = 1,000 Nm/s. According to Newton's law, 1 N is the force that accelerates the mass of 1 kg at an acceleration of 1 m/s^2:

$$F = \text{ma}\left(1.0\,N = 1.0\,\text{kgm/s}^2\right)$$

(1.6)

Equation 1.3 can therefore be written:
$P = 9.8\mu QH$ (1,000 Ns/m), or $P = 9.8\mu QH$ (kW)

$$P = 9.8\,\mu QH\left(1,000\text{ Ns/m}\right)\text{ or}$$
$$P = 9.8\,\mu QH\left(\text{kW}\right)$$

(1.7)

Considering the average values of the turbine and generator efficiency, we obtain the following formula:

$$P = 8.6QH\left(\text{kW}\right)$$

(1.8)

This formula is used in the preliminary evaluations of the power available at a given location. It should be noted that from Equations (1.5) and (1.7), we obtain directly the equivalence between the power units in the two systems of units: 1 kW = 102 kgf m/s.

1.6 COMPONENT STRUCTURES OF A HYDROELECTRIC

A typical hydroelectric consists of the following components:

- a dam that closes the river to create the reservoir (forebay);
- hydraulic conveyance facilities, the headrace, headworks, penstock, gates and valves, and tailrace;
- a spillway to discharge the floods, which normally includes the culverts, or bottom outlets;
- the powerhouse structure; the turbine-generator unit, including guide vanes or wicket gates, turbine, draft tube, speed increaser, generator and speed-regulating governor;
- systems of ventilation, fire protection, communications and bearing cooling water equipment;
- substation which includes transformer, switchgear, automatic controls, conduits, and grounding and lightning systems;
- transmission line.

The types of layouts vary depending on the topographical and geological characteristics of each site. They vary if the local is a plain or a "V" valley. It will be discussed in Chapter 3.

1.7 LARGEST HYDROELECTRICS IN THE WORLD

Tables 1.5–1.7 list the largest HPPs in the world, ranked by installed power (MW), gross head (m) and dam height (m).

Table 1.5 Largest Hydroelectrics in the World, Ranked by Power Output – Gulliver and Arndt[a] (1991)

Order	Name	Country	Hb[b] (m)	Output (MW)	Date
1	Three Gorges	China	181	22,400	2012
2	Itaipu	Brazil/Paraguay	196	14,000	2004
3	Belo Monte	Brazil	88	11,233	2019
4	Guri[c]	Venezuela	162	10,600	1968
5	Tucuruí	Brazil	93	8,400	2003
6	Grand Coulee	USA	168	7,460	1942
7	Sayano-Shushensk	Russia	245	6,400	1980
8	Krasnoyarsk	Russia	124	6,000	1968
9	La Grande 2	Canada	168	5,328	1979
10	Churchill Falls	Canada	32	5,225	1971
11	Bratsk	Russia	125	4,500	1961
12	Ust-Ilim	Russia	102	3,840	1977
13	Ilha Solteira	Brazil	74	3,200	1973
14	Brumley Gap	USA	78	3,200	1973
15	Xingó	Brazil	140	3,012	1987
16	Bennett, W.A.C.	Canada	183	2,730	1968
17	Mica	Canada	242	2,660	1976
18	São Simão	Brazil	120	2,680	1979
19	Volvograd	Russia	47	2,563	1958
20	Paulo Afonso IV	Brazil	33	2,460	1979
21	Cahora Bassa[b]	Mozambique	171	2,425	1975
22	Chicoasén	Mexico	261	2,400	1980
23	Gezhouba	China	47	2,340	1975
24	La Grande 3	Canada	93	2,304	1982
25	Volga-V.I. Lenin	Russia	45	2,300	1955
26	Iron Gates	Romania/Serbia	60	2,300	1970

[a] Updated.
[b] Dam height above the foundation.
[c] Additional capacity planned or in construction.

Table 1.6 Largest Hydro Powerplants in the World, Ranked by Gross Head

Order	Name	Country	Hb (m)	Output (MW)	Date
1	Bieudron[a]	Switzerland	1,883	1,269	2010
2	Reisseck-Kreuseck	Austria	1,772	24[b]	1953
3	Chandoline[a/c]	Switzerland	1,748	120[c]	1957
4	Portillon	France	1,420	45	-
5	Roselend	France	1,203	475	-
6	Nendaz[a]	Switzerland	1,008	390	1964
7	Fionnay[a]	Switzerland	874	290	1964
8	Mont-Cenis	France	843	350	1968
9	Danhim	Vietnam	800	160	-
10	Malgovert	France	750	290	-

[a] These plants are supplied by the reservoir of La Grande Dixence, 284 m height, cited in Chapter 6.
[b] Power output until 1953. The total expected output is 112 MW.
[c] Plant out of service since July 2013.

Design of Hydroelectric Power Plants – Step by Step

Table 1.7 Largest Hydro Powerplants in the World, Ranked by Dam Height[a]

No.	Name	Country	Type	H (m)	Date
1	Rogun	Tajikistan	Earth	335	2018
2	Jinping-1	China	Arch	305	2013
3	Nurek	Tajikistan	Earth	300	1980
4	Xiaowan	China	Arch	292	2010
5	Xiluodu	China	Arch	285.5	2013
6	Grand Dixence	Switzerland	Gravity	285	1962
7	Inguri	Georgia	Arch	272	1987
8	Chicoasen	Mexico	Rockfill	265	1981
9	Vaiont (fora de uso)	Italy	Arch	262	1961
10	Nuozhadu	China	Earth	261.5	2012
11	Tehri	India	Rockfill	261	1990
12	Kinshau	India	Earth-Rockfill	253	1985
13	Laxiwa	China	Arch	250	2009
14	Gilgel Gibe III	Ethiopia	CCR	250	2015
15	Deriner	Turkey	Double Arch	249	2012
16	Alberto Lleras (Guavio)	Colombia	Rockfill	243	1989
17	Mica	Canada	Earth	242	1972
18	Sayano-Sushensk	Russia	Gravity Arch	242	1980
19	Mihoesti	Romania	Earth	242	1983
20	Ertan	China	Gravity Arch	240	1999
21	Changheba	China	Rockfill Conc Face	240	2016
22	Mauvoisin	Switzerland	Arch	237	1957
23	Chivor (La Esmeralda)	Colombia	Rockfill	237	1975
24	Oroville	USA (Calif.)	Earth	236	1968
25	Chirkey	Russia	Arch	233	1978
26	Shuibuya	China	Rockfill Conc Face	233	2008
27	Goupitan	China	Double Arch	232.5	2009
28	Karun-4	Iran	Gravity Arch	230	2010
29	Bhakra (Gob. Sagar)	India	Gravity	226	1963
30	El Cajon	Honduras	Double Arch	226	1985
31	Hoover	USA (A./Nevada)	Arch Gravity	221	1936
32	Jiangpinghe	China	Rockfill Conce Face	221	2012
33	Contra	Switzerland	Arch	220	1965
34	Dabaklamm	Austria	Arch	220	1989
35	La Yesca	Mexico	Rockfill Conc Face	220	2012
36	Piva (Mratinje)	Montenegro	Arch Gravity	220	1976
37	Dworshak	USA (Idaho)	Gravity	219	1972
38	Glen Canyon	USA (Arizona)	Arch Gravity	216	1964
39	Toktogul	Kyrgyzstan	Arch	215	1974
40	Daniel Johnson (N. 5)	Canada	Multiple Archs	214	1968
41	San Roque	Philippines	Earth	210	2003
42	Ermenek	Turkey	Double Arch	210	2009
43	Luzzone	Switzerland	Arch	208	1963
44	Keban	Turkey	Rockfill Gravity	207	1974
45	Irapé	Brazil	Rockfill Conc Face	205	2005
46	Bakun	Malaysia	Rockfill Conc Face	205	2011
47	Karun-3	Iran	Arch Gravity	205	2005
48	Zimapán	Mexico	Arch Gravity	205	1993
49	Dez	Iran	Arch	203	1963
50	Almendra	Spain	Arch	202	1970
51	Campos Novos	Brazil	EFC	202	2006
52	Berke	Turkey	Arch Gravity	201	2001
53	K'olnbrein	Austria	Double Arch	200	1977

(Continued)

Table 1.7 (Continued) Largest Hydro Powerplants in the World, Ranked by Dam Height[a]

No.	Name	Country	Type	H (m)	Date
54	Shahid Abbaspour (Karun I)	Iran	Arch	200	1976
55	Altinkaya	Turkey	Rockfill	195	1988
56	New Bullards Bar	USA (Calif.)	Arch Gravity	194	1968
57	Lakwar	India	Gravity	192	1985
58	New Melones	USA (Calif.)	Rockfill	191	1979
59	Itaipu	Brazil	Earth/Rock/Gravity	190	1983
60	Aguamilpa	Mexico	Rockfill Conc Face	187	1993
61	Kurobe No. 4	Japan	Arch	186	1964
62	Swift	USA (Wash.)	Rockfill	186	1958
63	Kurogegawa N. 4	Japan	Arch	186	1964
64	Mossyrock	USA (Wash.)	Arch	185	1968
65	Oymopinar	Turkey	Arch	185	1983
66	Barra Grande	Brazil	Rockfill Conc Face	185	2005
67	Atatürk	Turkey	Rockfill	184	1990
68	Shasta	USA (Calif.)	Grav-Arch	183	1945
69	Bennett, W.A.C.	Canada	Earth	183	1967
70	Amir Kabir	Iran	Arch	180	1964
71	Dartmouth	Australia	Rockfill	180	1979
72	Emosson	Switzerland	Arch	180	1974
73	Tehchi	Taiwan	Arch	180	1974
74	Tignes	France	Arch	180	1952
75	Takase	Japan	Rockfill	176	1978
76	Ayvacik	Turkey	Rockfill	175	1981
77	Lijiaxia	China	Gravity	175	1996
78	Revelstoke	Canada	Gravity/Rockfill	175	1983
79	Alpe-Gera	Italy	Arch Gravity	174	1964
80	Don Pedro	USA	Earth	173	1971
81	Karakaya	Turkey	Arch	173	1987
82	Hungry Horse	USA	Arch	172	1953
83	Longyangxia	China	Arch Gravity	172	1992
84	Thissavros	Greece	Earth	172	1996
85	Cahora Bassa	Mozambique	Arch	171	1974
86	Daniel Palacios	Ecuador	Arch	170	1982
87	Charvak	Russia	Rockfill	168	1977
88	Gura Apelor	Romania	Rockfill	168	1986
89	La Grande 2 Barrage	Canada	Rockfill	168	1978
90	Grand Coulee	USA	Arch Gravity	168	1942

[a] *Sources:* Jansen (1983); WP&DC, nov/1979; ICOLD World Register of Dams.

Chapter 2

Planning hydropower generation

2.1 CATCHMENT AREAS AND MULTIPLE USES OF WATER

There are 12 catchments areas in Brazil, as shown in Figure 2.1 (Resolution 32, 10/15/2003, CNRH – National Water Resources Council). In seven of them, the name of its main rivers prevails: Amazonas, Tocantins-Araguaia, São Francisco, Parnaíba, Paraná, Paraguai e Uruguai. The others are grouping of several watercourses. There's no main river as an axis and therefore are called gouped basins, namely: South Atlantic Basin, Southeast Atlantic Basin, East Atlantic Basin, Eastern Northeast Atlantic Basin and Western Northeast Atlantic Basin.

Brazil has one of the most extensive, dense, diversified and extensive river networks in the world. Approximately 13% of all fresh water on the plane is in Brazilian territory. After China and Russia, Brazil has the third largest hydroelectric potential in the world. Catchment areas of Amazonas and Paraguay occupy predominantly extensive plains. Catchment areas of the Paraná and São Francisco rivers are typically plateau. The Urubupungá (Paraná river), Pirapora, Sobradinho, Itaparica and Paulo Afonso (São Francisco river) deserve special mention among the natural falls that were used for the implementations of developments. The Iguaçu falls (Figure 2.2) are among those preserved by its scenic, environmental and touristic value.

The Brazilian rivers present fluvial regime, that is, are supplied by rainwater. The average annual precipitation in South America is 1,600 mm (i.e. twice the other continents). In Brazil, it is 1,800 mm.

Due to the predominance of the tropical climate, in most of the Brazilian territory, floods occur during the summer (December–March), except for some northeastern rivers that flood during autumn and winter seasons (April–July). Since the Amazonian rivers have varying temporary contributions (precipitations and melting of the Andes), they attenuate ebb (the flow back from the flood stage). Due to the possibility of intense precipitations during the year (cold fronts combined with centers of low atmospheric pressure), the Southern rivers have low-precision ebbs in the time of lower intensity. With the exception of some rivers in the semi-arid region of the Northeast, most of the rivers are perennial.

As per Chapter 4, according to the Hydropower Inventory Studies, the first phase is to mandatorily evaluate all other possible uses when considering the use of the water resources of a basin for the purpose of generating electricity.

The main uses of water are classified in: use consumptives: urban and rural supply, animal husbandry, irrigation and industrial use; non-consumptive uses: power

DOI: 10.1201/9781003161325-2

26 Design of Hydroelectric Power Plants – Step by Step

Figure 2.1 Catchment areas. (Adapted of IBGE 2013.)

generation, navigation, flood control, tourism and leisure, aquaculture and fisheries, and ecosystem maintenance. To obtain data on water use, extensive and diligent research is required in the governmental agencies listed below:

- Energy Research Company – Federal Government (EPE);
- National Energy Electric Agency – Federal Government (ANEEL);
- National Water Agency – Federal Government (ANA);
- Brazilian Power Plants – Holding and State Energy Companies (ELETROBRAS);
- Brazilian Institute of Environment and Renewable Resources (IBAMA);
- National Council of the Environment – Federal Government (CONAMA);
- Mineral Resources Company – Federal Government (CPRM);

Planning hydropower generation 27

Figure 2.2 Iguaçu heads (Clebersander Rech 2007).

As seen in Chapter 1, the power output in a hydroelectric plant is estimated by Equation 1.8:

$$P = 8.6 Q H \tag{1.8}$$

where Q = flow (m³/s) and H = net head across the turbine (m), defined in Figure 1.17.

For the Inventory Studies, it is indispensable that the cartographic and hydrological studies are realized with level of detail that gives them the quality and the necessary precision.

Hydrological studies provide the floods for the design of permanent and temporary structures, along with the series of monthly average flows, used in the determination of power and in simulations of energetic studies.

Cartographical surveys must be carried out at a scale that allows for each alternative of division of the available head in the basin: identification and evaluation of all the interferences and impacts of the plants with the infrastructure and the environment; determination of curves elevation x area x volume of the reservoirs; and the determination of the total head of each plant.

From the environmental point of view, these studies of alternatives of division of the available head in the basin are complex. So, they need to be performed by an experienced team in such a manner that the associated measures and programs promote sustainable development with minimal damage. In this way, the problems in obtainingzalso get minimized. It is worth noting that the time spent on collecting quality data is long and involves extensive filed campaigns.

The author is of the opinion that the Integrated Environmental Assessment (AIA) should be carried out concomitantly with the Inventory Studies, though it has not happened systematically. In recent years, all works have been delayed in schedule, especially in the Amazon region, due to the problems involved in obtaining licenses.

Though a strong national and international political action is going on against hydroelectric dams and their reservoirs, in the author's opinion, it doesn't make much sense to be so critical about hydroelectric projects, as hydroelectricity is a renewable source of energy, has lesser of an adverse environmental impact than the technologies involved in electric energy generation, and is also much cheaper than the rest. As the work began, the owner companies have faced various contestation actions from the communities involved (indigenous nations, quilombolas, various ONGs, etc.). As a consequence, the works have often been shut down by the Public Prosecution. This confusing and complicated situation has not been adequately resolved by the involved parties, government, entrepreneurs and affected communities, and this may have negative repercussions on schedule of generation expansion program required to meet the growing demand.

2.2 GENERATION EXPANSION PLANNING

According to methodology of the Eletrobras Manual (2007), the planning of expansion of the generation of energy has been made by EPE, from the Inventory Studies. Figure 2.3 shows the institutional structure of electric sector in Brazil.

As mentioned in Chapter 1, after creation of Eletrobras in 1961 the implementation of hydroelectric plants intensified. The total installed capacity is 104,000 MW (2020 10 Year Plan). Brazil has an estimated potential of 249 GW, out of which 157 GW

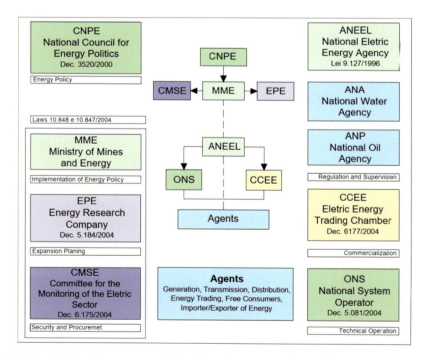

Figure 2.3 Institutional structure of the electric sector – Brazil (ONS 2013).

can be generated from Amazon and Paraná river basins. There are several projects to be implemented in the medium and long terms.

Detailed information in generation planning in Brazil can be found on the ONS-EPE website. The average annual growth is of the order or 4.6% which means 3,200 MWmed (2011–2020). It is important that the country keeps its energy matrix fundamentally hydroelectric because it is clean energy, renewable and cheaper than the energy from other sources.

2.3 PHASES OF STUDIES

Figure 2.4 shows a classic simplified matrix of field surveys and basic studies. Regardless of the type of Project (CGH, SHP, PCH<30 MW, or large plant, HPP>30 MW), the three phases of study are Inventory Studies, Feasibility Studies (EVE), Basic Project (FEED) and Executive Project. The level of studies deepens with the advancement of phases.

According to Aneel/Eletrobras standards (1997, 1999, 2000), the execution of these activities is essential to ensure quality of projects and their approval. The scope of activities, by disciplines and by stage, is listed in these references. The studies and projects of an HPP are performed according to the guidelines, standards and criteria already mentioned, and obeying the current legislation, including environmental legislation. These studies are carried out in several stages, in an orderly sequence, according to the prevailing practices in the respective countries. In Brazil, it is not possible to carry out engineering and environmental studies at their ideal times.

- Inventory Studies (EIH): these are the first studies that characterize the potential and identify the plants to be implemented in future. Depending on the power to be installed, these plants can be micro (MCH), small (PCH), and medium to large (HPP). The AAI, carried out since 2006, should be made concurrently with the Inventory Studies (see Section 2.3.1).
- Basic Design of a Micro Power Plant (MCH): $P < 1.0$ MW (Resolution 393, 4/12/1998 SGH/ANEEL). This project shall be treated as a PCH, and is exempt from concession, permission or authorization. Status of this project is only to be informed to ANEEL (Law 9,074/1995, art. 8°). They are not, however, exempted from environmental licenses, which must be obtained from environmental agencies, responsible for issuing respective terms of reference of the studies to be carried out.
- Basic Design of a PCH, 1.0 MW $< P \leq$ 30.0 MW (Resolution 652, 9/12/2003, da SGH/ANEEL). See also Resolutions 395/1998 e 343/1998. These resolutions are to be implemented only after necessary approval from the Basic Project (FEED) of the PCH by SGH and the grant issued by SCG for a period of 30 years (renewable). Depending on each case and decisions of Environmental Agency licensing body, it is necessary to have Environmental Impact Studies (EIA) and Environmental Impact Studies (RIMA) to obtain the Prior License (LP) which define the enterprise as viable. The Installation License (LI) of the Works must be obtained for its implementation, along with the approval of RDPA (Environmental Programs Detail Report) in the case of simplified licensing (associated with RAS) or approval of the PBA (Environmental Basic Project associated with EIA/RIMA). After 2009, Law 11943/09 established that projects with an output of more than 30 MW and less than 50 MW with an aim to independent production or self-production may be

30 Design of Hydroelectric Power Plants – Step by Step

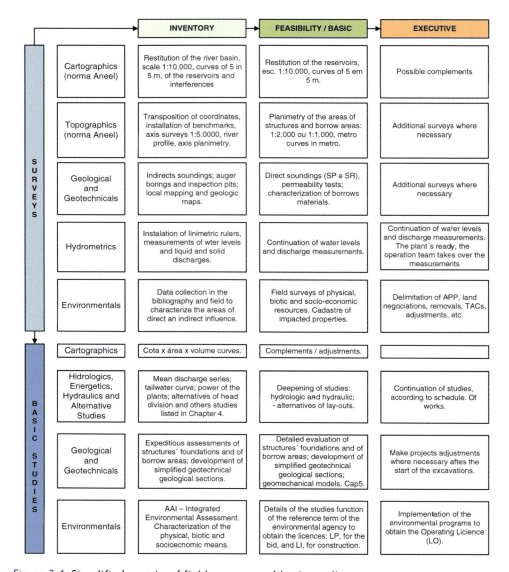

Figure 2.4 Simplified matrix of field surveys and basic studies.

granted through ANEEL's authorization, without any unnecessary dispute about the concessions in bids. A Consolidated Basic Project (FEED) that normally contains some revision of the project previously also needs to be filed with the agency after obtaining approval of the Basic Project (FEED).

- Studies of Technical, Economic and Environment Feasibility of Power Plants with a power greater than 30 MW which, after approval by SGH/ANEEL, are included in the bidding program for granting the concession (Power Auctions). Only those

ventures whose Feasibility Studies have been approved and the ones with respective environmental licenses (LPs) can enter into the auctions.
- Basic Design of the HPP: presents an optimization of the Studies of Technical, Economic and Environment Feasibility of Power Plants and must have the approval of ANEEL.
- Environmental Studies of the HPP: The Hydroelectric Inventory Study phase of the basin should be carried out jointly with the AAI. Both of these are the responsibility of the EPE, and the alternatives should be discussed broadly in terms of environmental issues. For each HPP selected for feasibility analysis, the Environmental Impact Studies (EIA) and the Environmental Impacts Report (RIMA) should be elaborated in detail, in an objective language, easily accessible to the interested parties. The EIA/RIMA must be approved by environmental licensing department (Ibama or state departments), and by other departments that have socio-environmental issues associated with the HPP (Funai, DNPM, FCP, IPHAN, Ministry of Health – if the plant is in the Legal Amazon). The approval of EIA/RIMA, after the Public Hearing, results in the edition of the Previous License (LP) confirming the environmental feasibility of the plant. The next stage is the elaboration of the Basic Environmental Project (PBA), detailing the specified environmental programs, whose approval will give rise to the Installation License (LI), which will authorize the beginning of the works. The implementation of the recommended measures in the PBA will allow obtaining the Operating License. When analyzing the impacts of the plant, if the environmental agency admits a simplified licensing, then the studies will be less detailed, replacing the EIA/RIMA by RAS and the PBA by RDPA.
- Engineering executive projects, carried out concurrently with the implementation of plants, deal with the phases of river diversion, excavation and treatment of the foundations. They also deal with detailed projects of all the definitive structures of the enterprise.

Figure 2.5 illustrates different phases of studies, the concession of which is obtained through auction, and includes the estimated average time to elaborate each study phase. The process for the implementation of an HPP in Brazil is extensive, complex and bureaucratic.

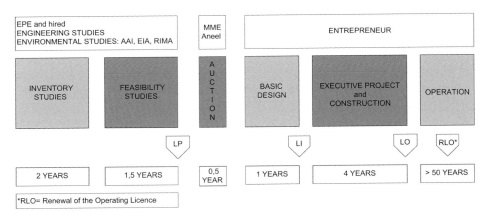

Figure 2.5 HPP – Phases of study for the auction (2014).

2.3.1 Inventory hydroelectric studies

The hydroelectric potential of a hydrographic basin corresponds to the estimated total that can be technically, economically and environmentally exploited considering the configuration of the electric reference system. It contemplates the plants in operation in the basin, under construction and planned, and takes into account multiple water use scenarios. This potential is determined in the Inventory Studies (IHE). Before the beginning of these studies, the planning is elaborated in the office using the cartographic base of IBGE.

The Inventory Studies are the basis for planning the expansion of the country's hydroelectric generation, and **identify the plants that will be implanted in the basin in the medium and long term**. They are developed in two phases. First is the, Preliminary Studies, where the alternatives of division of the head are investigated. The second phase is called Final Studies, consisting of details of two or three alternatives chosen as more attractive.

The identified plants are classified as follows (according ANEEL criteria):

- MCHs: $P < 1.0$ MW, according Resolution n° 395, 12/4/1998;
- PCHs: $1.0 < P < 30.0$ MW, according Resolution n° 652, 12/9/2003;
- HPPs: $P > 30$ MW.

The studies must be developed with base in topographic, batimetric, hydrometric, geologic, geotechnical and environmental data, collected in the river basin. It is important to observe the official guidelines for the elaboration of mapping services, topography and georeferencing of maps, drawings and electronic files.

The hydrological data must be that of the fluviometric stations of the basin, belonging to the ANA hydrometric network, whose consistency must be verified. Secondary basin data collected in the literature should also be used.

All the following basic studies should be carried out as recommended in the Inventory Manual whose compliance is carefully verified by the government:

- Cartographic, Topographic, Topobatimetrics and Hydrometric Studies;
- Hydrological and Hydroenergetics Studies;
- Geological and Geotechnics Studies;
- Environmental Studies.

To define the alternatives for dividing the head, one must use the existing cartographic plans (scales 1:50,000, if available, or 1:100,000, or even 1:250,000). Satellite surveys have been used, along with ArcGIS support, that allow the development of a preliminary assessment of the basin.

In planning the development of studies, it is possible today to fly over the basin without leaving the office and identify the possible axes of dams. If the cartographic base does not exist it must be surveyed to enable the studies to be carried out. A quality cartographic base is fundamental for the precision of the identification of future projects, since it conditions the height of the dams and, therefore, the gross head (H) of each plant.

It is noted that the alternatives for dividing the head are defined considering the minimization of the impacts on population and the existing infrastructure (towns, highways, bridges, etc.).

The layouts of the projects and the structures design, the evaluation of the impacts, the estimation of costs of each one, the comparison and selection of the alternatives, must be done as precisely as possible to reduce the risk of errors in the optimum use of hydroelectric potential in the final phase of the Inventory Studies.

In the final studies, a detailed cartographic base must be used, with scales 1:5,000 or 1:10,000. The alternatives are detailed and compared for choosing the most attractive head partition. The costs of interconnecting the power plants to the region's transmission system is also evaluated. The alternative that has the lowest cost-benefit ratio will be the one chosen to divide the basin head.

The water levels of the plants of the selected alternative can't be changed in the later phases of studies, unless authorized by the government in special cases.

As illustration, the selected alternative "D" for the division of the head of the Teles Pires river, Mato Grosso, is presented in Table 2.1 and Figure 2.6.

Table 2.1 Selected Alternative for the Head Division of the Teles Pires River, MT-Brazil (2005)

PLANT	Gross Head (m)	Inundated Surface (km^2)	Power (MW)
TPR 1230	17.0	60.0	53
TPR 775	31.5	329.6	461
TPR 680	23.8	123.3	342
TPR 329	59.0	123.4	1,820
API-006[a]	44.8	59.5	275
TPR 287	24.4	53.0	746

[a] Plant in the Apiacás river.

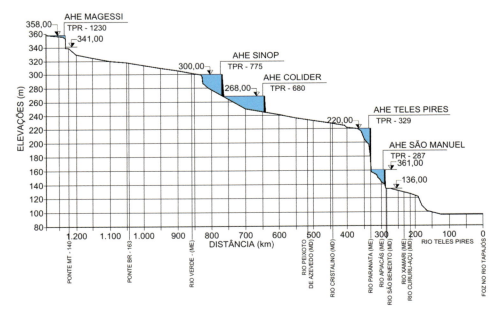

Figure 2.6 Profile of the river Teles Pires (ANEEL 2005).

2.3.2 Integrated environmental assessment

The implementation of development models on a sustainable basis has required integrative approaches to the management of environmental resources that allow to evaluate the cumulative and synergistic impacts of the interventions in a given area, aiming at understanding the interactions and the dynamics of the relevant processes that define or constitute the environment.

In the ambit of environmental licensing of hydroelectric power plants, since 2003, IBAMA has started to demand that studies of environmental impacts be reported to the river basin in accordance with Resolution Number 001 (CONAMA, 1986).

In 2006, the Ministry of the Environment (MMA) issued the document AAI, "Integrated Environmental Assessment" (EPE, 2006), which became the new instrument for planning and developing knowledge for environmental studies and management. It is a comprehensive document that is available on the internet and mandatory to use in the Brazilian electrical sector. The methodological structure of this AAI will not be covered in this book, but the steps are as follows:

- Characterization of the main ecosystems of the hydrograph basin;
- Characterization of the impacts of the plants according to their spatiality, that is, how they are distributed and affect local populations and the environment;
- Identification of the conflicts in the basin;
- Application of the information integration techniques;
- Recommendations for the products to be developed due to the uncertainties arising out of lack of knowledge.

Public participation, although it is one of the accompanying components for the development of studies, should not be seen as a step, but as a guarantee of a principle of transparency and of the effective contribution of society throughout the process, from the initial characterization to associated decisions.

2.3.3 Basic project of mini plants

The activities for the development of the Basic Project (FEED) of an MCH, Mini Hydroelectric Plant, are the same as those developed for the Basic Project of a PCH (see Eletrobras Guidelines). These projects are done together simultaneously.

2.3.4 Basic project of small plants

The Basic Project (FEED) must be done according the Eletrobras/ANEEL Guidelines of PCH, Small Hydroelectrics Plants (2000). These guidelines specify to perform field surveys (topography, geology and geotechnics, hydrometry and environmental), basic studies, alternative studies, selection of alternative and detailing, bill of quantities, standard Eletrobras budget, schedule of construction and evaluation of the economic attractiveness of the plant, as well as the design of the system of transmission and interconnection of the plant.

2.3.5 Feasibility studies

The Feasibility Studies must be done according the Eletrobras/ANEEL Guidelines (1997), which specify the same survey roadmap for surveys and studies planned for the projects of MCHs and PCHs.

Some chapters may be more detailed, such as flow, energy and sediment studies, especially if the plant is in the basin of sedimentary rocks. Each case should be analyzed in detail and considered in terms of environmental issues to support decisions.

2.3.6 Environmental impact studies

The Environmental Impact Studies (EIAs) are carried out by a multidisciplinary team according to IBAMA's instructions, and the instructions provided by the state environmental agencies, always obeying the CONAMA Resolutions.

The terms of reference to guide the preparation of the EIA and RIMA, Environmental Impact Report, as required by the legislation are issued, on the basis of the data and information about the basin, including the Inventory Studies and the Integrated Environmental Assessment, if available. For the case of MCHs and PCHs with less than 10 MW, these studies can be simplified (RAS).

Brazil has adopted three-phase environmental license, that is, it depends on the approval in three phases to complete the project:

- LP – Preliminary License: normally granted for a period of 5 years, based on EIA/RIMA; approves the site and the design of the plant, certifies the environmental feasibility and establishes the basic requirements and conditions to be met in the PBA, Basic Environmental Project (PBA) and, preliminarily, during the future phases of construction and operation;
- LI –Installation License: authorizes the installation of the works according to the specifications of the approved plans, programs and projects, components of the PBA, including environmental control measures and other conditions;
- LO – Operating License: authorizes the operation of the plant after verifying the fulfillment of requirements of the previous licenses (LP e LI) and the construction without harming the environment.

The Feasibility Studies and the EIA/RIMA are documents that also subside obtaining the water availability reserve declaration (DRDH) for the enterprise issued by the water authority (ANA).

2.3.7 Consolidated basic engineering project

The detailed or executive project is followed by consolidated basic engineering project (PBC), prepared to record some revision of the project after approval of ANEEL.

This project, with a more detailed bill of quantities and more precise budget, subsidizes the contracting of wall te works, including the reservoir. In this phase, a detailed evaluation of the risks involved takes place, such as topographic, geological and environmental risks, which may imply errors in the bill of quantities.

36 Design of Hydroelectric Power Plants – Step by Step

This is an important point because it influences the costs of contingencies normally included in the budget. In spite of the criteria of each investor, the basic Eletrobras/ANEEL Hydroelectric Power Plant Design Guidelines (1999) must be observed.

2.3.8 Environmental basic project

The environmental basic project of each plant is elaborated so that all the environmental programs characterized in the EIA, or those that are added by environmental agencies in the LP, are detailed.

These programs should be implemented during the executive project throughout the works of the plant (HPP). In the case of smaller plants, such as PCHs or MCHs, a more simplified document should be elaborated to obtain LI.

During the construction, the environmental agency accompanies the implementation of the programs to finally issue the operating license (LO).

2.3.9 Detailed project

The detailed project of all permanent structures is developed, concomitantly with the works of construction in the job site.

The environmental programs are adjusted according to the dates of progress of the construction works in the job site.

Calculation memories of the phases of river diversion and of all the definitive structures are made, as well as detailed drawings for the construction, such as:

- location drawings of the civil works;
- river diversion phase drawings;
- excavation and treatment drawings of the foundations of the permanent structures;
- shape and frame drawings of concrete structures, including embedded parts;
- drawings of all installations and drawings of the cross-sections of the earthworks and rockfill; and
- finishing and architectural drawings of the structures, including permanent access routes.

It should be emphasized that the drawings "as built" are elaborated concomitantly with the development of the construction.

2.4 BUDGET AND EVALUATION OF PLANT'S ATTRACTIVENESS

An HPP budget is usually attached with a large project, and so is not an easy task. Estimating the cost of foundations and respective treatments of soil and rock embankments, of concrete structures, of electromechanical equipment including manufacturing, transportations and assembly, the costs of the substation and the transmission line, is always a heavy activity. Any estimation error in any of these items may compromise the project's viability. In Brazil budgeting has two phases:

- at the time of state enterprises; and
- after the privatization in 1993 of the Electrical Sector.

As described below, it was mandatory for the state-owned companies to use the OPE-Eletrobras standard budget. After privatization each company started using its own budget, which are usual and known to everyone who works with these ventures. As expected, with privatization, the prices of the energy, in US$/MWh, experienced a significant reduction.

2.4.1 Standard budget

When the country started its development, the Eletrobras was founded in 1962. After Eletrobras, the evolution of installed capacity in Brazil increased significantly (Chapter 1, Figure 1.15). Since that time, it has been implemented on a standard budget that had the merit of creating a culture of budgeting a large project. This budget has the following listed accounts.

- Account 10: is the environmental costs account. It includes land, relocations and other social and environmental actions. In an EPC Contract, the risk of this account is very high;
- Account 11: includes structures and other improvements. It also includes powerhouse and the operator's village;
- Account 12: is dams and hydraulic conveyance facilities account. It includes dams (earth, rockfill and concrete), the spillway, the headrace canal, the intake, the tunnels and penstocks, surge tank, tailrace channel, as well special constructions (lock, for example); included also the river diversion
- Account 13: includes main equipment – turbines and generators;
- Account 14: includes electrical equipment accessories;
- Account 15: includes other various equipment of the plant;
- Account 16: includes highways, railways, airport, bridges;
- Account 17: includes indirect costs: such as the construction of the job site and the owner's administration.

2.4.2 Budgets after privatization

After the privatization in 1993, this type of budget has fallen into disuse. At that time, the energy auctions started and to participate in the auctions each company has its own budgeting and risk assessment methodology, including insurance. It was at that time that EPC Contracts, Engineering, Procurement and Construction appeared.

It should be noted that account ten is considered as environmental cost, reaching a high value in relation to other accounts. The cost of the reservoir, for example, is generally high and must be expended regardless of the mitigating or compensatory activities of the impacts. This account also includes land. Account ten sometimes reaches about 20% of the total cost of the enterprise. All engineering accounts have their costs estimated in the traditional way: volumes of services x unit prices. In the standard budgets all accounts have a forecast for eventual costs. In Brazil, the cost of

38 Design of Hydroelectric Power Plants – Step by Step

interconnecting the plant to the national system (SIN), substations and transmission lines, must be considered.

For example, as for the cost, base reference should be made, among others, to the report prepared by Georgian (Gross Energy Group) and Norconsult AS consultants in cooperation with the Ministry of Energy of Georgia and NVE, Norwegian Water Resources and Energy Directorate, in 2016, that contains relevant cost data for hydropower development in Georgia: "Cost Base for Hydropower Plants in Georgia > 13 MW". The manual may serve as a useful guide to potential investors in renewable energy anywhere. It may also serve as a tool for the government for assessment and prioritization of the hydropower development. This document is also available on the internet.

2.4.3 Assessment of plant's attractiveness

The technical and economic evaluation of the HPP is carried out using the following expression. It is a classical method and gives a good estimate of the energy cost.

$$CE = \frac{(Ci \times FRC + O \& M \times EG)}{EF \times 8.760} \tag{2.1}$$

CE cost of energy generated (US$/MWh);
Ci value of total investment, including interest rate during construction;
FRC capital recovery factor; $FRC = = 0.12414$ for $n = 30$ years and $i = 12\%$ per year;
$O\&M$ annual cost of operation and maintenance (US$/kWh);
EG annual generated energy (MWh);
EF firm energy/guaranteed energy during the critical period (MW med);
8.760 number of hours of the year.

The attractiveness of the plant is evaluated by comparing the cost of energy it generates with the costs of the alternative energy sources such as nuclear, solar, wind, petroleum, coal, wood chips and sugarcane bagasse.

Chapter 3

Types of power plants and layouts

3.1 INTRODUCTION

The characteristics of the catchment area and that of the environment, condition the development of the studies and projects to make use of the available water resources. The government did the first phase of studies, called the Inventory Hydroelectric Studies (EIH), that defined not only how many plants can be implemented in the basin, but also their operational WLs, which will characterize the types of each plant. They addressed multiple uses of water, which could be considered as a part of first Integrated Environmental Assessment (AAI) of the basin. It should be noted that this evaluation was also made in the Studies of Canambra Engineering Consultants in the 1960s, a consortium formed by Canadian and American specialists who were hired by Eletrobras to perform the studies.

These studies usually determine that reservoirs should address multiple uses of water, such as flood control, urban, rural and industrial supply (consumptive uses), irrigation, navigation, tourism and leisure, aquaculture and fisheries, animal husbandry and ecosystems (non-consumptive uses). Uses for irrigation and navigation may conflict with the use for generation due to the priorities defined for rational use of the basin's resources. Therefore, all uses must be well evaluated.

3.2 TYPES OF POWER PLANTS

3.2.1 Function of the type of operation

Depending on the type of operation, the hydroelectric power plants are characterized with or without regularization reservoir. Plants without regularization reservoir are those whose operating water level (WL) does not vary (or varies very little), and are denominated by "run of river". These include daily or weekly regularization plants that provide peak power at peak periods. WLs vary greatly in the periods of flood and drought in the plants with accumulation/regularization reservoir. The levels that characterize the useful and standby volumes of the reservoir are shown in Figure 3.1. There are also reversible plants, with more than one reservoir. Water is pumped from these plants to the higher reservoir during the hours of lower energy demand of the system, so that it can be used later and generation can take place during peak hours (Plant of Pedreira, at the Billings dam). This type of plant is not in the scope of this book.

DOI: 10.1201/9781003161325-3

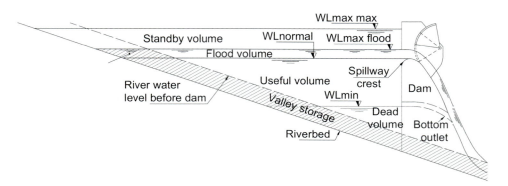

Figure 3.1 Reservoir. Characteristic levels.

Due to environmental constraints, since the 1990s, plants with regularization reservoirs are less in use, and instead, larger reservoirs are used more. At first, reservoirs surely has an impact on ecological balance. But, as said earlier, there is much evidence of success in adapting to the new, and better, ecological environment. The reservoirs are indispensable, not only for generation of energy, but also for the management of the risks associated with extreme hydrological events, and for the storage of water and energy (Gomide, 2012).

The reservoirs are classified according to the duration of filling. In this respect, we distinguish them under three categories:

- daily or weekly regulating reservoirs, with volumes of the same order of magnitude as the volumes of affluent flows (daily or weekly flows);
- seasonal reservoirs, whose volumes are of the order of magnitude of the contributions of the rainy season. These reservoirs, in general, are located in the superior course of the rivers, and allow to store water during this season for use in the period of heavy consumption;
- interannual reservoirs, whose volumes are greater than the volume of annual contributions; they allow to store water during a wet year to use in a dry year.

The choice of reservoir volume is a function of the characteristics of the site of use and is subject to environmental constraints and existing activities in the region (agglomerations, infrastructure, existing uses, etc.). This question is not part of the scope of this book.

3.2.2 Function of type of use

Hydroelectric plants are characterized by the type of "base" and "peak" use. A base plant is one in which the energy generated by it is used to continuously supply the load demanded by the System (SIN). HPPs are by nature basic plants. A peak plant is designed and built especially, to meet the peak times. It uses the reservoir to provide the necessary flow rates for generation of energy regardless of the tributaries. Energy from these plants is available at almost all the time and is referred to as "firm energy" (available energy 95% of the time). However, the energy demand varies during the day, above or below the base energy. These load peaks can usually be served by power plants with regulation reservoirs, for greater flexibility in supplying consumption variations.

3.2.3 Function of the head

Depending on the head the hydroelectric plants can be classified as follows:

* low-head plants with bulb turbines (H <20 m) or Kaplan (20 m <H <60 m);
* medium to high power plants with Francis turbines (40 m <H <400 m);
* plants of very high head, with Pelton turbines (350 m <H <1,100 m).

The main types of turbines will be presented in summary in Chapter 10.

3.3 TYPES OF LAYOUTS

The layouts of the hydroelectric plants vary according to the topographic and geological-geotechnical conditions of the site. Basically, there are two types of layouts that are adopted routinely:

* dam layouts with all structures positioned along the axis of the dam, including the powerhouse;
* canal drop layouts for places with a large natural drop, where the dam is positioned upstream of the drop and the hydraulic circuit of adduction and generation on another axis, at one of the banks, with the powerhouse placed at downstream of the main dam; the location of the spillway is variable: it can be included in the body of the dam, it can be a chute spillway in the abutment, or a culvert spillway, etc.

Each case has its peculiarities and the most economical, technical and environmental alternatives are routinely researched in the studies and projects of the enterprises.

The main projects of HPPs are illustrated in the books of the Brazilian Committee of Dams, Topmost Dams of Brazil (CBDB, 1978) and "Main Brazilian Dam" (CBDB, 1982, 2000, 2009). These documents are of utmost importance and indispensable consultation for those who work for the Electric Sector in Brazil. They contain, in some detail, the general layouts of the main Brazilian plants. The layouts are the same for places with similar characteristics, since they already correspond to the technical evolution on the subject.

In the 17-volume Norwegian Institute of Technology (NIT) books are some examples of Norwegian plants. Volume 14, written by Edvardsson and Broch (2002), specifically discusses underground powerhouses and tunnels under high pressure. Some of these jobs are available on the internet.

In the book by Kollgaard and Chadwick (1988), *Development of Dam Engineering in the United States*, there is a summary of the American dams concluded until that year. Similarly, in JNCLD (1988), there is a summary of Japanese dams up to that year.

3.3.1 Dam layouts

In the dam layouts, the intake/powerhouse and spillway are incorporated into the dam axis. The head (H) is created by the dam. When the sites are in plains in low stretches of rivers, one has a dam axis with some extension and the lateral structures, usually extensive, are of earth fill or rockfill dams.

This is the case of the Balbina HPP, 3.2 km long, on the Uatumã river, Amazonas (Figure 3.2), Tucuruí Plant, 7.0 km long, on the Tocantins river, Pará, Itaipu Plant, 7.8 km long, on Paraná river border between Paraguay and Brazil, and the Sobradinho HPP, 8.5 km long, on the São Francisco river, Pernambuco, among others, as will be presented later in this chapter (Figure 3.2).

When these sites are located in steep-walled valleys ("V"), the main dam may be a rockfill with an earth core, or concrete face rockfill, or asphalt core, or even a concrete dam.

This is the case of the Itapebi HPP, 583 m long, Jequitinhonha river (Figure 3.3), Barra Grande HPP, 670 m long, Pelotas river, and Campos Novos HPP, 590 m long, Canoas river, among others. In the case of Foz do Chapecó HPP the dam, 548 m long, has an asphalt core. In the case of the Funil HPP, the dam is arched concrete, 360 m long (CBDB, 1978).

3.3.2 Canal drop layouts

Where there is a marked natural fall (H), in a short section of the river, the layout is usually bypass, with the dam positioned upstream of the fall, and the power generation circuit positioned on one of the shoulders, with the powerhouse downstream, far from the main dam.

This is the case of the Itá HPP works, Uruguay River (Santa Catarina/Rio Grande do Sul), 880 m long, Santa Rosa SHP, Rio Grande (Rio de Janeiro) and Colino SHP, on the Jucuruçu do Sul river (BA) (dam with 200 m extension), among others. Similar layout is there of the Conde d'Eu SHP Basic Project (8.2 MW), in Paquequer river on the municipality of Sumidouro (Rio de Janeiro). At the site, there is a natural drop of 270 m (Figure 3.4). The layout includes a 50 m long, 8 m high concrete spillway dam, a power circuit with low pressure tunnel 276 m long and

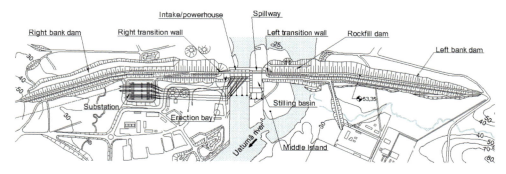

Figure 3.2 Balbina HPP, Uatumã river (Amazonas, Brazil) (CBDB, 2000).

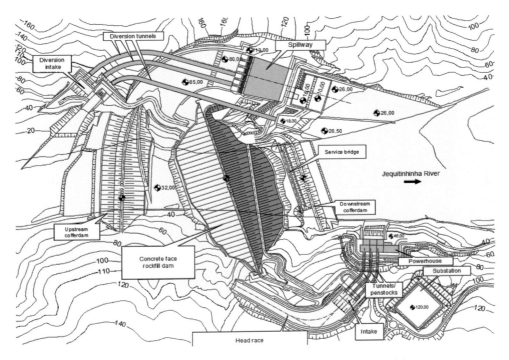

Figure 3.3 Itapebi HPP, Jequitinhonha river (Bahia, Brazil) (CBDB, 2009).

a high pressure stretch 570 m of extension and slope of 10%. The beginning of the high-pressure segment coincides with the position of the vertical shaft, 221 m high, and the surge tank (Figure 3.5).

Special mention should be made of the "La Grande Dixence" plant, 2,069 MW (Figures 3.6 and 3.7), Valais, Switzerland, implemented between 1950 and 1965. This dam, 285 m high, on the small river Dixence, forms Lake Dix, with an area of 4.0 km². The lake receives water from other rivers, pumped at Z'Mutt, Stafel, Ferpècle and Arolla stations, transported by 100 km of tunnels, as well as Lake Cleuson dam (87 m from the height), 7.0 km to the northwest. The rest of the water comes from melting glaciers during the summer. The lake has its maximum capacity in September and its minimum capacity in April.

As related previously, the Grande Dixence reservoir supplies four hydroelectric plants (Table 1.3):

- Chandoline (120 MW; $Hb = 1,748$ m; $Q = 6.25$ m³/s);
- Fionnay (290 MW; $Hb = 874$ m; $Q = 45$ m³/s),
- Nendaz (390 MW; $Hb = 1,008$ m; $Q = 45$ m³/s);
- Bieudron (1,269 MW; $Hb = 1,883$ m; $Q = 75$ m³/s).

With the exception of the Fionnay Power Plant, whose turbine flow supplies Nendaz, all the plants flow into the Rhône river.

Figure 3.4 (a–b) Conde d'Eu SHP, downstream of a natural fall. Municipality of Sumidouro, Rio de Janeiro, Brazil (Watermark 2009).

Chandoline, the plant for the original Dixence dam, was commissioned in 1934. The Grande Dixence dam submerged the original dam, but Chandoline's powerhouse remained. Its turbines were rebuilt, after the new Dixence, with the dimensions, 5 Peltons of 24 MW each. The plant has been out of service since July 2013.

Types of power plants and layouts 45

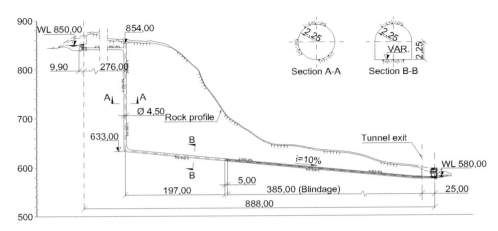

Figure 3.5 Profile of the SHP Conde d'Eu. Basic Design (Watermark 2009).

The Fionnay plant is supplied through a 9.0 km long pressurized tunnel, excavated with an average slope of 10%. From the surge tank, it turns into a forced conduit that descends 800 m with a high gradient, 73%, to the powerhouse cavern, where it divides into the 6 turbines.

The Nendaz plant receives the design flows at Fionnay, which travel 16 km in a tunnel pressure to the equilibrium chimney, where it turns into a forced conduit that plunges 1,000 m into the upstream region of the powerhouse, where it goes for the six turbines.

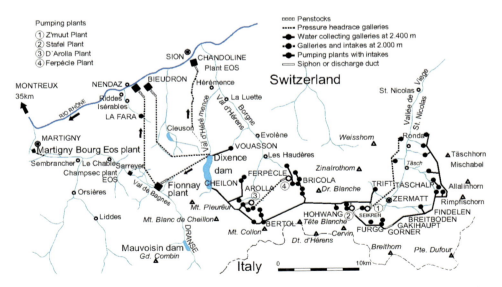

Figure 3.6 (a) Scheme of La Grande Dixence dam (Ginochio et al., 2012). (Note: After Martigny, down the river, it is 35 km to Montreux on Lake Geneva.) (b) La Grande Dixence dam (Wikipedia).

Figure 3.6 (Continued) (a) Scheme of La Grande Dixence dam (Ginochio et al., 2012). (Note: After Martigny, down the river, it is 35 km to Montreux on Lake Geneva.) (b) La Grande Dixence dam (Wikipedia).

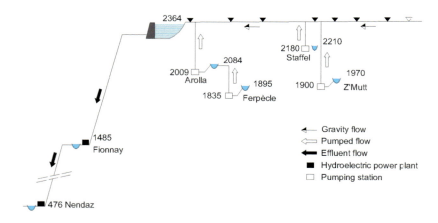

Figure 3.7 La Grande Dixence HPP – Profile (Ginnochio et al., 2012).

The Bieudron HPP has a long-pressurized tunnel until the surge tank. Here it transforms into a penstock that dives to the powerhouse; with an upstream of the powerhouse having penstock tifurac to the three Pelton turbines of 433 MW each. This plant was put into operation in 1988. It has the net head of 1,869 m, the flow is of 75 m³/s and the efficiency is 92%. Pelton turbines have five jets, with a diameter of

193 mm and an output velocity of 192 m/s $\left(\sqrt{2\times 9.81\times 1.869}\right)$. For information, it should be noted that the kinetic energy of each of the five jets has an output of 80,367 kW (Output $= kQH = 8.6 \times 5\,m^3/s \times 1,869\,m$).

The penstock broke on December 12, 2000, over 1,000 m head, at El. 1,234 m (see Chapter 13 and Ginocchio and Viollet, 2012). The reconstruction was completed in December 2009 and the plant started operations in January 2010.

3.4 NOTES ON THE SPILLWAY POSITON IN THE LAYOUT

The position and types of spillways may vary greatly from project to project, depending on the site and on the type of the dam. In layout with earth/rockfill dam, the spillway is incorporated to the main axis. Its structure usually forms a gravity dam with Creager profile, with energy dissipation in ski jump (e.g. Tucuruí HPP, Figure 3.8) or in a hydraulic jump in a stilling basin (e.g. Balbina HPP, Figure 3.2). There are other examples of the same presented in this book.

Figure 3.8 Tucuruí Spillway. Eletronorte Magazine (1998). Side view: end of the powerhouse, spillway and the earth-rockfill dam.

48 Design of Hydroelectric Power Plants – Step by Step

Figure 3.9 Karakaya HPP. Wikemedia.Ski jump spillway. $Q = 18{,}000\,m^3/s$.

If the dam is made of concrete, the spillway can be over the dam, with control of gates and dissipation in ski jump, such as that of Karakaya (Turkey), Figure 3.9, or in crossed jets, such as the Cahora-Bassa-HPP, Figure 8.34 (Moçambique). It can be without control of gates with partial dissipation in steps, such as that of the HPP Dona Francisca (Rio Grande do Sul), Figure 8.37, among other examples presented in this book. In Karakaya's case, the bucket of the spillway is the roof of the powerhouse (see Chapter 8).

It is often used as a side spillway as in the Monjolinho HPP followed by a restitution channel (Figure 3.10), or as in the Hoover HPP followed by a restitution tunnel (Figures 3.11 and 3.12).

Morning glory spillways are also used, as in the Paraitinga and Euclides da Cunha HPPs, and in the Colino SHP, whose owner is Renova Energy. The Colino I SHP (11 MW) is located in the Colino river, in the Jucuruçu do Sul river basin (Bahia). The plant has a gross head of 141 m. The layout has a bypass with the outlet and powerhouse away from the main dam (Figures 3.13 and 3.14).

In many cases, there is a chute spillway in the abutments, as in the Itaipu, Barra Grande, Itá, Foz do Areia, Emborcação, Campos Novos, Nova Ponte and Serra da Mesa HPPs, among others. Some of these spillways are illustrated in Annex A2. It should be noted that these last three spillways have part of the chute over the rocky mass on the abutment.

Types of power plants and layouts 49

Figure 3.10 Monjolinho HPP. (CBDB, 2011). Side spillway. $L = 210$ m; $Q = 6,755$ m^3/s.

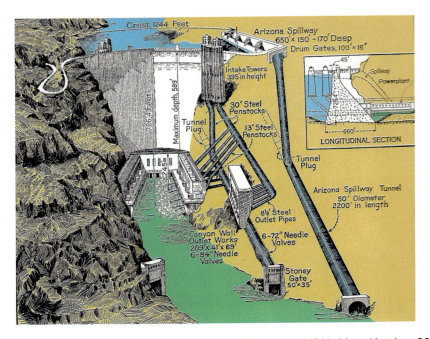

Figure 3.11 Hoover Dam, Colorado River (Arizona/Nevada, USA). Max. Height = 221.4 m. Sketch Schematic (Magazine of the Plant, 1988).

Figure 3.12 (a–b) Hoover Dam. (a) Observe the two side spillways; (b) Spillways operating (1998). Maximum flow capacity: 11,325 m³/s. (c) Hoover Dam. Observe the bridge downstream of the dam.

(*Continued*)

Figure 3.12 (Continued) (a–b) Hoover Dam. (a) Observe the two side spillways; (b) Spillways operating (1998). Maximum flow capacity: 11,325 m^3/s. (c) Hoover Dam. Observe the bridge downstream of the dam.

Figure 3.13 Colino SHP. Morning glory spillway. Maximum capacity ~143 m^3/s.

The Sobradinho HPP, on the São Francisco river, has two spillways: one of surface with gates and a stilling basin. The other is of bottom spillway, with dissipation in a terminal sill and in the riverbed downstream. The bottom spillway was used to pass water during construction to maintain power generation at the Paulo Afonso HPP, downstream.

Figure 3.14 Colino SHP. Morning glory spillway in operation.

Chapter 7 provides more information on spillway projects. The structures of the generation hydraulic circuit are presented in more detail in Chapter 8. In addition to the references cited above, it is recommended to consult CBDB's Large Brazilian Spillways (2002).

Several other examples of layouts are presented in the Appendix.

Chapter 4

Hydrological studies

4.1 INTRODUCTION

This chapter presents hydrological and energetic studies for small hydroelectric power plants projects. The studies for large power plants (>30 MW) follow the same lines, but with a greater degree of complexity and with particular aspects. As seen in Chapter 1, the power to be installed in a hydroelectric plant is directly proportional to the product of the "average flow" by "average net head". These parameters, together with the tailwater key-curve, are the main input data of the project. Cartographic information of quality is also essential.

The definition of the series of medium flows should be made with the support of a hydrology consultant with proven experience in studies of this nature. Regarding the series of extreme outflows, it is important to note that their importance is due to the fact that a large part of the accidents with power plants are due to insufficient discharge capacity of the spillways (>20% – ICOLD, 1995), which means that the floods remain undersized (see example of the El Guapo dam accident in Chapter 12).

The tailwater key-curve characterizes the variation of tailwater levels (head variation) as a function of the plant's disfluent flow rates (equal to the sum of the turbine flows with the spillways discharges). The determination of this curve is made on the basis of the data history of the fluviometric station installed in the tailwater channel of the plant itself. If the readings history is long, the key-curve can be obtained with good precision. As this is rare, data in the current projects have been used from other stations in the basin, transposed to the site of the plant. These data, together with those obtained locally through systematic measurements, make it possible to obtain a key-curve with reasonable accuracy.

This curve can be improved as the data series becomes more extensive throughout the studies and also during period of operation of the plant. It should be noted that in the current model of the Electric Sector, for SHPs has an annual verification of generation history that is carried out during the month of July, with the result being issued by SRG, ANEEL in August. The average energy marketed can be increased or decreased depending on the history of the operation. Cartographic surveys in the reservoir area, as well as topographic surveys of the site of the plant, should be carried out obeying the cartographic standards of SGH-ANEEL, in order to minimize any errors (which are not uncommon).

DOI: 10.1201/9781003161325-4

54 Design of Hydroelectric Power Plants – Step by Step

The hydrological studies are presented below, taking as reference the Basic Project (FEED) Report of a plant, the SHP Poço Fundo, that are prepared by Watermark Engenharia (Rio de Janeiro). For this plant, with an estimated 73 m gross head and a net head of 69.44 m (4.88% loss), an installed capacity of 14 MW was estimated to be generated for a turbine flow of 22.55 m^3/s. The hydrological studies, performed as recommended in the Eletrobras Manual (2000) are presented in Section 4.1. The energy studies are presented in Section 4.2.

4.2 HYDROLOGICAL STUDIES

4.2.1 Basin characterization

In the physiographic characterization the river and its basin (or watershed), along with its location are identified. The river spring should be identified in terms of municipality, the altitude, the extension of the river to the mouth, and its main tributaries. From the spring to the mouth, it is necessary to characterize the relief, whether flat, wavy or mountainous, and the number of hydroelectric plants foreseen in the basin (look for information in ANEEL). It is also necessary to characterize the possibility of floods in the basin during the rainy season and, as a consequence, estimate its capacity of natural regulation of the defluxions.

The shape of the basin, whether elongated in the longitudinal direction of the river or circular, etc., is also of importance for the studies. In order to establish relationships and comparisons, the characteristics listed below must be determined for both the integral basin and the part from the spring to the site of the future plant.

4.2.1.1 Drainage area

The drainage area of a basin is the projection of the surface contained between its topographic dividers (water dividers) in a horizontal plane. It is obtained through planimetry on maps and is usually expressed in km^2.

4.2.1.2 Shape of the basin

For the characterization of the shape of a basin, indexes are used that seek to associate it with known geometric forms. The index or coefficient of compactness (K_c) shows the relationship between the perimeter of the basin and a circle of area equal to that of the basin, i.e.:

$$K_c = 0.28 \frac{P}{\sqrt{A_d}} \tag{4.1}$$

where P is the perimeter of the basin (km) and A_d is the drainage area (km^2).

The compactness index (K_c) is the measure of the degree of irregularity of the basin. It is recorded that for a circular basin, it would be equal to 1.0. The potentiality of occurrence of high flood spikes in the basin as the compactness index gets closer to the unit.

The conformation index (K_f), or shape factor, is the ratio of the area of the basin to the square of its axial length, measured along the main water course, from the mouth to the farthest fall near the divider of waters of the basin.

$$K_f = \frac{A_d}{L^2} \tag{4.2}$$

where L is the axial length, km.

This index of conformation relates the shape of the basin to a rectangle. In a narrow and long basin, the possibility of intense rainfall covering its full extent is smaller than in wide and short basins. For basins of the same area, the one that has a lower K_f will be less subject to flooding.

4.2.1.3 Mean bed slope

The mean bed slope is obtained by dividing the difference between the spring and the mouth quota by the total length of the main watercourse. The flow velocity of the river depends on its slope. The higher the slope, the greater the flow velocity and consequently the more acute the hydrographs of the floods.

4.2.1.4 Time of concentration

The time of concentration corresponds to the time required for the entire basin to contribute to the runoff in a given section. In other words, it is the time taken by a drop that plunges into the farthest point of the considered section of a basin to reach this section. In the case of floods, it is the minimum possible time for this maximum course of this drop.

There are several formulas, depending on the region and the characteristics of the basin, for the calculation of the time of concentration, such as the quite usual formula of the Natural Resources Conservation Service (NRCS, former Soil Conservation Service), of the US Department of Agriculture.

$$t_c = 57 \cdot \left(\frac{L^3}{H}\right)^{0.385} \tag{4.3}$$

where:

t_c = time of concentration, minutes;

L = main water course length, km;

H = the difference between the elevations of the farthest point and the point considered, m.

These characteristics should be summarized in table as the Table 4.1.

56 Design of Hydroelectric Power Plants – Step by Step

Table 4.1 Summary of the Physiographic Characteristics of the Basin

Characteristics	Unit	Value
Drainage area	km^2	
Perimeter	km	
Axial length	km	
Compactness coefficient	-	
Form factor	-	
Maximum altitude	M	
Minimum altitude	m	
Average declivity	m/m	
Time of concentration	min	

4.2.2 Hydrometeorology

In order to characterize the basin climatically, it is necessary to first look for the list of existing stations (Table 4.2) and to complement it with neighboring stations if there are not enough climatological stations within the basin.

The station with the longest data period should be chosen, with its characteristics presented in the table, depicting Temperature, Relative humidity and Precipitation, as shown in Tables 4.3–4.6.

Table 4.2 Climatological Stations

Station	Code	Latitude (S)	Longitude (W)	Historic	Operator

Table 4.3 Average Monthly Temperatures According to Climatological Atlas

Month	O	N	D	J	F	M	A	M	J	J	A	S
T (°C)												

Table 4.4 Extreme Normal Temperatures (Climatological Atlas)

Temperature	T (°C)
Minimum absolute	
Maximum absolute	
Absolute thermal amplitude	
Average minimum	
Average maximum	

Table 4.5 Relative Humidity in the Basin (Climatological Atlas)

Month	O	N	D	J	F	M	A	M	J	J	A	S	Ano
Relative Humidity (%)													

Table 4.6 Average Monthly Precipitation

Months												Year (mm/ano)
O	N	D	J	F	M	A	M	J	J	A	S	

4.2.2.1 Temperature

The average monthly temperatures, estimated from the isotherms of the normal climatological maps, should be presented in tables like Table 4.3, where the coldest quarter and the warmest quarter are usually the most important.

The extreme temperature values and the absolute thermal amplitude should also be presented as in the model of Table 4.4.

4.2.2.2 Relative humidity

Table 4.5 shows the relative humidity that prevails in the area of the basin. The high relative humidity (above 70%) indicates that there is more possibility of the conversion process in precipitation.

4.2.2.3 Precipitation

The analysis of the historical series of the selected fluviometric stations, presented in Table 4.6, should allow to characterize the annual distribution of mean precipitation in the basin.

The seasonal distribution of rainfall characterizes the rainy season and the dry season. The percentage with which the rainy season contributes to the annual total should be reported. It is important to identify the rainier quarter and the total precipitation in the period. In Table 4.6, it was considered the most common hydrological year in Brazil, which runs from October to September.

4.2.2.4 Climate classification

The climate of the region can be characterized using the traditional classification proposed by German climatologist Köppen.

4.2.3 Fluviometric measurements

ANA is responsible for supplying the data series (data contained in the Hydroweb database) of the fluviometric stations in the stretch of the river and in the river basin (Figure 4.1).

Figure 4.2 has been extracted to depict the recommended discharge measurement procedures practiced by the United States Geological Survey

58 Design of Hydroelectric Power Plants – Step by Step

Figure 4.1 (a) Limnometric rule. (b) Meteorological station. (c) Balbina HPP. Meteorological station. (d) Balbina HPP. Evaporimetric tank. Totaling anemometer. Floating thermometer. Hook micrometer.

Figure 4.2 Discharge measurement using ADCP equipment-acoustic doppler current profiles. USGS-United States Geological Survey.

Hydrological studies 59

Table 4.7 Available Fluviometric Stations

Code	Station	River	Installation	Entity

These data are usually presented in a standardized table (Table 4.7), according to the periods of observation.

The river discharge curve (or flow curve), that is, the $Q \times WL$ in the tailrace, is of extreme importance for the evaluation of the head available for power generation at the site of the project. Therefore, it is fitting to make some observations about the process of its determination. This relationship is established by direct on-site measurements at the on-site fluviometric station at various flows and river levels, and may be univocal or not, constant or variable over time, depending on local conditions.

The installation of the station consists essentially of water-level meter devices, limnometric rulers or lingraph, duly referred to a known and materialized quota on the ground. Flow measurements should be made by specialized companies. The fluviometric station of the plant must be positioned in a suitable place, with stable and sensitive characteristics, in order to obtain a well-defined relation.

This place must meet the following requirements:

- be in a rectilinear section of the river, with well-defined margins and without any singular points that can significantly disturb the flow;
- cross section as much as possible symmetrical and with accentuated slopes;
- regularly distributed speeds; and
- average speed greater than 0.3 m/s.

The most important characteristics of the stations are stability and sensitivity. It is evident that if the systematic flow is obtained from the observed water levels, the constancy of the ratio quota-flow assumes a preponderant value.

However, rivers do not always present the necessary conditions for this stability. The natural physical characteristics, such as the configuration, the vegetation cover of the banks and flood zones and the nature of the bed, which constitute the control of the hydrometric station, can undergo more or less rapid changes over time, causing variations in the ratio of flow.

The sensitivity of a hydrometric station, translated by greater or lesser variation of the water level for a given flow change, is another important aspect in the characterization of its qualities.

It is evident that the records of a more sensitive station can be converted to flow with greater precision than those of sections of lower sensitivity.

For details on the characteristics of the section, regarding stability and sensitivity, hydraulic controls and their changes, it is recommended to consult Chow (1959), Linsley (1959) and Pinto et al. (1973).

4.2.4 Tailwater elevation curve

For the determination of the tailwater elevation curve, discharges measurement campaigns must be carried out, normally, for at least one complete hydrological cycle.

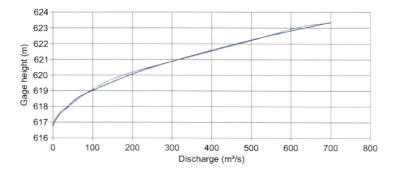

Figure 4.3 Tailwater elevation curve. Poço Fundo SHP (FEED).

The author thinks that a complete hydrological cycle, however, is a very short period. Ideally, all river basins should be routinely monitored for long and quality data records.

Special attention should be given to direct discharges measurements and the range of actual discharge rates, which ensure representativeness of the tailwater elevation curve.

The accuracy of the tailwater curve depends, among others, on the number of measurements. The more data that is available and with large amplitude of the levels and flows, in the high branch of the curve, the better the precision.

However, the observed data range is often insufficient for the project definitions. Measurements often cover only small flows in the low stretch of the curve. So, we need to extrapolate the flows to the high stretch of the curve, a task always subject to errors that may be significant.

The extrapolation is done by means of hydraulic calculation, with the aid of the HEC-RAS 4.0 model of the US Army Corps of Engineers (USACE).

This software is freely distributed and allows one-dimensional analysis of both permanent and non-permanent flows through the Standard Step Method.

More details about this software can be found at www.heo.usace.army.mil.

This extrapolation process should not be carried out without a careful study of the local conditions with respect to the possibility of changes of the type of control for the higher flows

The quota-flow relationship cannot always be expressed by a simple equation and the tailwater elevation curve can have an inflection point (Pinto et al., 1973).

Figure 4.3 presents an example of a typical tailwater elevation curve of the Poço Fundo SHP project already carried out.

Polynomials defined with a high degree of arithmetic precision do not, per se, guarantee the accuracy of the tailwater elevation curve. It is very important to indicate the values actually measured for the judgment of the quality of the tailwater elevation curve (showing the points in the curve).

4.2.5 Flow-duration curves

Based on the data available at the stations, a consistency analysis should be carried out, illustrated by comparative hydro grams as shown in Figure 4.4. In these graphs, one can observe the relative behavior of the flows of the stations to verify the consistency between them.

Hydrological studies 61

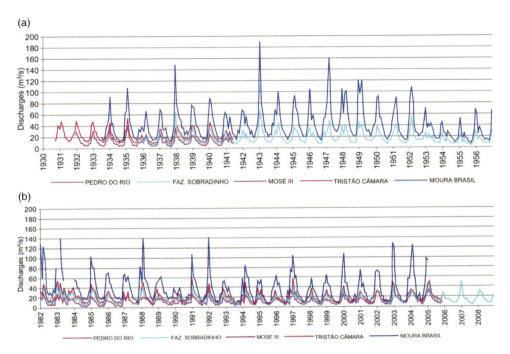

Figure 4.4 (a–b) Hydrograph of monthly average discharges of the stations of the Piabanha river basin (RJ) (SHP Poço Fundo Basic Design (FEED)).

It is expected that when there is a peak at the upstream station there is a similar peak at the downstream also.

In order to fill the existing flaws in the records, mean monthly flow correlations are established between them, thus allowing a common continuous series of monthly average flows for all stations, relative to the longest observation period. These correlations are presented in typical graphs as indicated in Figure 4.5.

The equations used and the correlation coefficients obtained are presented in Table 4.8. The closer to 1 (one) the coefficient, the stronger the correlation.

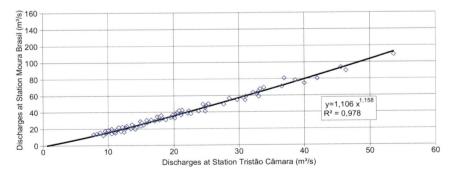

Figure 4.5 Correlation of monthly average discharges (SHP Poço Fundo Basic Design).

Table 4.8 Equations Used and the Correlation Coefficients Obtained

Station	Moura Brasil	Morelli	Pedro do Rio	Areal	Faz. Sobradinho
River	Piabanha	Preto	Piabanha	Piabanha	Rio Preto
Drainage area (km²)	2.052	927	412	510	720
Pedro do Rio Rio Piabanha				$Y = 1.2615X^{0.9547}$ $R^2 = 0.9752$	
Areal Rio Piabanha			$Y = 0.7788X^{1.057}$ $R^2 = 0.9935$		
Faz. Sobradinho Rio Preto		$Y = 1.3822X^{0.9358}$ $R^2 = 0.9208$	$Y = 0.6635X^{0.9902}$ $R^2 = 0.8867$		
Morelli Rio Preto	$Y = 0.8742X^{1.2426}$ $R^2 = 0.9006$				$Y = 0.8968X^{0.984}$ $R^2 = 0.9208$
Tristão Câmara Rio Preto	$Y = 1.1065X^{1.1581}$ $R^2 = 0.9783$		$Y = 0.2633X^{1.2408}$ $R^2 = 0.9401$		
Moura Brasil Rio Piabanha		$Y = 1.4505X^{0.7248}$ $R^2 = 0.9006$			

Hydrological studies 63

The discharges at the site should be obtained from the series consisting of the stations, transposed to the location of the axis by correlations of drainage areas, as expressed in Equation 4.4:

$$Q_{local} = \left(\frac{Q_{station}}{A_{station}} \right) \times \left(A_{local} \right) \qquad (4.4)$$

The series thus obtained should be presented in a standardized table as Table 4.9. The example (Figure 4.4) shows a series with data from 1931 to 2008.

The year 1931 is the standard initial year for the energy simulations of the Brazilian Electricity Sector.

The flow duration curve (permanence of monthly average discharges) at the Poço Fundo SHP site is depicted in Figure 4.6.

Table 4.9 Medium Flows Series on Site of the Poço Fundo SHP. AD = 730 km² (Basic Design)

	Jan	Fev	Mar	Abr	Mai	Jun	Jul	Ago	Set	Out	Nov	Dez
1931	26.5	34.6	30.0	23.3	17.5	14.4	12.6	11.6	13.0	17.1	20.8	25.7
1932	35.0	28.9	23.6	17.1	17.1	14.7	11.8	12.5	11.0	13.3	16.0	31.5
1933	34.4	22.9	22.2	16.5	15.7	12.6	11.2	9.3	10.0	16.0	17.9	22.8
1934	33.3	18.6	19.9	15.0	11.3	9.6	10.1	7.8	8.9	8.8	12.0	25.2
1935	26.4	37.2	21.3	18.3	14.3	11.6	10.5	9.8	12.1	16.5	14.7	19.9
1936	11.2	15.3	22.4	17.2	11.3	9.2	7.0	6.3	6.8	7.1	7.7	16.7
1937	30.1	28.4	14.4	15.6	15.2	9.3	7.7	5.9	5.6	10.8	20.6	65.1
1938	37.1	32.6	25.2	24.8	16.3	16.0	10.8	14.3	10.7	11.5	16.2	33.0
1939	35.6	26.0	17.1	16.6	11.7	8.6	7.7	6.4	7.2	6.4	10.5	25.8
1940	37.4	28.7	24.5	15.1	11.8	9.2	7.7	6.7	7.1	10.6	17.8	26.8
(*)												
2000	38.5	17.8	20.1	16.4	10.0	8.2	8.7	9.7	13.6	9.1	11.9	19.0
2001	25.4	21.2	17.2	14.8	11.9	8.3	7.3	6.0	6.6	7.7	10.0	23.5
2002	23.0	22.8	16.1	11.8	10.3	7.7	7.2	6.0	8.4	6.0	16.9	43.1
2003	32.6	16.1	17.2	12.5	10.7	8.7	7.5	8.5	8.7	11.2	21.1	23.4
2004	28.3	28.3	21.2	17.3	12.9	11.1	13.5	8.7	6.9	10.8	16.2	26.7
2005	38.1	43.0	30.3	17.7	14.4	12.5	12.4	9.3	10.0	8.4	16.1	29.1
2006	13.7	12.5	13.0	15.5	10.2	8.1	6.5	6.3	6.6	8.8	16.7	23.0
2007	49.4	22.1	14.1	13.9	12.0	9.1	7.9	6.3	5.5	6.5	13.2	18.3
2008	20.1	29.4	27.2	21.3	13.7	10.7	8.4	7.7	7.8	9.7	18.3	22.4
Min	9.8	9.6	8.5	8.5	7.5	6.0	4.6	3.8	3.5	4.4	6.2	6.9
Aver.	30.0	26.5	24.1	18.4	13.4	10.6	9.1	8.0	8.8	10.9	16.8	25.7
Max	69.5	55.6	54.6	38.0	28.8	32.5	17.4	14.3	31.5	37.2	32.7	65.1
		MLT = 16.9 m³/s						Q_{mpc} = 15.4 m³/s				

AD, drainage area; Q_{MLT}, average long-term flow (MLT). Data for each month of the 1931/2008 series of this example; Q_{mpc}=average flow of the critical period (Jun. 1949–Nov. 1956), used in the energy simulations of the Electrical Sector.
(*) Data for the years 1941–1999 were omitted.

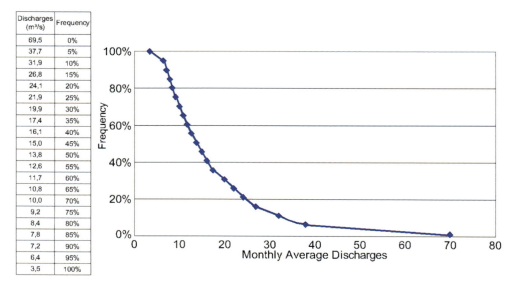

Figure 4.6 Poço Fundo SHP. Flow duration curve – example. Basic Design (Watermark Engineering).

4.2.6 Extreme flows

In the study of extreme flows, the daily data series of the fluviometric station that is located closest to the utilization is used. These flows are used for spillway and river diversion projects.

For studies of river diversion, it is common to study the frequency of the maximum flows during the dry season (between May and October) when this is well defined to highlight the lower risk of flooding in this period. Floods should be transferred to the site of recovery using the correction factor. Floods related to the hydrological year and not to the calendar year should be considered, and the correction for instantaneous peak values should be applied to the calculated values by Fuller's formula (1914):

$$Q_{máx} = Q\left(1 + 2.66\, A^{-0.30}\right) \tag{4.5}$$

where:
A_{max} = peak flow (m³/s);
Q = average daily flow of the peak day (m³/s);
A = drainage area (km²).

In statistical studies, traditional distributions should be used for two Gumbel and Exponential parameters. Gumbel's is preferred when the asymmetry of the distribution is ≤1.5. For larger asymmetries, the Exponential distribution is shown to be more robust (Eletrobras, 1987). The results of this study for the fluvial station Fazenda Sobradinho (closest to the SHP site) are presented in Tables 4.10–4.14.

Hydrological studies   65

Table 4.10 Maximum Flow Series, Station Fazenda Sobradinho

Year	$Q_{máx}$ (m^3/s)	Year	$Q_{máx}$ (m^3/s)
	Nov–Oct		May – Oct
1936	61.6	1936	27
1937	207	1937	40
1938	116	1938	60
1939	97.4	1939	22
1940	164	1940	24
1941	108	1941	30
(*)			
2000	125	1999	35.7
2001	183	2000	62.2
2002	186	2001	45.2
2003	157	2002	30.4
2004	262	2003	55.7
-	-	2004	42.7
-	-	2005	45.6
Total of years	60	Total of years	62
Maximum	266.0	Maximum	234.0
Minimum	61.1	Minimum	17.5

Source: Poço Fundo SHP Basic Design.
(*) Data from 1942 to 1999 were omitted in this example.

Table 4.11 Design Flows for the Distributions Studied (November - October).

Adjustment TR^a (years)	Exponential – 2 Parameters Flows (m^3/s)		Gumbel – 2 Parameters Flows (m^3/s)	
	Calculated	Instant	Calculated	Instant
2	134	183	141	193
5	182	249	187	257
10	218	299	28	299
25	266	365	257	352
50	303	414	286	391
100	339	464	314	431
500	424	580	381	521
1.000	460	630	409	560
10.000	581	796	503	689

[a] TR, Recurrence time.

Transferring to the site of SHP Poço Fundo you get the results presented in Table 4.15.

Spillway design flood (or dam design flood)

According to Eletrobras and CBDB (2003), for dams with more than 30 m, or whose collapse involves the risk of loss of human life (existence of permanent dwellings downstream), the existence of spillway design flood will be the maximum likelihood (VMP).

66 Design of Hydroelectric Power Plants – Step by Step

Table 4.12 Design Flows for the Distributions Studied (May – October)

Adjustment TR^a (years)	Exponential – 2 Parameters Flows (m^3/s)		Gumbel – 2 Parameters Flows (m^3/s)	
	Instant	Calculated	Instant	
2	39	53	44	60
5	70	96	74	101
10	94	129	94	129
25	126	172	120	164
50	149	205	138	190
100	173	237	157	215
500	229	313	201	275
1,000	253	346	219	300
10,000	332	454	281	385

[a] TR, Recurrence time.

Table 4.13 Parameters of Applied Distributions (November–October)

Adjustment	Exponential – 2 Parameters	Gumbel
Formula	$Q = \beta_0 - \beta_1 \times \ln (1/TR)$	$Q = \mu - \alpha \times \ln [(TR\text{-}1)/TR]$
Parameters	$\beta_0 = 97.11$	$\alpha = 40.97$
	$\beta_1 = 52.52$	$\mu = 125.99$
Asymmetry	0.70	
Recommended Distribution:	Gumbel	

Table 4.14 Parameters of Applied Distributions (May–October)

Adjustment	Exponential – 2 Parameters	Gumbel
Formula:	$Q = \beta_0 - \beta_1 \times \ln (1/TR)$	$Q = \mu - \alpha \times \ln [(TR\text{-}1)/TR]$
Parameters	$\beta_0 = 14.84$	$\alpha = 26.84$
	$\beta_1 = 34.41$	$\mu = 33.76$
Asymmetry:	3.66	
Recommended Distribution:	Exponential	

Table 4.15 Instant Flow (m^3/s) - SHP Poço Fundo

TR (years)	November–October[a]	May–October[b]
2	196	54
5	260	98
10	303	131
25	357	174
50	397	208
100	437	241
500	528	318
1,000	568	351
10,000	699	461

[a] Full period (spillway design).
[b] Dry period (river diversion studies).

Hydrological studies   67

For dams with a height <30 m, or with a reservoir with a volume of less than $50 \times 10^6 \, m^3$, and with no risk of loss of human life (no permanent dwellings downstream), the flood shall be defined through a risk analysis, with respect to the minimum recurrence of 1,000 years.

4.2.6.1   Powerhouse design flow

It will be defined through a risk analysis of the average daily natural flows, with respect to the minimum recurrence of 1,000 years.

4.2.6.2   Diversion flows

For each phase of river management, the diversion flows will be defined by the recurrence times resulting from a risk analysis, comparing the cost of the diversion works with the expected value of the cost of damages resulting from the respective floods. In calculating the damage, local damages will be considered, along with the costs due to delay of schedule and eventual damages upstream and downstream.

4.2.6.3   Risk analysis

For each phase of river management during construction, the flows of diversion works should be defined according to the risk of flooding of the land area, taking into account the time of exposure to this risk, as defined in Chapter 11.

4.2.7   Minimum flows

The minimum flows are characterized by the lowest daily values of each year of the annual series. The minimum flow rate is associated with a duration t. The minimum flow of any one year, with a duration of 30 days, indicates that it is the lowest value of the year of the average flow of 30 consecutive days.

In practice, the durations of 7 or 30 days are of greater interest, since the sequence of low flows is the most critical condition in the use of water. The minimum flow probability curve allows the estimation of the risk of flow rates lower than a chosen value.

4.2.8   Regularization of discharges

Nowadays, the only known way to accumulate electrical energy in large quantities is to store water in reservoirs (potential energy) to transform it, when necessary, into electrical energy.

However, the issue of environmental impact has taken such a limiting power in infrastructure works that hydroelectric facilities are currently run of river, in which the reservoir does not have the capacity to regularize.

The generation of energy is at the mercy of the tributary flows to the reservoir and suffers all the influence of the seasonality. The benefit of regularization is lost, and in many projects there is a possibility of attending to the loading point of the plant in the dry season.

Basically, the study of regularization of flows of an isolated plant determines the flow rate that can be maintained during the critical period, according to the available useful volume. The Electrical Sector established the period from June 1949 to November 1956 as the critical period of the Interconnected System to be considered by all in projects.

For these studies, see Pinto et al. (1973, 1976), Gomide (1970, 1981, 2012), Lanna (1993) and Souza et al. (2009). The first reference presents the example of the Iguaçu river. The latter presents, in a very didactic way, the Conti-Varlet method of flow rate regulation and presents the case of the Pai Querê HPP in the Pelotas river (RS).

If the plant is interconnected to a power transmission system the calculations become more complex and specialized and are not covered in this book (in Brazil there is one of the largest interconnection systems in the world, the SIN).

4.2.9 Determination of sanitary flow

During the filling of the reservoir, or in the case of plants that the powerhouse is downstream, a hydraulic device must be provided to ensure the sanitary or ecological flow necessary for the maintenance and survival of the ecosystem in the river downstream of the dam or between the dam and the power house. The layout must also provide for a fish ladder if there are migratory species in the basin. Thus, the sanitary or ecological flow can be complemented by the flow in the fish ladder.

In Brazil, each state has its own rule regarding sanitary flow. In Rio de Janeiro, the Environmental Institute (Inea) determines that the sanitary flow should be equal to 50% of the reference flow equal to Q7,10- minimum flow with 7 days of duration for a return period of 10 years (Ordinance No. 591 of 08/14/2007 of the SERLA – Rio de Janeiro).

4.3 CURVES QUOTA × AREA × VOLUME

The curves quota x area x reservoir volume (Figure 4.7) should be defined from the plant elaborated in the cartographic surveys of the reservoir, at a scale of 1: 5,000,

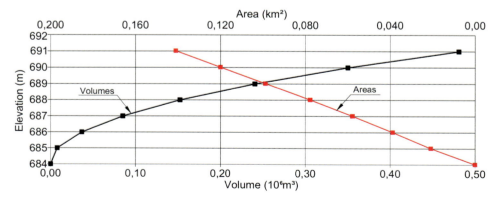

Figure 4.7 Curves quota × area × reservoir volume - example. Poço Fundo SHP Basic Design (Watermark Engineering).

Table 4.16 Curves Quota × Area × Reservoir Volume

Quota (m)	Area (km²)	Volume (10^6 m^3)
684	0.00	0.00
685	0.02	0.01
686	0.038	0.03
687	0.06	0.08
688	0.078	0.15
689	0.09	0.24
690	0.12	0.34
691	0.14	0.47

usually with contour lines of 5–5 m. This survey can be performed by laser or by interferometric radar, or by conventional aerial photogrammetry. It provides fairly fast and accurate methods and are suitable even for densely forested sites. Table 4.16 corresponds to those curves shown in Figure 4.7.

4.4 RESERVOIR FLOOD ROUTING

Flood routing can be done by raising the WL of the reservoir after complete opening of the spillway gates. Basically, it consists of the calculation of the difference between flows affluents and effluents with the repercussion of the volume retained in the NA of the reservoir. The elevation of the WL increases the load on the spillway, which results in a new diffluent flow. It is an iterative calculation. At any given moment, the effluent flow will be greater than the affluent and the WL will begin to decrease until it returns to its normal maximum value.

When using the reservoir below its normal operating WL, and for flood routing, there is a so-called "standby" volume provided for this purpose. The form of calculation is the same as the calculation of superelevation. The flood routing studies can be carried out according to the methodology of the routinely used hydrological books cited at the end of this chapter.

4.5 BACKWATER STUDIES

The backwater studies should be carried out by applying established methodology in projects of this nature, as recommended in the document Eletrobras (1997) "Economic Feasibility Studies of Hydroelectric Power Plants".

As mentioned in Section 4.1.3, one of the most used tools is the HEC-RAS model, v. 4.0 (USACE). For backwater studies, instantaneous waterline profiles, topometameric sections topographically tied, distances between these sections and flows on the days on which the instantaneous profiles are surveyed are required. With these data, the model allows to calculate the average roughness of the stretches to repeat the observed levels. This is, therefore, the calibration of the model. After this calibration, the extrapolations of the natural conditions to the reservoir conditions are made.

4.6 FREE BOARD

The free board studies should be performed using the methodology established in projects of this nature as recommended in Section 3.1.1 of CBDB/Eletrobras (2003).

The recommended method is that of Saville Jr. (1962), which is also presented in USBR (1981). Savile recommended that, in landfills and rockfill, the normal freeboard should be defined to absorb the effect of the waves caused by the maximum wind of 100 km/h and a wave with a 2% probability of being overcome (Table 4.17).

The normal free board is defined as the difference in elevation between the crest of the dam and the normal level of the reservoir. The minimum free board is defined as the difference in elevation between the dam crest and the reservoir WL (for the maximum design flood of the spillway).

The dimensioning of the free board needs determination of the height and action of the waves. The height of the waves in the reservoir depends on the wind speed, wind duration, fetch, water depth and reservoir width.

The fetch is defined as the overwater distance from the shore opposite of the dam in the direction the wind is blowing. It is the "lane" in the body of the water (surface of the reservoir) on which the wind will act. This parameter conditions the height of the wave.

When the waves approach the dam upstream slope, the height can be altered by increasing the depth or by reducing the width of the reservoir. After the wave comes in contact with the slope, the effect is influenced by the inclination of the wave train in relation to the axis of the dam, the inclination of the slope and the texture of the face of the slope (roughness). Since there is no data on wave height and wave runup, free board determination requires experience and consideration of local factors.

Information in Table 4.18 is extracted from the book USBR (1974), which records a summary of results obtained by empirical formulas proposed for the determination of wave heights.

As a criterion, the USBR suggests that 100 mph (160 km/h) winds are adopted for the normal freeboard and winds of 50 miles/h (80 km/h) are adopted for the minimum freeboard. Based on these speeds, they recommend the free boards, as specified in Table 4.19.

The methodology for determining wind-generated waves according to Saville Jr. (1962) is presented below. The effective fetch is determined as the average distance in the direction of the wind within a 90° sector centered about the wind direction:

$$F = \sum_{n=-45}^{n=+45} Ri \, \cos^2 \alpha i \, / \sum_{n=-45}^{n=+45} \cos \alpha i \tag{4.6}$$

Table 4.17 Free Board

Structure	Free Board Normal (m)	Free Board Minimum (m)
Earth and rockfill	Minimum of 3.0 m	1.0 m above the WLmax. maximorum
Concrete dam	Minimum of 1.5 m	0.5 m above the WLmax. maximorum
Power house	Minimum of 1.0 m	-

Source: Eletrobras and CBDB (2003).

Hydrological studies 71

Table 4.18 Fetch, Wind Velocity and Wave Height

Fetch		Wind velocity	Wave Height
Mile	km	km/h	M
1	1.609344	80.47	0.82
1	1.609344	120.70	0.91
2.5	4.02	80.47	0.98
2.5	4.02	120.70	1.10
2.5	4.02	160.93	1.19
5	8.05	80.47	1.13
5	8.05	120.70	1.31
5	8.05	160.70	1.46
10	16.09	80.47	1.37
10	16.09	120.70	1.62
10	16.09	160.93	1.86

Source: USBR (1974).

Table 4.19 Free Board (m) (USBR)

Fetch (km)	Free Board Normal (m)	Free Board Minimum (m)
< 1.60 (~1 mile)	1.22	0.91
1.60	1.52	1.22
4.02	1.83	1.52
8.05	2.44	1.83
16.1	3.05	2.13

Source: USBR (1974).

where the parameters defined in Figure 4.6 are:

F = fetch;

Ri = radial distance from the contour line of the reservoir to the dam (m);

αi = angle between the wind direction and the radial direction radial (°) (Figure 4.8).

The non-dimensional formulae for calculating fully developed wave characteristics are presented below, according to Kjaernsli et al. (1992):

$$\frac{gHs}{U_*^2} = 0.0506 \left(\frac{gF}{U_*^2} \right)^{0.5} ; \text{Saville's formula is}: \frac{gH}{U_2} = 0.0026 \left(\frac{gF}{U_2} \right)^{0.47} \tag{4.7}$$

$$\frac{gTp}{U_*} = 0.903 \left(\frac{gF}{U_*^2} \right)^{0.333} ; \tag{4.8}$$

$$\frac{gT}{U_*} = 23 \left(\frac{gF}{U_*^2} \right)^{0.667} \tag{4.9}$$

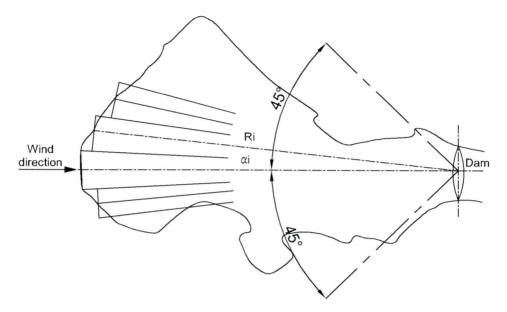

Figure 4.8 Determination of effective fetch (Saville 1962).

where:
Hs = significant wave height (m);
Tp = peak wave period (min);
T = minimum wind duration (min);
g = acceleration of gravity (m/s^2);
U_* = wind friction speed (m/s).

The friction velocity is calculated from wind speed, U_{10}, 10 m above the earth's surface by the empirical form:

$$U_* = C_D U_{10}^2 = \left(0.0008 + 0.000065\, U_{10}\right) U_{10}^2 \qquad (4.10)$$

where C_D is the coefficient.
Adopting $U_{10} = 100$ km/h, equal to the CBDB design criteria, we have:

$$U_* = C_D U_{10}^2 = (0.0008 + 0.000065 \times 100)\, 10{,}000 = 73\, \text{m/s}$$

Replacing this value in Equation 4.7 of the significant wave, we have:

$$Hs = 0.14 F^{0.50}\ (\text{m}) \qquad (4.11)$$

According to Saville (1962), this same equation would be $0.10\, F^{0.47}$.
    Therefore, for a fetch of 16 km (10 miles) $Hs = 0.56$ m, much smaller than that recommended by the USBR (1974) for a wind speed of 80 km/h (see Table 4.19).

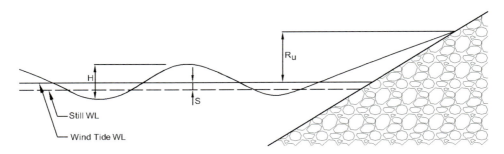

Figure 4.9 Wave characteristics (Kjaernsli et al. 1992).

According to Kjaernsli et al. (1992), the waves hitting the upstream slope will run up to a height (*Ru*) above the surface of the reservoir as shown in Figure 4.9.

The wind tide, *S*, needs to be taken into account for shallow reservoirs. For deep reservoirs it is insignificant when compared to run up (*Ru*), being estimated by Equation 4.12:

$$S = \frac{U^2 Fe}{4,800 D} \qquad (4.12)$$

where:
  S = wind tide (m);
  U = wind velocity (m/s);
  Fe = fetch (km);
  D = average depth of the reservoir along the fetch (m).

The run-up exceeded by 2% of the waves in a storm is given by:

$$Ru = r\,Hs \qquad (4.13)$$

where:
  Ru = vertical height above the WLres (m)
  r = attenuation factor; and
  Hs = significant wave height (m).

The attenuation factor, r, depends on the number of Iribarren for the wave geometry:

$$Ir = \tan\beta \sqrt{\frac{Lo}{Hs}} = Tp \tan\beta \sqrt{\frac{g}{2\pi Hs}} \qquad (4.14)$$

where:
  Ir = number of Iribarren;
  β = slope angle;
  Lo = wave length in deep water (m);
  Tp = peak period of the wave (s); and
  g = acceleration of gravity (m/s^2).

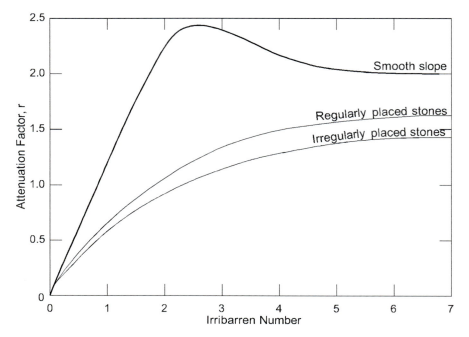

Figure 4.10 Number of Iribarren × Attenuation factor "r" (Kjaernsli et al. 1992).

The Iribarren number is an expression for each type of wave breaking on a slope, as specified below:

$Ir < 0.5$ spilling wave;
$0.5 < Ir < 2.0$ plunging wave;
$2.0 < Ir < 2.6$ plunging and collapsing wave;
$2.6 < Ir < 3.1$ collapsing or surging wave;
$3.1 < Ir$ collapsing wave.

The graph of the number of Iribarren × Attenuation Factor is shown in Figure 4.10. According to Kjaernsli et al. (1992), most dams have $0.5 < Ir < 2.0$.

4.7 RESERVOIR FILLING STUDIES

The reservoir filling studies shall be performed using the average daily or monthly flows at the site of the dam, based on the curve × area × volume of the reservoir.

The simulation should characterize the evolution of the WL during the filling, with the times necessary until the levels of operation of the reservoir are reached.

It is the calculation of the flow that reaches the reservoir, minus the flow that leaves the reservoir, resulting in volume accumulated in the time interval adopted.

4.8 RESERVOIR USEFUL LIFE STUDIES

Aggradation problems, or accretion, reduction of reservoir volume due to sediment deposition throughout its useful life, are frequent. The most famous case is the Mascarenhas HPP in the Doce river, State of Espirito Santo (Almeida and Carvalho, 1993).

Aggradation could affect the useful life and also their generation capacity. This is an issue that should be monitored by concessionaires or authorized hydroelectric power generation, according to Joint Resolution No. 3 of August 10, 2010 (ANEEL, ANA, 2010).

Sediment production involves many factors: regional geology and geomorphological characteristics of the basin, rainfall, winds, and anthropogenic actions that generally promote deforestation and degradation of the region, increasing solid flow.

The issue has received increasing attention from designers, researchers and funding agencies around the world. World Water (1988) cites a World Bank study in 1974, indicating that the average life expectancy of reservoirs had dropped from 100 to 24 years. This issue is addressed in several references, among which are USBR (1974), Mays (1999) and Carvalho (1994, 1995, 2007, 2008) which should be consulted for details.

The following is a summary of the evaluation of the useful life of a reservoir, the PCH Poço Fundo (Rio de Janeiro), the basis for this chapter. For this evaluation, one should choose the station with sediment metric measurements closest to the project, whose data are available at Hidroweb/ANA.

Using these data, it is necessary to calculate the solid flow in suspension obtained by the product of the liquid flows by the concentrations, determined from the analysis of the collection of water samples made during the field measurements, through the following expression:

$$Q_{ss} = 0.0864 Q_L C \tag{4.15}$$

where:

Q_{ss} - suspended load, t/dia;
Q_L - net flow, m^3/s;
C - sediment concentration in suspension, mg/l.

In order to establish the curve of the solid flow, or sediment transport curve, the solid flow in suspension must be correlated with the liquid flow through an expression type $Q_{ss} = a\, Q^b$, where a and b are constants obtained as a function of this graph (example Figure 4.11).

Based on this transport curve, it will be possible to calculate the annual sediment contribution to the reservoir, applying the average daily (if available) or monthly liquid flows to the formula. To the result, a percentage of sediment retention, referring to the bed of the water course, should be added, not contemplated by the measurements of the station.

For the evaluation of the silting time, the recommendations of the USBR for reservoir sedimentation may be followed. In the case of Poço Fundo SHP, the calculations performed using ANEEL methodology are presented in Table 4.20, and due to the results obtained, it can be verified if sedimentation problems are expected in the reservoir over its useful life.

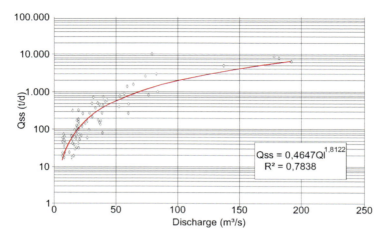

Figure 4.11 Sediment transport curve at the station considered. Poço Fundo SHP R^2 = Correlation coefficient.

Table 4.20 Poço Fundo HPP. Calculation of the Predicted Silting Time for the Reservoir

	Description/Formula	Un.	Value
AD	Drainage area	(km^2)	730
Q_{mlt}	Long-term average net flow	(m^3/s)	16.86
L	Length of the reservoir	(m)	5.191
WL_{res}	Maximum normal reservoir water level	(m)	691.00
A_{res}	Reservoir surface at WLmax	(10^3 m^2)	140
V_{res}	Reservoir volume at WL	(10^6 m^3)	0.47
Q_{st}	Average long-term suspended load	(t/year)	43,518
IS	Sedimentation Index = V_{res}^2 / ($Q^2 \times L$)	-	1.83 E+05
ER	Retention efficiency[b] (given by Churchill curve)	(%)	20%
Pc	Percentage of clay contained in the sediment	(%)	25%
Pm	Percentage of silt contained in the sediment	(%)	25%
Os	Percentage of sand contained in the sediment	(%)	50%
Wc	Coefficient of compaction of clay[a]	-	0.416
Wm	Coefficient of compaction of silt[a]	-	1.121
Ws	Coefficient of compaction of sand[a]	-	1.554
Kc	Constant dependent on the type of reservoir operation (clay)[a]	-	0.256
Km	Constant dependent on the type of reservoir operation (silt)[a]	-	0.091
Ks	Constant dependent on the type of reservoir operation (sand)[a]	-	0.000
K	Constant dependent on the type of reservoir operation (total)[a]	-	0.087
g_i	Apparent specific weight = Wc × Pc + Wm × Pm + Ws × Os	(t/m^3)	1.161
T'	Time of silting (initial calculation) = $V_{res} \times g_i$ / ($Q_{st} \times ER$)	(years)	69.22
g_T	Apparent specific weight in T years (compacted) = Kc × Pc + Km × Pm + Ks × Ps	(t/m^3)	1.319
T	Time of silting (corrected) = $V_{res} \times g_T$ / ($Q_{st} \times ER$)	(years)	78.63

[a] Valid values for the case of "always or almost always submerged sediment".
[b] The retention efficiency obtained by the Churchill curve for suspended sediment was zero. It was considered a 20% retention, referring to the estimated percentage for bed sediment.

It is verified that no sedimentation problems are expected in the reservoir of SHP Poço Fundo throughout its useful life. The reservoir is narrow and shallow, and the flow velocity is not very low, so the retention efficiency obtained from the Churchill curve is zero due to a low sedimentation rate. 20% retention was considered, referring only to bed sediment.

Frame 4.1 shows a list of reservoirs of hydroelectric plants with some type of problem. This list, drawn from Carvalho (1994), is being republished here to showcase the importance of the subject matter to young engineers early in their careers.

As recorded by the author, the Doce river basin is one of the most prolific in sediment production in the country. Heavy rains and large basin slopes are the main natural causes that favor soil erosion. As anthropogenic causes, the exploitation of open-pit ores and extensive deforestation for the production of steel coal are highlighted. A simple visual inspection allows the observation of numerous areas of potential erosion

Frame 4.1 Some Reservoirs in Brazil, Partially or Totally Silted

Reservoir	River	Owner	Power Output
Tocantins-Araguaia basin	Itapecuruzinho	CEMAR	1.0 MW
Itapecuruzinho			
São Francisco basin	Velhas	CEMIG	10 MW
Rio de Pedras	Paraúna	CEMIG	30 MW
Paraúna	Pandeiros	CEMIG	4.2 MW
Pandeiros			
Atlantic-East Basin	Contas	CHESF	30 MW
Funil	Contas	CHESF	23 MW
Pedras	Santa Bárbara	CEMIG	9.4 MW
Peti	Piranga	-	10.5 MW
Brecha	Piracicaba	BELGO-MINEIRA	-
Piracicaba	Piracicaba	ACESITA	50 MW
Sá Carvalho	Tanque	-	2.41 MW
Dona Rita	Santo Antônio	CEMIG	104 MW
Salto Grande-Madeira Lavrada	Tronqueiras	-	7.87 MW
Tronqueiras	Suaçuí Pequeno	-	-
Bretas	Doce	ESCELSA	120 MW
Mascarenhas	Paraitinga	CESP	85 MW
Paraitinga	Jaguari	CESP	27.6 MW
Jaguari			
Parana basin	Pardo	CESP	80.4 MW
Caconde	Pardo	CESP	108.8 MW
Euclides da Cunha	Atibaia	CPFL	34 MW
Americana	Paranapanema	CESP	120 MW
Jurumirim	Paranapanema	SCFL	22 MW
Piraju	Tibaji	Klabin	22.5 MW
Presidente Vargas	Itiquira	ITISA	156 MW
Itiquira	Coxim	ENERSUL	7.5 MW
São Gabriel	São João	ENERSUL	3.2 MW
São João			
Uruguay basin	Caveiras	CELESC	4.3 MW
Caveiras	Chapecozinho	CELESC	5.76 MW
Celso Ramos			
Atlantic-South basin	Jacuí	CEEE	1.0 MW
Ernestina	Jacuí	CEEE	125 MW
Passo Real			

in the basin. The uses of the Doce river basin, such as Brecha, Peti, Piracicaba, Sá Carvalho, Salto Grande, Bretas and Dona Rita, present silting problems.

The Mascarenhas HPP (Figure 4.12) of 120 MW capacity was inaugurated in 1974. This project did not include sedimentological studies (Almeida and Carvalho, 1993). Despite the existence of so many signs, studies and projects did not address the issue properly. The work itself does not have a structure to mitigate the effects of the probable silting. Corresponding to 88% of the basin, the dam is 33 m high at the 63.50 m height and the drainage area is 74,300 km². The volume of the reservoir, in the WL-max, is $42 \times 10^6 \, m^3$ and, in the WLmin, it is $31 \times 10^6 \, m^3$. The volume stored at the 38.00 m height, corresponding to the water outlet threshold, is only $8 \times 10^6 \, m^3$.

The operation of the plant required permanent dredging which was not enough. This plant had its reservoir almost totally silted in 1979, 5 years after its inauguration.

The studies could have been conducted on the basis of data from nearby basins. In 1971 Cemig began to make daily measurements at the Ferros station in the Santo Antônio river. This post changed later to Ouro Fino, near Ferros. It is recorded that the first measurements of solid flow in the Doce river at the Tumiritinga station, upstream of Mascarenhas, date back to 1974 and were carried out by DNAEE. Existing information showed that the sedimentological aspects could not have been neglected.

These authors report in the same paper the case of the shutdown of the three generating units of the Funil HPP (30 MW) in the Contas river, Bahia. These generating units were owned by Chesf, from January 1992 to March 1993. The plant was inaugurated in 1962. In order to restore the plant to operation, it was necessary to dredge a volume of approximately 33,000 m³ of the reservoir and remove 1,000 m³ of sediment from the penstocks. It should be noted that the dam is of concrete-gravity, with a length of 292.70 m in the ridge, incorporating eight gates; with the maximum height at 60 m. The gross head is about 40 m. The total turbine flow is 8.850 m³/s. The width of the block of intake and power house is 54 m. The reservoir has 4.1 km² of surface, total volume of 46.4 hm³ and useful volume of 27 hm³.

Figure 4.12 Mascarenhas HPP.

According to Roig et al. (2003), the reservoir of Funil HPP, of Light in Rio de Janeiro, silted 23% in 23 years, or 1% per year, consistent with those data presented by Mahmood (1987) in a technical report to World Bank.

Power (1988), in the article "Siltation is threat to whole world's storage dams", records that: the reservoir of the Hoover HPP, which came into operation in 1935 (Table 6.7), is silting up 0.3%, and therefore, it has already lost 24% of its reservoir in approximately 79 years; Tarbela is stocking 1.5% per year; and Three Gorges, inaugurated in 1970, is silting 1.7% per year; and that the Warsak reservoir on the Kabul River in Pakistan lost 18% of its capacity in its first year of operation.

White (2010), in the work "World water resources, usage and the role of man-made reservoirs", presented data of 2,300 reservoirs in 31 countries, World Register of Dams. He cited the annual storage loss of 0.5 per year. He also mentioned the half-life of the reservoirs of 12 regions of the world; for South America this half-life is 500 years.

Coelho (1993), in his Master's thesis, shows that the Americana reservoir (São Paulo State) silted 8.9% in 40 years, corresponding to 0.22% per year.

According to Carvalho (1994), in Brazil our loss is also 0.5% per year. It is recommended to consult the author's other works, for example, on the Itaipu HPP (Carvalho, 1995), for which it was estimated that, in 130 years, the volume up to the intake threshold will be silted.

Miranda (2011) shows that the Três Irmãos HPP reservoir, on the Tietê river (SP), silted 14% between 1975 and 2008, corresponding to 0.42% per year. Based on this result, it evaluated the energy losses of 1993, year of operation of the first turbine, to 2008 at 377 MWh/month. According to him, this energy loss could supply the service of 1,508 residences that had a monthly consumption of 250 kWh.

It is worth mentioning the researches of the State University of Campinas (Unicamp) in the backwater section of the reservoir of Barra Bonita HPP (Figure 4.13) coordinated by Professor Gireli.

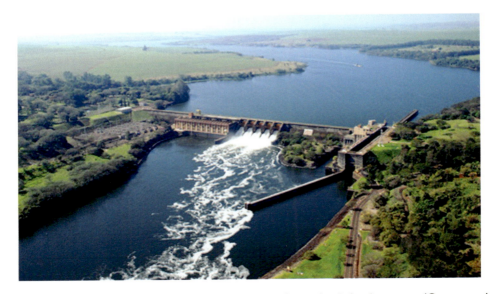

Figure 4.13 Barra Bonita HPP. See navigation lock in the left abutment (Governo do Estado de São Paulo).

In addition to the damage to power generation, there is a risk of a stoppage of freight transportation along Tietê-Paraná, the stretch of the most important waterway in the country,. In less than 5 years, some points were silted up to 12 m as shown in the bathymetric surveys performed.

According to the survey it is no longer possible to sail with vessels with a draft of 2.9 m. In this section the waterway fully works, only for six months a year and the tendency is that this will get worse with the increase of the silting. For details it is recommended to consult Gireli et al. (2011, 2012).

Chapter 5

Power output

5.1 AVAILABLE HEAD

As seen in Chapter 2, the operating WLs are defined in the hydroelectric Inventory Studies of the hydrographic basin. These studies define WLs for each site and consequently the available head (H) of each enterprise as:

- maximum headwater level (WLmax);
- normal headwater level (WL);
- minimum normal headwater level (WLmin); and
- tailwater level (TW).

These levels in principle cannot be changed, except in very special cases. To change the WL in Brazil, authorization is required from SGH-ANEEL/MME.

5.2 POWER OUTPUT

The estimated power output should be made in accordance with the international recommendations. In this book, the instructions for the Feasibility Studies of hydropower plants of Eletrobras (1997) are provided.

Between the 1970s and early 1980s, the electric sector was used as a concept of expansion of supply to serve the consumer market, the so-called firm energy criterion, or deterministic criterion, which established that the electric system, through a simulation of the hydrothermal operation of the generating park, should be able to meet the projected market without energy deficits in the event of any of the flow sequences existing in the historical registry.

For the given system configuration, a set of existing or planned plants, along with the following concepts, are applied:

- Firm system energy: The highest continuous load (critical load) that this configuration can meet considering the aforementioned criterion, throughout the critical period;
- Critical period: The sequence of months of the historical record for which the flow meets the firm's energy of configuration without having leftovers and energy deficits; i.e., it is the time interval in which the system moves from the maximum

DOI: 10.1201/9781003161325-5

storage situation to the minimum storage situation and then returns to the initial situation (maximum storage of the reservoirs).

The average generation of each plant in this critical period is called local firm energy. When this plant is considered in the configuration under analysis, the firm energy gain of a plant corresponds to the increase of the firm energy of the system. The firm energy gain is evaluated by simulating the configuration with and without the plant in question.

This deterministic criterion was replaced by the probabilistic criterion, or criterion of guaranteed energy, compatible with the stochastic nature of the supply to the consumer market. In spite of this, the evaluation of the energy benefit of a plant, for sizing purposes, is still based on the concept of firm energy gain. In addition to this, there is a concept of guaranteed power gain under deterministic criterion, which corresponds to the maximum power generation capacity in the plant. As can be observed from the history of natural flows, this plant functions with a permanence of 95% in its operation in an integrated way to the system.

Another parameter that quantifies the energy benefits of a plant is the secondary energy gain. It corresponds to the energy generated in excess of the firm energy in the months of favorable hydrological inflows. It is evaluated by the difference between average long-term generation of the system and its energy with and without the inclusion of the plant.

Considering these concepts, the determination of power to be installed in the plant can be made in the following way. Increasing the level of installed power at a plant can provide steady energy gains, especially from the generation of secondary energy and by increasing guaranteed energy. When the number of machines of the plant is enough to turbine the regularized flow, it ensures firm energy gain that quickly saturates from a given power range. Therefore, higher motorizations are justified, almost exclusively, due to the gains of secondary energy and guaranteed energy.

In the determination of installed capacity, the motorization is increased, while, in incremental terms, the expression used is:

$$\Delta Ef \cdot \text{CME} + \Delta Pg \cdot \text{CMP} + \Delta Es \cdot \text{CMS} \geq \Delta C \rightarrow \tag{5.1}$$

where:

ΔEf, ΔPg, ΔEs: incremental benefit of firm energy, guaranteed tip and secondary energy, respectively, resulting from the increase of installed power (MW);

ΔC: Incremental cost corresponding to the change from one motorization level to the next;

CME: Marginal cost of energy, US$/MW year;

CMP: Marginal cost of peak, US$/kW/year; and

CMS: Marginal cost of secondary energy, US$/MW year.

As the peak marginal (CMP) and secondary energy (CMS) costs are zero in the period in the Southeast/Center-West region, and with the annualized incremental values of economic benefit and of costs involved, the power increase should be pursued while checking Equation 5.2:

Power output 83

Table 5.1 Marginal Expansion Costs

Year	CME Expansion (R$/MWh)
2022	178.89
2023	189.50
2024	214.96
2025	271.23
2026	232.30
Média	217.38

Source: EPE

$$\left[\left(\frac{8760 \cdot \Delta Ef \cdot CME}{FRC(n,i) \cdot \Delta C}\right)\right] \geq 1 \tag{5.2}$$

where:

$$FRC\ (n,\ i) = \frac{i \cdot (1+i)^n}{(1+i)^n - 1} \tag{5.3}$$

being (n) the useful life of the enterprise and (i) the discount rate.

In recent studies, it has been adopted $n = 30$ years and $i = 12\%$ p.a. It should be emphasized that market values should be used instead of Marginal Costs CME, CMP and CMS.

When the project is an SHP, the energy considered must be the average, obtained from a history of flows with a minimum of 30 years of data. If the data history includes the critical period of the system, the energy considered will be the average of the entire period.

The CME, marginal cost of expansion, is provided by EPE. The average value between 2022 and 2026 is in the order of 217 R$/MWh.

More recently, considering private ventures, it was possible to choose value of energy from the prices practiced in the market (Table 5.1).

5.3 TURBINE TYPE SELECTION

The selection of the turbine type is carried out and expressed as a function of the head and the flow, as presented in Chapter 10. Francis turbine was selected on the basis of Poço Fundo SHP with a gross head of 73 m and a nominal flow of 22.55 m³/s.

5.4 ENERGY SIMULATION

The energy simulation of the plant becomes a simpler problem in the case of an SHP with run of river reservoir, and without the presence of another upstream undertaking, that regulates the affluent flows. Thus, for the energy simulation of the plant, an individualized plant model was elaborated, which considers operation of the plant on month to month basis, According to the history of monthly average flows generated for the site, and allowing for the determination of average energy of the nominal flows and of the spilled flow rates, energy simulation of the plant can be determined.

Data used in the simulation:

- reservoir WLmax (m): 691.00 m;
- WLtw (m): 618.00 m;
- hydraulic losses (m): 4.88% of gross head = 3.56 m;
- nominal overall efficiency of the turbine-generator group (%): 90.6%;
- sanitary flow (m^3/s): 1.96 m^3/s;
- forced unavailability: 2%;
- programmed unavailability: 3%.

Based on the data presented and considering several levels for the installed capacity, the energy simulation was performed for the Poço Fundo SHP, using the mean flow series (Table 4.9) and the flow duration curve (Figure 4.4). The results are recorded in Table 5.2 and Figure 5.1, and are used in the calculations presented in the following items.

Table 5.2 Poço Fundo SHP – Energetic simulations (Basic Design)

PO (MW)	Medium Energy (MW med)	Capacity Factor	Reference Flow (m^3/s)
12.00	7.29	0.61	19.05
13.00	7.52	0.58	20.78
14.00	7.69	0.55	22.55
15.00	7.83	0.52	24.37
16.00	7.94	0.50	26.24

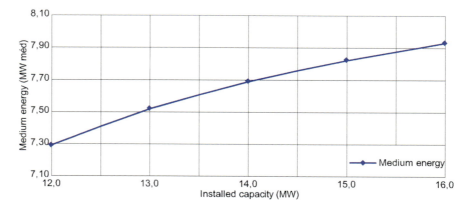

Figure 5.1 Poço Fundo SHP. Installed capacity × medium energy (Basic Design).

5.5 ENERGY-ECONOMIC DIMENSIONING

Taking into account the economic parameters of reference, for the energy-economical design of the plant, an incremental analysis should be performed by gradually increasing the plant's installed power until the economic benefits resulting from the additional energy are exceeded at the cost of this increase.

The annual cost was calculated for an economically useful life of 30 years, at an interest rate of 12% per year and an annual cost of operation and maintenance of R$ 10.00/MWh. The investment costs of the motorization alternatives used in the energy-efficient Basic Design are estimated as a function of the work arrangement, conceived as previously described, are shown in Table 5.3.

The energy benefits, valued by market parameters, as well as the analysis of benefits/costs at an incremental level are presented in Tables 5.4 and 5.5.

The analysis of these data allows to conclude that the power of the Poço Fundo SHP should be 14 MW, since, from this motorization, the benefit/cost ratio is less than 1 (one), i.e. the energy cost is higher than the benefit. This power output corresponds to the nominal flow of $22.55\,m^3/s$, with a permanence of 22%. For comparison purposes, the MLT flow rate is $16.86\,m^3/s$; the average flow rate of the critical period is $15.20\,m^3/s$, and the steady flow (with 95% probability of occurring or being exceeded) is $6.4\,m^3/s$.

5.6 NUMBER OF GENERATING UNITS

The Poço Fundo SHP, with 14 MW of installed capacity, should provide an average energy of 7.69 as an average MW, and with a capacity factor of 55%. Two generating units were adopted, which will allow good operational flexibility and make it possible

Table 5.3 Poço Fundo SHP – Investment Cost of the Alternatives

PO (MW)	Investment Cost (R$)	Annual Cost (R$/ano)	Incremental Cost (R$)
12	59.168.870	7.984.044	
13	60.952.189	8.225.580	241.536
14	62.437.950	8.424.919	199.340
15	63.980.630	8.628.697	203.778
16	65.493.610	8.826.160	197.463

Reference date: April 2010.

Table 5.4 Poço Fundo SHP – Energy Benefits (Basic Design)

PO (MW)	Medium Energy (MW Medium)	Annual Benefit (R$)	Incremental Annual Benefit (R$)
12	7.29	8,940,456	-
13	7.52	9,222,528	282,072
14	7.69	9,431,016	208,488
15	7.83	9,602,712	171,696
16	7.94	9,737,616	134,904

Reference date: April 2010.

Table 5.5 Poço Fundo SHP, Incremental Analysis – Benefit/Cost (B/C)

PO (MW)	Incremental Benefit (R$)	Incremental Cost (R$)	B/Ct
12	-	-	-
13	282,072	241,536	1.17
14	208,488	199,340	1.05
15	171,696	203,778	0.84
16	134,904	197,463	0.68

Reference date: April 2010.

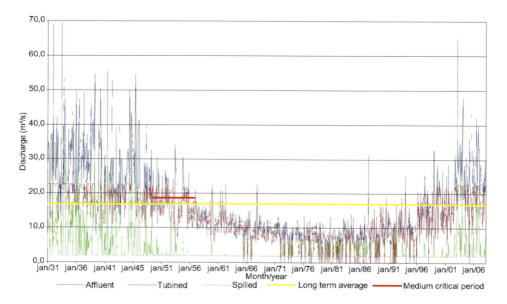

Figure 5.2 Hydrograms – Poço Fundo HPP (Basic Design).

to supply up to 75% of the average energy with only one unit when the other machine is unavailable for maintenance.

The temporal utilization of the water availability at the place of the enterprise is presented below, as per the graphs depicting the expectation of inflows to the reservoir. This explains the transformation of the hydraulic energy into electric energy and the turbination characteristics of these inflows by the machines.

Graphs depicted in Figures 5.2 and 5.3 and Table 5.6 show the hydrograms of the Preto river in the SHP Poço Fundo, the energy simulation of the operation of this SHP, as well as the analysis of the unavailability of the turbines as a function of the fluvial regime. It is observed that the possibility of total shutdown of the machines in the adopted solution as a function of insufficient affluence is around 8%.

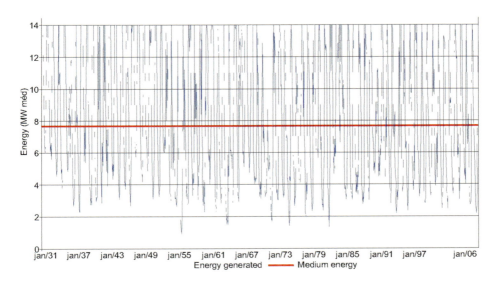

Figure 5.3 Energy – Poço Fundo HPP (Basic Design).

5.7 DETERMINATION OF PHYSICAL GUARANTEE

A value is defined that represents the physical guarantee for concessionaires of hydroelectric plants that are interested in participating in the benefits of hydrological risk sharing through the Energy Reallocation Mechanism (MRE).

The owner of the plant may establish power supply contracts, whose sum is less than or equal to this amount, and register them at CCEE (see the list of acronyms).

For these contracts, both buyer and seller have the guarantee that they can count on this energy for the monthly accounting of their consumption/generation, without purchasing any additional amounts and eliminating exposure to the hydrological risk.

If this is the purpose for which the plant is being designed, it is suggested that in the final energy-economic design, the optimum power is determined from the net annual average energy that will be homologated later in the definition of the physical guarantee.

In order to find the value of the average annual net energy, the ANEEL Ordinance No. 463/2009 (2009) defines the use of the following parameters as necessary:

I. Generator output (kW);
II. Output by generator and its power factor (kVA)/fp;
III. Output by turbine and its minimum flow (kW]/[m3/s);
IV. nt: Nominal turbine yield (%);
V. ng: Nominal generator yield (%);
VI. TEIF: Equivalent rate of forced unavailability (%);
VII. IP: Scheduled unavailability (%);
VIII. h: Nominal hydraulic losses (m);

Table 5.6 Analysis of Number of Machines – Poço Fundo HPP (Basic Design)

WLres = 691.00 m	Francis Turbine	Qmin. Observed = 1.62 m³/s
WLtw = 618.00 m	Output = 14 MW	Qmax. observed = 67.52 m³/s
Href. = 73 m	Efficiency $T+G$ = 90.6%	Q 95% = 4.50 m³/s
Losses = 5.0%	Qmax. turb. = 22.55 m³/s	Ecological flow = 1.93 m³/s
Hnet = 69.44 m	Qmin. = 45%	

Characteristics 0.0 0.3 0.8 1.0 2.1 6.5 10 19.5 26.0 32.4 42.7 44.3 48.3 53.9 58.5 62.7 65.3 67.6 69.4 71.6 72.8 75.4 77.4 **Permanence**

UnID	Unit Qmax (m³/s)	Unit Qmin (m³/s)	% time stop	Und	1.6	2.1	2.7	3.2	3.7	4.8	5.3	6.5	7.9	8.9	10.6	11	12.1	13.1	14.2	15.2	16.3	17.3	18.4	19.4	20.5	21.5	22.6	Discharges (m³/s)
1	22.55	10.04	41%	1								X	X	X	X	X	X	X	X	X	X	X	X	X	X	X	X	Operating
2	11.28	5.02	8%	2								X	X	X	X	X	X	X	X	X	X	X	X	X	X	X	X	range
				1					X	X	X	X	X	X	X	X	X	X	X	X	X	X	X	X	X		X	
3	7.52	3.35	1%	3					X	X	X	X	X	X	X	X	X	X	X	X	X	X	X	X	X	X	X	
				2				X	X	X	X	X	X	X	X	X	X	X	X	X	X	X	X	X	X	X	X	
				1	X	X	X	X	X	X	X X	X	X	X	X	X	X	X	X	X	X	X	X	X	X	X	X	

IX.	Hb: Nominal gross head [m];
X.	Losses: losses to the point of connection with the system (%);
XI.	Cint: Intern consumption (MW médios);
XII.	qr: Remaining flow of the plant (m3/s);
XIII.	qu: Flow of consumptive uses (m3/s);
XIV.	History of average monthly flows (m3/s), not less than 30 years, generated in such a way that it is as extensive and up to date as possible, and must be in conformity, when applicable, with the flow history presented in the approved Basic Project (FEED);
XV.	Detailing of the methodology for obtaining the flow history specified in the previous paragraph, as well as all the information necessary for the reproduction of said history.

Applying the calculations defined in Section 5.4 and considering the following equation is found the probable value of the physical guarantee that always is defined by ANEEL and MME for the enterprise to participate in the MRE.

$$\text{Physical Guarantee} = (\text{Average energy} - \text{Consumption}) \times (1 - \text{Losses}) \tag{5.4}$$

where:

- Average energy: is the calculated energy (MWmed) according to Section 4.3.4, considering the unavailability indices (TEIF and IP);
- Consumption: is the sum of energy (MWmed) consumed in the plant's own facilities;
- Losses: is the sum of electrical losses in percentage from the elevation of the voltage in the plant itself to the point of connection with the transmission/distribution

Chapter 6

Geological and geotechnical studies

6.1 INTRODUCTION

This chapter will present the topic of geological and geotechnical investigations and studies based on the experience of dam breaks by foundations that have already occurred for young engineers and engineering geologists to understand the issue in depth. They need to clearly keep in mind that knowledge of the sub-soil conditions of the dam sites is essential for the development of a project suitable for the safety of dams and power plants.

The layout, the size of the structures and the estimated loads in each project will condition the extension of the program of investigations of the foundations and characterization of the materials that will be used in the construction. The program should include detailed surveys of the foundations, including the abutments, and, in addition, must include detailed researches to characterize the materials of the earth borrow areas, the quarries, the sand and gravel deposits closest to the site, likely locations for disposal areas, areas for industrial facilities, as well as areas for construction sites and support facilities.

Lessons learned from dam accidents, ruptured by the foundations, and ruptured by the abutments and also by overtopping, many of which lead to loss of life, corroborate the need for studies and geological and geotechnical investigations to be made and planned by engineers and engineering geologists with proven experience in projects of this nature, in all phases of studies.

It should be remembered that the dams have already exceeded 300 m of height, which means columns of water also greater than 300 m, that is, very high pressures. Examples: Nurek dam (304 m) and Rogun dam (335 m) in Tajikistan.

To sensitize young engineers to the importance of the theme, special mention is made of the 5th Lauritz Bjerrum Memorial Lecture, presented in Oslo, Norway, 05/05/1980, by Peck, R. B.: "Where Has All the Judgment Gone?". This 40-year-old Peck lecture continues and will continue to apply today and into the future. The lecture discussed, among other accidents, the rupture of the Teton dam that occurred on 05/06/1976 in Rexburg, Idaho (USA). This 100 m high earth dam, built by the USBR, ruptured by tubular erosion (piping) during the first filling of the reservoir. It caused the loss of 11 human lives and a high financial loss.

According to Peck, the collapse began at the right abutment, where a key trench with abrupt walls had been excavated on the rock of the canyon slope.

The rocky massif at the site was heavily fractured and the discontinuities on both sides of the trench were open and untreated. The sealing material used in the core and trench of the dam, a clay silt of wind origin, was thrown directly against the rock,

DOI: 10.1201/9781003161325-6

without placing any transition. A single-line injection curtain was flanked on both sides by shorter hole lines, which deepened from the center of the key trench. This curtain consisted of the injection of cement slurry into open holes from a concrete casing, shaped into a small groove in the rock, along the key trench.

Irrespective of the weakness that existed at the initial concentration site of the flow, the research groups, the Independent Panel (Chadwick et al., of which Peck participated, 1976) and the Interagency Review Group (1980), concluded that the project was flawed, since it allowed the conjugation of several unfavorable factors, such as: (1) highly erodible core materials; (2) heavily fractured rock without dental concrete or surface treatment; (3) absence of transition zones between core and rock; (4) unfavorable stress conditions associated with the narrow width of the abrupt-walled key trench; and (5) potential condition of leakage through curtain of injections below the mortar capping.

The placement of the thin and erodible core material with highly declassed foundation was fatal. Due to the experience with other accidents, described in the same lecture, it was possible to conclude that all collapses occurred by the conjugation of two or more defects. As quoted in the lecture, this conclusion supports the principle of projecting itself with deep defenses, which corresponds to the principle of the "belt and suspensory" advocated by Casagrande, A. (apud Peck, 1982). "This principle postulates that if any element of defense in the dam or its foundation fails to perform its function, there must be one or more forms of defense to take its place".

The Teton dam is a clear example of disrespect to this principle. The only lines of defense were represented by the nucleus and the curtain of injections, both poorly designed. Photographs of this disaster are presented in Chapter 13, on dam accidents.

Another paper by Peck, "The influence of non-technical factors in the quality of dams" (1972), remains current and must be observed by the technicians involved with the projects and construction of dams. These factors are (1) unrealistic customer, (2) insecure designer, (3) unfair designer, (4) super optimistic designer, (5) designer who misuses his consultants, (6) super busy consultant, (7) inexperienced designer in construction, (8) abuser of the observational method, (9) inefficient inspector, (10) loophole contractor, (11) financially weak contractor and (12) non-qualified contractor.

6.2 INVESTIGATIONS/STUDY PHASES

The text that follows was prepared based on the research program carried out to characterize the foundations of the structures of the SHP Poço Fundo recently built in the Rio de Janeiro, as well as to characterize the borrow areas for construction materials.

Is a practical text and had as references the traditional Design of Small Dams (USBR, 1973) and the most recent Rockfill Dams, Design and Construction book (Höeg, 1993). The book *Reliability and Statistics in Geotechnical Engineering* by Baecher and Christian (2003) should be mentioned here.

This book presents the subject in depth in four parts as described below. Part I: (1) Introduction – uncertainty and risk in geotechnical engineering; (2) Uncertainty; (3) Probability; (4) Inference; (5) Risk, decisions and judgement; Part II: (6) Site characterization; (7) Classification and mapping; (8) Soil variability; (9) Spatial variability within homogenous deposits; (10) Random field theory; (11) Spatial sampling; (12) Search theory; Part III: (13) Reliability analysis and error propagation; (14) First order second moment (FOSM) methods; (15) Point estimate methods; (16) The Hasofer-Lind approach (FORM); (17) Monte Carlo simulation methods; (18) Load an resistance factor design; (19) Stochastic finite elements; Part IV: (20) Event tree analysis; (21) Expert opinion; (22) System reliability assessment. And concluding the book the authors present an appendix with the Primer on probability theory. This is a mandatory reference. These details are clearly not part of the scope of this book.

Going back to the SHP Poço Fundo project example, the existence of a Geological Map of the region (Figure 6.1), with the characteristics, faults and joints of the region that can condition the project, greatly facilitated the planning of geological-geotechnical investigations to subsidize the project.

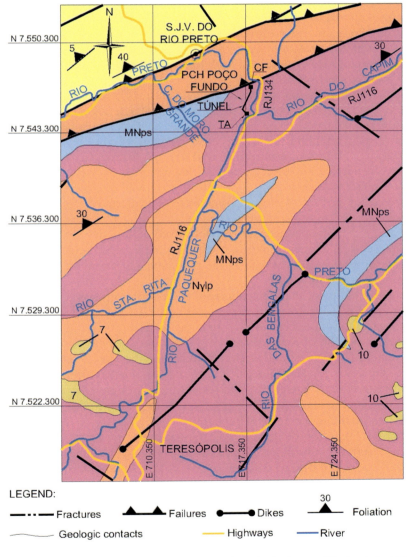

Figure 6.1 Poço Fundo SHP. Basic Design. Regional geological map. São José do Vale do Rio Preto. Rio de Janeiro. Watermark Engineering (2007).

It's worth remembering that in the Inventory Studies the works are limited to the execution of geophysical surveys. Some auger holes, test pits and exploratory trenches may also be performed. Analyzes of aerial photographs of the region (photointerpretation) are carried out, which is also extremely important in the definition of dam sites and for planning the program of field exploratory work. The program of geophysical investigations usually includes the following methods: refractive seismic, vertical electric survey (SEV), very low frequency (VLF) and more recently ground penetration radar (GPR).

All these investigations due to their character (indirect) measure physical properties that should be interpreted and associated with geological-geotechnical features of interest, that is, features such as: layers of soil; rocky top (sound, fractured, altered, etc.); fault zone, dykes, water table and so on. In this phase, the characteristics of the materials, the support capacity and the permeability of the foundations, as well as the identification of possible erosive processes and scars of landslides, the local vegetation and their state of conservation, accesses, etc. are diagnosed. Their results will subsidize the program of field exploratory work that should be prepared for the next phase of studies, the Feasibility Studies.

In the Feasibility Studies a program of field subsurface exploration (include testing) must be prepared. This program should consist of a detailed field reconnaissance and mapping by engineers and geologists and the execution of subsurface explorations, and soil and rock sampling, including auger borings, boreholes, test pits, trenches, audits, in situ soil and rock permeability, and strength testing. In addition, laboratory tests should be carried out to characterize the materials, as well as triaxial tests to determine the shear strength.

The execution of this program, as well as the collection of samples in the region of the foundation of all the permanent structures and borrow areas of soils, deposits and quarries, should be carried out by specialized companies, according to the international practice. The information obtained should be sufficient to characterize, in detail, the subsoil profile, as well as the parameters of strength, permeability and deformability of the foundation.

The determination of the permeability of the foundation should be done by means of infiltration tests (EI) in the stretches in soil in the drillings, at each meter, and the tests of loss of water under pressure (EPA) in the sections of rock in the drillings.

The determination of the parameters of the shear strength and deformability of the earth materials is made based on the SPT tests and the triaxial tests on undeformed or molded samples in the laboratory. For rock masses, the determination can be made in laboratory tests or, if possible, in situ tests (which are time-consuming and costly). These parameters can also be defined based on the experience obtained in works with similar masses, using semiempirical methods.

During the execution of the surveys, the geologists receive the survey bulletins and begin designing subsoil profiles. The geological-geotechnical sections of the foundations are elaborated to characterize in detail the foundations of the several structures of the dam. What is pursued is to know, in detail, the subsoil to define the geological-geotechnical parameters of the foundation mass of each structure of the use with the "maximum possible precision".

Of course, the spacing between the drill holes introduces limitations that require technicians to "infer" what happens between holes. The combination of direct and indirect methods can greatly improve the results of the foundation characterization process.

Geological and geotechnical studies 95

The Regional Geological Map of the Poço Fundo SHP (14 MW) in the Preto river, is presented as an example (Figure 6.1). This SHP was implemented in the granitic-gneiss massif of the Organ Saw, where the famous "Finger of God" is shown in Figure 6.2b.

In geological maps, it is possible to observe/identify units and geological structures, such as:

- Geological Units

    MNps – Paraíba do Sul Complex - São Fidélis Unit: Granada-biotite-sillimanite quartz-feldspar gneiss (metagrauvaca), with pockets and veins in situ or injected granitic composition. Intercalations of frequent calcissilictic and quartzite gneiss. Varieties with cordierite and sillimanite (kinzigito) with transitional contacts with the biotite gneiss. Horizons of graphite schists are common. Calcissilictic, metacarbonate and quartzite also occur. In rare domains with low deformation rate the turbidite structures are preserved;

Figure 6.2 (a) Organ Saw. (b) Rocks belonging to the Organ Saw. It is the largest granite batholith exposed in the Rio de Janeiro.

Nylp - leucogranite gneiss, paquequer saw: muscovite-biotite leucogranite, sillimanite-granada-biotite, S-type granite with thick granulation, with strong tangential foliation, rich in paragneiss remains;
- Geological Structures:
  Fractures, Faults, Dykes, Foliation, Geological Contacts, etc.
  Follow a satellite photo of the site of the Poço Fundo SHP on which the plot of the works of the plant was inserted (Figure 6.3). After it, three more photos of the rocky massif on the site are shown (Figures 6.4–6.7). Looking at the photos, even without expertise, quality of the rock mass as a foundation for any kind of engineering work can be observed.

At the site several mixed drills (percussion and rotary) were carried out that proved the quality of the sound massif, with poorly developed fracturing, but with excellent geomechanical qualities. The power tunnel will be excavated within a single lithology and most likely without any problem of stability and watertightness. Figure 6.7 shows its examples.

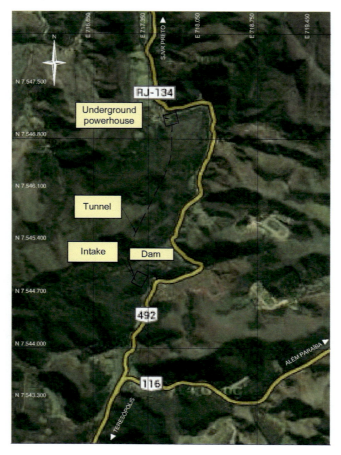

Figure 6.3 Poço Fundo SHP. Satellite photo of the site. Basic Design – Watermark Engineering (2007).

Geological and geotechnical studies 97

Figure 6.4 Poço Fundo SHP. Basic Design. Granite-gneissic rock massif along the stretch of the power tunnel. (Photo taken from the right bank of the river.)

Figure 6.5 Poço Fundo SHP. Basic Design. Granitic-gneiss massif downstream of the tailrace tunnel.

98 Design of Hydroelectric Power Plants – Step by Step

Figure 6.6 Poço Fundo SHP. Basic Design. Preto river valley with rocky, granitic-gneissic, and colluvial soils.

Figure 6.8 shows the Aimorés dam, Doce river (Minas Gerais), 18 m high. Figure 6.9 shows a typical example of a log extracted from Costa (2012). In Brazil, bulletin templates for all types of surveys and trials can be found in the Survey Manual – Bulletin Nº 3, 5th Edition (ABGE, 2013).

Some traditional tables for the classification of soils and rock masses are presented below, along with some tables containing the resistance and deformability properties of these materials, that are routinely used in the elaboration of geotechnical dam projects.

These tables were extracted from Höeg (1992). Of course, this information should be used by young engineers only under the supervision of consultants with proven experience in projects of this nature.

For the classification of fine soils, i.e., soils with more than 50% passing through the sieve # 200 $D = 0.074$ mm, use the SUCS-USBR plasticity chart (Figure 6.10).

Frame 6.1 is presented with the defects that can occur in rock masses (Fell et al., 2017). Afterwards, is presented as:

- Frame 6.2 with the soil classification in the SUCS – USBR groups, which is based on laboratory tests;
- Frame 6.3 with the expected properties of soil groups;
- Frame 6.4 with the classification of the rocky massif as a function of specific geotechnical parameters;
- Frame 6.5 with the rock massif classifications with hydraulic decomposition, consistency, fracturing and conductivity.

Geological and geotechnical studies 99

Figure 6.7 Drill cores. Poço Fundo SHP Basic Design. Watermark Engineering Rio de Janeiro (2007).

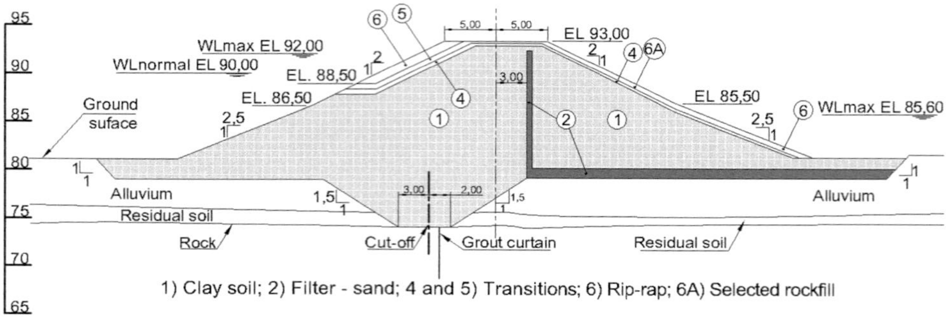

100 Design of Hydroelectric Power Plants – Step by Step

1) Clay soil; 2) Filter - sand; 4 and 5) Transitions; 6) Rip-rap; 6A) Selected rockfill

Figure 6.8 Aimorés earth dam. Minas Gerais. (CBDB – MBD. Vol III. 2009).

Follow the tables:

- Table 6.1, Classification of Soils by the Consistency of Clays and Sand Compactness;
- Table 6.2 – E and Cc values for some materials;
- Table 6.3 – Units and Geomechanical parameters (Itaipu HPP).

Geomechanical classifications are systems developed on the basis of the characteristics of rocky massifs. The definition of different classes of massifs is important for the excavation and treatment of foundations. The main geomechanical parameters are:

- resistance of intact rock;
- resistance of the discontinuities (joints, planes of structural weakness of the massif);
- orientation of unfavorable discontinuities in relation to excavations;
- density of compartmentation of the mass (frequency of joints);
- groundwater influence;
- tension conditions around the excavation.

The situations in practice vary from place to place and, similarly, vary the treatment. If one of the goals of the treatment is to improve the tightness of the foundations, obviously the most permeable and fractured masses are the most critical. The treatment based on injections will be heavy.

There are two more known geomechanical classification systems:

- the Rock Mass Rating (RMR) – Bieniawski (1984);
- the Q System – Barton (1993).

The details of these geomechanical classification systems can be found in several specific publications listed in the references. The construction of dam foundation

Geological and geotechnical studies 101

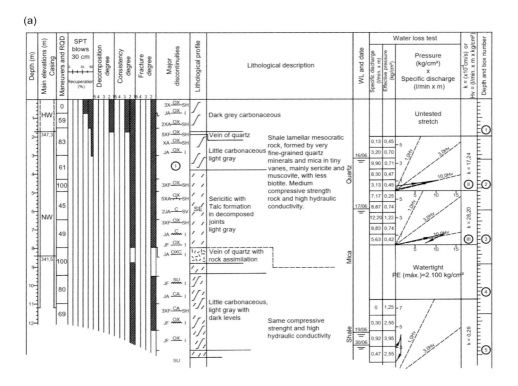

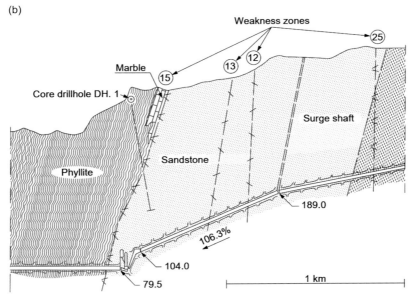

Figure 6.9 Borehole or drillhole log: (a) example from Costa (2012) and (b) example from Nielsen and Thidemann – Meräker HPP (1993).

102 Design of Hydroelectric Power Plants – Step by Step

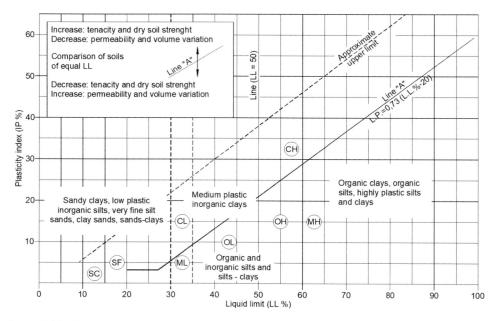

Figure 6.10 Plasticity Chart – SUCS-USBR (Vargas, 1978; ABGE, 1998).

models aims to characterize the prototypes for geological and geotechnical constraints based on information from all previous phases of research, showing the relationship between these constraints and the needs of the project (Costa, 2012). The model can be geomechanical or geo-hydrological.

According to Costa (2012), the geomechanical model can indicate the quantitative relationships between different geological components and the various design alternatives, provided that such components are characterized geomechanically in the text. Figure 6.11 shows an example of the Itaipu HPP.

Figure 6.12 presents the geomechanical model of a studied alternative for the Irapé dam, in the Jequitinhonha river – Minas Gerais State.

According to Costa (2012), in the geo-hydrological model, it is sought to show the generalized behavior of the rocky massif of the foundations due to its influences related to the infiltration of rainwater, the percolation and the consequent rock alteration. This model is fundamental in the analysis of the deformability, stability and watertightness of the foundations (see, for example, Figure 6.13).

It should also be remembered that the shear strength of soils and rockfills is given by the Mohr-Coulomb equation (Equation 6.1), whose parameters are shown in Figure 6.14:

$$\tau_f = c' + (\sigma - u)_f \cdot \tan\phi' \tag{6.1}$$

Defect Name		*Description*	*Typical Extent/Occurrence*	*Formed by*
	Joint	Almost planar surface or crack, across which the rock usually has little tensile strength. May be open, air or water-filled, or filled by soil or rock substances which acts as cement. Joint surfaces may be rough, smooth or slickensided.	From less than 1 m, up to 10 m.	Shrinkage on cooling (igneous rocks). Extension or shear (any rocks) due to tectonic stresses, or unloading.
	Sheared zone -1 (fault)	Zone with roughly parallel almost planar boundaries, of rock substance cut by closely spaced (often<50 mm) joints and or cleavage surfaces. The surfaces are usually smooth or slickensided and curved, intersecting to divide the mass into lenticular or wedge-shaped block.	More than 10 m, up to kilometers.	Faulting, during which small displacements occurred along slickensided surfaces distributed across the width of the zone.

(Continued)

Defect Name		Description	Typical Extent/Occurrence	Formed by
	Crushed seam (1)	Seam with roughly parallel almost planar boundaries, composed or usually angular fragments of the host rock Substance. The fragments may be of gravel, sand, silt or clay sizes, often mixtures of these. The finest fragments are often concentrated in thin layers or slivers next to and parallel to the adjacent rock. Slickensides are common in the sliver zones and along the rock boundaries.	More than 10 m, up to kilometers.	Faulting, during which relatively large displacements occurred within the zone.
	Soil infill seam	Seam of soil substance, usually with very distinct roughly parallel boundaries. This seams usually CH; thick near-surface seams sometimes GC or GC-CL (2).	Less than 10 m, within near-surface mechanically weathered zone, or in rock disturbed by creep or land sliding.	Migration of soil into an open joint or cavity.
	Extremely altered seams	Seam of soil substance, often with gradational boundaries. The type of soil depends on the composition of the host rock.	Varies widely, but extremely altered seams are usually near the surface weathered zone.	Weathering (or alteration) in place, of rock substance, usually next to or within a pre-existing defect.

Notes: (1) In the geological sense; (2) Group symbols as defined in Frame 6.2; (3) A marker feature, showing the nature of the displacements. Such features are observed only rarely.

Geological and geotechnical studies 105

Frame 6.2 Unified Soil Classification System (SUCS-USBR)

Field Identification Procedures (Excluding Particles Larger than 0.074 mm   Soil Classification and Basing Fractions on Estimated Weights)

Coarse-grained soils More than half of material is larger than 0.074 mm sieve size	Gravels More than half of coarse fraction is larger than 4.8 mm sieve size	Clean Gravels (% passed 0.074 mm sieve < 5%; little or no fines)	$Cu = D_{60}/D_{10} > 4$ $1 < Cc = (D_{30})^2 l$ $D_{10} \times D_{60} < 3$	GW	Well grade gravels
			$Cu = D_{60}/D_{10} < 4$ 4	GP	Poorly grade gravels
			$1 > Cc = (D_{30})^2 l$ $D_{10} \times D_{60} > 3$		
		Gravels with fines (% passed 0.074 mm sieve > 12%; appreciable quantity of fines)	Fine classified as	ML MH GC	GM Clayey gravels	Silty gravels
	Sands More than half of coarse fraction is smaller than 4.8 mm sieve size	Clean sands (% passed 0.074 mm sieve > 5%; little or no fines)	$Cu = D_{60}/D_{10} > 6$ $1 < Cc = (D_{30})^2 l$ $D_{10} \times D_{60} < 3$	SW	Well grade sands
			$Cu = D_{60}/D_{10} < 6$ $1 > Cc = (D_{30})^2 l$ $D_{10} \times D_{60} > 3$	SP	Poorly grade sands
		Sands with fines (% passed 0.074 mm sieve > 12%; appreciable quantity of fines)	Fine classified as	ML MH CL CH	SM SC	Silty sands Clayey sands
Fine-grained soils More than half of material is smaller than 0.074 mm sieve size	Silts and clays LL < 50%	Inorganics	IP > 7, points on or above line A	CL	Inorganic clays
			IP < 4, points below line A	ML	Inorganic silts
		Organics	LL dry < 0.75 LL natural	OL	Organic clays Organic silts
	Silts and clays LL > 50%	Inorganics	Points on/above line A	CH	Inorganic clays high plasticity
			Points below line A	MH	Inorganic silts Elastic silts
		Organics	LL dry < 0.75 LL natural	OH	Organic clays Organic silts
Highly organic soils			Mainly organic matter, dark color and smell	PT	Peat

$Cu = D_{60}/D_{10}$, coefficient of uniformity; $Cc = (D_{30})^2/D_{10} \times D_{60}$, coefficient of curvature; LL, liquidity limit; IP, plasticity index.
On the right in the soil classification column the other acronyms are defined. See also Figure 6.9

106 Design of Hydroelectric Power Plants – Step by Step

Frame 6.3 Expected Properties of Soil Groups SUCS-USBR. (ABGE, 98)

Gr.	Workability	Permeability[a]	Shear strength[b]	Compressibility[b]	$\gamma_{d\ máx}$ t/m^3 (PN)	Foundation Value	Drainage charact.
GW	Excellent	Pervious	Excellent	Negligible	2.0–2.2	Good to	Excellent
GP	Good	Negligible	Good	Negligible	1.8–2.0	excellent	Excellent
GM	Good	Semip to perv.	Good	Negligible	1.9–2.2		Reg to poor
GC	Good	Impervious	Reg. to good	Very low	1.85–2.1		Poor
SW	Excellent	Pervious	Excellent	Negligible	1.75–2.1		Excellent
SP	Regular	Pervious	Good	Very low	1.6–1.9	Poor to good	Excellent
SM	Regular	Semip to perv.	Good	Low	1.75–2.0	Poor to good	Reg to poor
SC	Good	Impermeável	Reg. to good	Low	1.7–2.0	Poor to good	Poor
ML	Regular	Semip to perv.	Regular	Medium	1.5–1.9	Poor to good	Reg to poor
CL	Reg. to good	Impervious	Regular	Medium	1.5–1.9	Poor to good	Poor
OL	Regular	Semip to perv.	Low	Medium	1.3–1.6	Poor	Poor
MH	Poor	Semip to perv.	Low/reg.	High	1.1–1.5	Poor	Reg to poor
CH	Poor	Impervious	Low	High	1.2–1.7	Reg. to poor	Poor
OH	Poor	Impervious	Low	High	1.1–1.6	Very poor	Poor
PT	Compaction extremely hard. Not used as earthfills. Must be removed from foundations. Excessive settlements. Very low resistance.						

Obs: $\gamma_{d\ máx}$ = maximum dry bulk density (t/m^3); PN, Proctor Normal.
[a] When compacted.
[b] Compacted and saturated.

where:

τ_f' = shear strength at rupture in the plane of rupture;
c' = effective cohesion;
σ' = $\sigma-u$ = effective normal tension (at the points of contact of the particles);
σ = normal stress in the plane of rupture;
u = pore pression in the plane of rupture (water pressure confined to the voids between the particles);
ϕ' = angle of effective friction.

It can be seen that:

$$a' = c'\tan\phi'$$

(6.2)

The shear strength of poorly permeable soils, clays and silts in which water in the pores does not drain rapidly, depends on the neutral pressures developed during the loading process. In order to predict the load and the pore pressures that develop in a time-dependent earth core, the neutral pressure parameter, B, and the consolidation coefficient, Cv, are measured in triaxial tests, which are not the subject of this book. It is recommended that you consult the specific references of Soil Mechanics listed at the end of this chapter, p. ex. Lambe and Whitman (1969).

Geological and geotechnical studies   107

Frame 6.4 Classification of the Rocky Massif as a Function of Specific Geotechnical Parameters. Ilha Solteira HPP (Paraná river) (Monticeli and Costa, 2012)

Geological Classification		Geotechnical Classification		
	Alteration	Fracture		
		Description	Fractures/ meter	Rock Classification
Clay basaltic breccia	Extremely altered	Irregular fractures		V
Compact basalt vesicle	Heavily altered	Extremely	> 20	IV
amygdaloidal	Virtually sound	fractured rock		
		Very fractured	11–20	III
		Few fractured	2–10	II
		Occasionally	1	I
		fractured		

Rock Class	Deformability Modulus (kg/cm^2)	Design Shear Parameters (kg/cm^2)
I	~ 200,000	-
II	125,000	$\tau = 7.8 + \sigma tg56°$ (*)
II	35,000	$\tau = 6.1 + \sigma tg48°$
III*	-	$\tau = \sigma tg34° \left(0 < \sigma < 6\,kg/cm^2\right)$
		$\tau = 1.6 + \sigma tg34° \left(0 < \sigma < 16\,kg/cm^2\right)$
IV	9,000	$\tau = 1.0 + \sigma tg35°$
V	4,000	$\tau = 0.5 + \sigma tg30°$

(*) Note: used: $\tau = 8.0 + \sigma tg45°$ for foundations project.

In summary, it is recorded that neutral pressure at a point in the landfill prior to any consolidation is defined by:

$$u = B\gamma z \tag{6.3}$$

where:

B = neutral pressure parameter measured in triaxial tests;
γ = specific soil weight above the point (t/m^3);
z = depth of point (m).

The coefficient of consolidation, Cv (cm^2/s), characterizes the rate of reduction of neutral pressure during the consolidation process and is proportional to the product of permeability by the modulus of deformability of the material. As previously stated, this coefficient is determined in oedometrics tests, or triaxial drained tests, by measuring the change in volume over time of a sample subjected to constant external load (Lambe and Whitman, 1969).

All such studies, investigations and tests are desirable and necessary to characterize in detail the foundations of dams, including their parameters of resistance, permeability and deformability, and allow the preparation of a safe and reliable design of the foundations and also of the dam structures.

Frame 6.5 Rock Massif Classifications with Hydraulic Decomposition, Consistency, Fracturing and Conductivity (ABGE, 1983; Costa, 2012)

Decomposition

Degree	Denomination	Characteristics
D1	Sound rock	The rock presents its constituent minerals without decomposition. Eventually, it presents rusty joints.
D2	Little decomposed rock	The rock presents incipient decomposition in its matrix and along the planes of fractures.
D3	Medium-decomposed rock	The rock presents about 1/3 of its decomposed matrix. The decomposition along the fractures is accentuated.
D4	Very decomposed rock	The rock has about 2/3 of its matrix or its totally decomposed minerals. All fractures are decomposed.
D5	Extremely decomposed rock	The rock presents your whole body fully decomposed.

Consistency

Degree	Denomination	Characteristics
C1	Very consistent rock	Rock with metallic sound, it breaks with difficulty to the blow of the hammer. Its surface is scratched by steel.
C2	Consistent rock	Rock with weak sound, breaks with relative ease to hammer blow. When scratched by steel leaves superficial grooves.
C3	Moderately consistent rock	Rock with hollow sound, easily breaks the blow of the hammer with brittle fragments at the pressure of the fingers. Light furrow to scratch of steel.
C4	Poorly consistent rock	Rock breaks very easily to hammer blow, edge of fragments easily brittle manually Deep grooves to scratch of steel.
C5	Rock without consistency (crumbly)	Rock crumbles to the blow of hammer, disintegrating with the pressure of the fingers. It can be cut with steel, being scratched with the nail.

Fracturing

Degree	Denomination	Fractures per meter	Space between fractures (m)
F1	Occasional. fract.	≤ 1	≥ 1.0
F2	Little fractured	1.1–5	0.20–0.50
F3	Med. Fractured	5.1–10	0.10–0.17
F4	Very fractured	10.1–20	0.05–0.09
F5	Extr. fractured	> 20	<0.05

Hydraulic conductivity

Degree	Denomination	Specific water loss $(l/min.m.kg/cm^2)$
H1	Very low	$CH < 0.1$
H2	Low	$0.1 \leq CH < 0.5$
H3	Medium	$0.5 \leq CH < 5.0$
H4	High	$5.0 \leq CH < 25.0$
H5	Very high	$CH \geq 25.0$

Geological and geotechnical studies   109

Table 6.1 Classification of Soils by the Consistency of Clays and Sand Compactness (Monticeli and Costa, 2012)

Material	Standard Sampler Terzaghi-Peck (SPT)			
	No. of Blows	Classification	Approx. Cohesion (kg/cm^2)	Allowable Pressure qa (kg/cm^2)
Clay	< 2	Very soft	<.125	< 0.30–0.22
	2–4	Soft	0.125–0.25	0.30–0.60
				0.22–0.45
	4–8	Medium	0.25–0.50	0.60–1.20
				0.45–0.90
	8–15	Stiff	0.50–1.00	1.20–2.40
				0.90–1.80
	15–30	Very stiff	1.00–2.00	2.40–4.80
				1.80–3.60
	>30	Tough	>2.00	>4.80
				>3.60
Sand	0–4	Very cute	-	Compaction is
	4–10	Cute	-	required
	10–30	Medium	-	0.70–2.50
	30–50	Compact	-	2.50–4.50
	>50	Very Compact	-	>4.50

Obs.: (1) The permissible pressures of the upper line refer to the isolated foundations and those of the lower line to the continuous foundations. (2) In the value of qa, only for the clays was considered a safety factor equal to 3 – rupture pressure = 3 qa.

Table 6.2 E and Cc Values for Some Materials (Cruz and Costa, 2012)

Material	E (kg/cm^2 × 10^3)	Cc (l/kg/cm^2 × 10^{-6})	Tension Range (kg/cm^2)
Residual soil	1.70	350	0–2
Basalt saprolite soil	0.60	100	0–6
Residual gneiss soil	0.50	1,200	0–2
	0.40	1,500	0–4
	0.35	1,700	0–6
Rockfill from basalt	0.80	875	0–4
	0.40	1,750	0–8
Crushed stone	0.73	820	0–4
Artificial sand	0.28	2,100	0–4

E, deformability module; Cc, coefficient of compressibility.

6.3 MATERIAL PARAMETERS

In cases of projects with few investigations and tests, it is necessary to use the bibliography to obtain data and parameters of works executed with similar materials. For this reason, data and parameters of several materials obtained in the specific bibliography of the experience with dam works are presented in Tables 6.4–6.9. As already mentioned, in view of the risks involved, the use of this information should only be done with the support of engineers and consultants with proven experience in projects of this nature.

Table 6.3 Geomechanical Units/Parameters. Itaipu HPP (Monticeli, 1986; apud Costa, 2012)

Geomechanical Units/Parameters	c	ϕ	E	K_n	K_t	μ	γ	K_{eq}
	kg/cm²	°C	10³kg/cm²	Kg/cm.cm			t/m³	cm/s
(1) Soil	0.2	28	0.3	-	-	0.30	1.7	10^{-4}
(2) Boulder of basalt	2.0	32	2	-	-	0.25	2.2	10^{-4}
(3) Dense basalt	15	45	200	-	-	0.18	2.9	10^{-5}
(4) Contact	0	45	-	200	8	-	-	10^{0}
(5) Sandy breach	8	30	80	-	-	0.22	2.3	10^{-4}
(6) Clayey breach	0	27	40	-	-	0.22	2.2	10^{-3}
(7) Basalt vesicular-amygdaloidal	10	37	150	-	-	0.20	2.6	10^{-4}
(8) Discontinuity sub-horizontal	0	38	-	200	70	-	-	10^{0}
(9) Dense basalt	12	40	180	-	-	0.19	2.7	10^{-5}
(10) Open contact	0	45	-	200	80	-	-	10^{0}
(11) Weathered sandy breach	6	30	60	-	-	0.18	2.2	10^{-4}

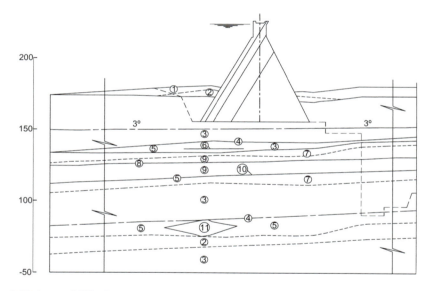

Figure 6.11 Itaipu HPP. Geomechanical model. Monticeli (1986, apud Costa, 2012).

The classical Leps curves (1970) on shear strength of rockfills are shown in Figure 6.15 (Kjaernsli et al., 1992).

Table 6.9 shows the influence between the directions of discontinuities and the dam axis with respect to stability and tightness of the foundations. Frames 6.6 and 6.7 show systems and conditions of discontinuities (see also Frame 6.1).

In order to separately evaluate the influence of the parameters chosen in the three approaches (classification regarding deformability, stability and watertightness), it is recommended to follow the guidelines proposed by Costa (2012) shown in Frames 6.8–6.10.

Geological and geotechnical studies 111

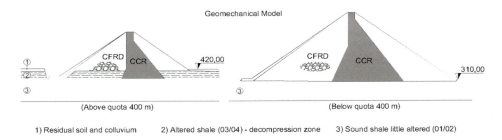

Figure 6.12 Irape. Dam alternative geomechanical model (Costa, 2012).

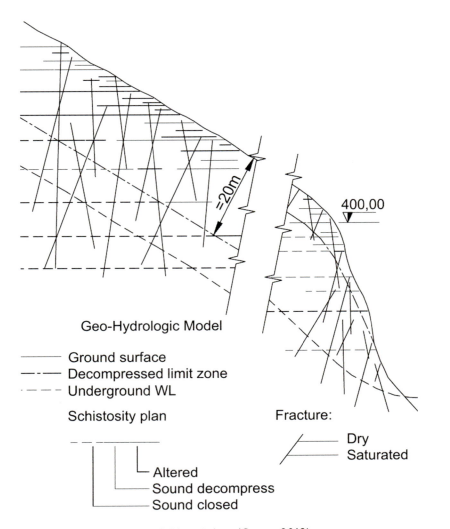

Figure 6.13 Geo-hydrological model Irapé dam (Costa, 2012).

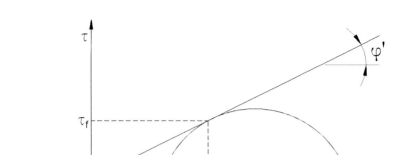

Figure 6.14 Shear strength Mohr-Coulomb circle.

Table 6.4 E Values and Poisson Coefficient for Rocks (Costa, 2012)

Rock Group	Rock	Young's Module ($E \times 10^{-5}$ kgf/cm^2)	Poisson Coefficient (ϑ)
Igneous	Basalt and gabbro	6.0–12.0	0.15–0.20
	Amphibolite	6.0–7.0	0.25–0.30
	Granite and granodiorite	5.0–9.0	0.10–0.30
	Diabase	3.0–9.0	0.15–0.20
	Andesite	1.2–3.5	0.11–0.20
	Riolite	1.0–2.0	0.10–0.20
Metamorphics	Marble	6.0–9.0	0.11–0.20
	Quartzite	4.0–10.0	0.15–0.20
	Gneiss	2.5–6.0	0.08–0.20
	Quartz-xist	1.2–3.0	0.15–0.20
	Micaxist	1.0–2.5	0.10–0.15
Sedimentaries	Limestone	4.0–8.0	0.10–0.20
	Sandstone	1.5–5.0	0.07–0.15
	Dolomite	2.0–3.0	0.08–0.20
	Argilite	1.5–3.0	0.10–0.25

6.4 FOUNDATION TREATMENT METHODS

The development of the foundation treatment project should be done considering the limits defined in the structure foundation excavation project.

Based on the geomechanical model of foundations, treatments are defined that aim to increase the safety of the work and, consequently, reduce the level of contingencies in the budget. In summary, the objectives of the treatments are:

- improve and guarantee the conditions of contact structure-foundation;
- improve the qualities of strength and deformability of the foundation mass;

Geological and geotechnical studies 113

Table 6.5 Parameters of Residual Soil Resistance (Cruz and Costa, 2012)

Dam (Brazil)	Origin Rock	c' (kgf/cm^2)	$\phi'(o)$
Porto Colômbia	Basalt	020–0.25	17–24
Marimbondo		0.10	15
Tucuruí		1.00	24
	Metabasite	0.30	25
	Filito	0.36	24
	Quartzite	0.31	22.5
Corumbá	Chloritaxist	1.20	29
Cana Brava	Metagabbro	0.40–0.80	20–22
Serra da Mesa	Micaxist	1.80–2.85	30
Simplício	Migmatite	0.20–0.30	23–27
Sapucaia	Gneiss	0.90	24
Itaocara		0.30	24

Table 6.6 Mean Values of Resistance of the Main Rocks (Costa, 2012)

Rock Group	Rock	Compression Resistance (kgf/cm^2)	Shear Strength (kgf/cm^2)	Tensile Strength (kgf/cm^2)
Igneous	Basalt and gabbro	800–4,000	50–400	60–200
	Amphibolite	1,700–2,800	150–300	10–150
	Granite and granodiorite	1,200–2,800	100–300	100–250
	Diabase	1,200–2,500	100–150	100–200
	Andesito	500–3,000	80–150	50–150
	Rhyolite and phonolite	1,000–3,000	80–200	80–120
Metamorphics	Marble	600–1,800	100–250	60–160
	Quartzite	2,800–3,000	150–200	150–200
	Gneiss	800–2,500	50–100	40–70
	Quartz-xist	1,300–2,500	120–150	100–150
	Micaxist	500–1,500	40–80	30–70
Sedimentaries	Limestone	600–1,800	100–180	50–120
	Sandstone	300–1,500	100–200	30–100
	Dolomite	200–1,200	80–150	25–100
	Argilite	400–1,000	40–70	30–50

- reduce the permeability of the foundation and, consequently, reduce the underpressure; and
- prevent the carrying of soils by the foundation.

The treatment project is developed over hydrogeological model of the foundation, which defines the conditions of the masses to be treated, taking into account the foundation requirements of each structure. Conventional treatment techniques should be used and have already been used, checked and accepted in other similar works.

The surface treatment aims to prepare the foundation to receive the final structure. The sequence of surface treatments is similar for the various types of

114 Design of Hydroelectric Power Plants – Step by Step

Table 6.7 Mean Values of Cohesion and Angle of Friction of the Main Rocks (Costa, 2012)

Rock	Cohesion – c (kgf/cm^2)	Friction Angle – ϕ (^{oo})
Granite	140–500	45–60
Basalt	200–600	50–55
Sandstone	80–400	35–50
Shale	30–300	15–30
Limestone	100–500	35–50
Quartzite	200–600	50–60
Marble	150–300	35–50

Table 6.8 General Classification of Rock Massifs (Costa, 2012)

Aspect Involved	Parameter	Variation of Indices				
V	Module E (GPa)	>200	100–200	50–100	10–50	<10
V		10	8	6	4	2
V	Resist to uniaxial	>100	50–100	20–50	5–20	<5
Deformability	compression (MPa)	10	8	6	4	2
V	Rock alteration	D1	D2	D3	D4	D5
V	(Frame 6.4)	16	12	8	4	2
V	Discontinuity systems	A	B	C	D	E
∧	(Frame 5.5)	**8**	**6**	**4**	**2**	**0**
∧	Direction of	a	b	c	d	e
∧	discontinuities	**8**	**6**	**4**	**2**	**0**
Stability∨	(Table 6.9)					
V	Conditions of	a′	b′	c′	d′	e′
V	discontinuities	**8**	**6**	**4**	**2**	**0**
V	(Frame 6.6)					
V	Spacing of	>3.0	1.0–3.0	0.3–1.0	0.05–0.3	<0.05
∧	discontinuities (m)	**10**	**8**	**6**	**4**	**2**
∧	Hydraulic Conductivity	H1	H2	H3	H4	H5
∧	(Table 6.9)	**15**	**10**	**7**	**4**	**0**
Watertightness∧	Direction of discontinuities	a	b	c	d	e
∧	(Table 6.9)	**15**	**10**	**7**	**4**	**0**
∧						
Sum of indexes		81–100	61–80	41–60	21–40	<20
Massive class		**I**	**II**	**III**	**IV**	**V**
Description of the rocky massif		Very good	Good	Regular	Poor	Very poor

Table 6.9 Influence between the Directions of the Discontinuities and the Dam Axis with Respect to the Stability and the Tightness of the Foundations (Costa, 2012)

Class	Type of Influence	Angle between Discontinuity and Dam Axis (°)	
		Regarding Stability	Regarding Watertightness
A	Null	71–90	0–10
B	Reduced	51–70	11–30
C	Regular	31–50	31–50
D	High	11–30	51–70
E	Very high	0–10	71–90

Geological and geotechnical studies   115

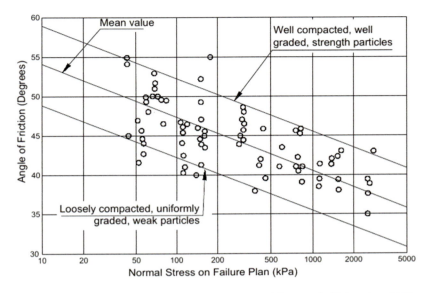

Figure 6.15 Shear strength of rockfills (Leps, 1970; apud Kjaernsli, Valstad and Höeg 1992).

Frame 6.6 Discontinuity Systems. Costa (2012)

Class	Characteristics	Denomination
A	A system of the types: V ou MF	Very favorable
B	A system of the types: JF ou Mf	Favorable
C	Two systems of the types: H/MF, H/Mf, V/JF, MF/JF Two systems of the types: H/JF, H/V, JF/Jf, V/Jf, V/MF, MF/Mf Three systems of the types: H/JF/Jf, H/V/JF, V/JF/Jf	Regular
D	A system of the types: H ou Jf Two systems of the types: H/Jf, V/Mf Three systems of the types: H/V/Jf, H/V/MF, H/MF/Mf, V/MF/JF	Unfavorable
E	Two systems of the types: Jf/MF, Mf/Jf Three systems of the types not included in classes C e D More than three systems	Very unfavorable

Symbology: H, horizontal; V, vertical; JF, joint dipping downstream with strong angle (> 50°); Jf, joint dipping downstream with weak angle (<30°); MF, joint dipping upstream with strong angle; Mf, joint dipping upstream with weak angle.

Frame 6.7 Conditions of Discontinuities (Costa, 2012)

Class	Surface Description of Discontinuity Plan
a'	Rough surface, discontinuous, strong walls
b'	Slightly rough surface, separation <1 mm, resistant walls
c'	Slightly rough surface, separation <1 mm, weak walls
d'	Continuous surfaces, grooved or filled <5 mm or with apertures of 1–5 mm
e'	Continuous surfaces with soft fill > 5 mm or with opening > 5 mm

116 Design of Hydroelectric Power Plants – Step by Step

Frame 6.8 Application of the Classification for Deformability (Costa, 2012)

Sum of Indices	26–36	16–24	< 16
Classif. of the massif	Good	Regular	Poor
Recommended dam types	Concrete	-	-
	Rockfill	Rockfill	
	Earth	Earth	Earth

Frame 6.9 Application of Stability Classification (Costa, 2012)

Sum of Indices	42–60	26–40	< 26
Classif. of the massif	Good	Regular	Poor
Recommended dam types	Concrete	-	-
	Rockfill	Rockfill	
	Earth	Earth	Earth

Frame 6.10 Application of the Classification for Watertightness (Costa, 2012)

Hydraulic Conductivity	H1	H2–H3	H4–H5
Sum of indices	>40	20–40	<20
Classif. of the massif	Good	Regular to bad	Bad to poor
Recommended procedures	None	Injection over two or three lines	Injection over three or more lines

structure-foundation interface. The treatments will be differentiated according to their intensity, the conditions of the foundations in soil and rock, and the type of structure that will cover the foundation: concrete and embankment (earth or rockfill).

Once the stage of excavation and removal of all the undesirable materials is finished, the foundation regularization phase begins. Existing topographic irregularities can cause problems of stress concentration in structures and/or executive difficulties.

The design and technical specifications should define the type of treatment to be carried out: removal by excavation and remodeling of slopes or regularization with concrete. It should also be defined whether the regularization should be done in a localized manner (only in some parts of the foundation) or generalized (throughout the foundation area). These services include the cleaning and filling of possible geological discontinuities.

Following are the cleaning services, which consist of removing any loose material in the foundation in two stages: thick cleaning and fine cleaning, the latter including washing with water. In the technical specifications, the limitations regarding the use of equipment and the water and/or air pressure of the washing/cleaning process should be defined, considering the erodibility and the disintegrability of the foundation materials with the necessary rigor.

The requirement to cover, or lining, the clean surface must be defined as a function of the possibility of carrying material from the embankment to the foundation, in the case of soil foundation, or vice-versa.

Lining may be required to protect materials with weathering disintegration characteristics. The materials for protection include: granular filters, porous concrete, mortar, cast concrete, swept concrete, asphalt emulsion and geosynthetics. The latter must be carefully assessed with regard to their durability and maintenance of their physical characteristics throughout the useful life of the dam (filtration, waterproofing, possibilities of fouling, etc.).

The control of the waters during the construction, placing of the materials, must be contemplated in the drainage projects of the foundation.

Localized and superficial injections may be necessary to seal detonated areas, regularization concrete contacts with the foundation, temporary drains, specific geological discontinuities, etc. In this case, the types of drilling, their orientation and depth should be defined, depending on the features to be injected.

The deep treatment of the foundation aims to consolidate the mass, improving its properties of resistance, deformability and permeability and still provide means for its better drainage. In addition, it may aim to homogenize the foundation, eliminating areas with high flow concentration, especially when there is a risk of internal erosion. Any reduction in the permeability of the foundation leads to decrease of the flow affluent to the drainage system. The intensity of the injection program depends on the interest of each project. In general, there is a practical limit to the effectiveness of the injection, and it is not possible to obtain a complete seal of the foundation, which would greatly hamper the project without efficiency gains.

The deep treatments must be the object of a geotechnical project, justified technically and economically, based on the experience acquired in other projects. The final decisions on this subject, however, will always be left to the executive stage, depending on the actual conditions found in the field, during the work. The execution of the treatments must be accompanied by the design team and the consultants. The most used method is to inject mortar or cement slurry into one or more parallel lines of holes, forming an injection curtain. The design normally defines the area to be treated and the type of drilling, the spacing between holes, the type of injection, the use of special shutters, the type of cement slurry, criteria for closing the lines and changing the cement slurry, among other things. It also defines the criteria for direct verification of the finished product.

This subject is exposed in several works listed in bibliography. In Brazil, it is recommended to consult the cases of treatment of the foundations of the plants: Itapebi on the Jequitinhonha river, Peixe Angical on the Tocantins river and Porto Primavera on the Paraná river, reported in CBDB-MDB, Vol. III (2009).

Figure 6.8 shows the foundation sealing device and the drainage system of the Aimorés dam on the Doce river, Minas Gerais. Figures 6.16–6.19 show treatment steps of the foundations of the Belo Monte HPP, Xingu river, Pará, and Barra dos Coqueiros HPP, Claro river, Goiás.

6.5 DRAINAGE SYSTEMS

On the drainage systems of dam structures, only the basic aspects will be presented in this book. It is recommended that young engineers consult the book *Seepage, Drainage and Flow Nets* by Cedergren (1967) for details on the subject.

6.5.1 Drainage system of earth and eockfill dams

The internal drainage system for an earth dam shall consist of a vertical or inclined filter and a combined sub-horizontal drainage mat beyond the foot drain (see Figure 6.8). The chimney drain was used for the first time in 1951 at the Vigario dam

Figure 6.16 Belo Monte HPP – Dyke 19B Foundation of the diversion gallery (2013) Manual cleaning of granitic rock (Geologist Paulo Guimarães).

Figure 6.17 Belo Monte HPP. Foundation of the powerhouse (2013). Final cleaning (Geologist Paulo Guimarães).

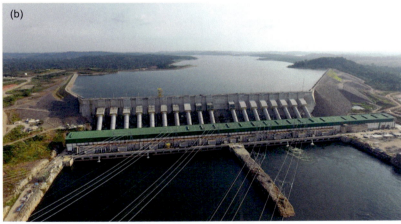

Figure 6.18 (a–b) Belo Monte HPP (2013). (a) Powerhouse at Pimental site. (b) Preparation of the foundation on the left bank to receive the earth dam: applied treatment over the granite rock: dental concrete and cement injection.

(or Terzaghi dam), Figure 6.20, Piraí river, Rio de Janeiro (Sherard et al., 1963). The vertical or inclined filter shall be at the top elevation of the reservoir, WLmax max. Figures 7.5 and 7.6 show examples of rock fil dams.

The internal drainage system must consist filtering layers, filter and transition, both in the upstream and downstream backrest. Filters must meet the consolidated criteria worldwide (see Section 7.3.1).

The thicknesses and distribution of the draining layers should be checked for the flow obtained in the percolation analysis, considering a safety coefficient of equal to

Figure 6.19 (a–b) Barra dos Coqueiros HPP (Goiás). Foundation of spillway in basaltic rock. Clean and marked for geological mapping. (2008). Source: Geologist Paulo Guimarães (DF).

or more than 10. For constructive purposes, the minimum dimensions of the drainage devices are given below. The hydraulic dimensioning of the sub-horizontal filter should be done by applying Darcy's law, or Dupuit's theory. The MEF, Finite Element Method, can also be used, incorporating the filter itself in the percolation analysis. The minimum dimensions of drainage devices are depicted in Frame 6.11.

Geological and geotechnical studies 121

(a)
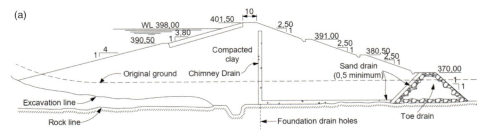

(b)

(c)

Figure 6.20 Vigário dam. Drainage system: (a) cross section ($H = 40$ m), (b) lateral view and (c) aerial view.

122 Design of Hydroelectric Power Plants – Step by Step

Frame 6.11 Minimum Dimensions of Drainage Devices

Device	Minimum Dimensions (m)
Vertical or inclined filter	0.60
Sub-horizontal filter – thickness	0.25
Foundation drainage trench – width	0.60
Relief wells – diâmetro	0.10
Transitions downstream of the core – width	0.60
Transitions in foot drains – width	0.40

6.5.2 Drainage system of the concrete dams

The main drainage system of the concrete dams consists of galleries within the structures in which the drain holes were drilled (roto-percussive or rotating). There are many cases where drainage galleries have been used at great depths. The galleries should be designed to be kept always drained, preferably by gravity or by pumping. Flooded galleries should only be adopted in special cases.

6.6 INSTRUMENTATION OF FOUNDATIONS

The theme related to the instrumentation of foundations and the observation of the behavior of structures is of fundamental importance for the continuous evaluation of the dam safety.

The deformability of the rocky foundation masses is a complex matter, which was practically not realized even until the 1950s. The knowledge of deformability, that is, the relationship between the stresses and deformations resulting therefrom, is of fundamental importance in the mechanics of rocks applied to the study of foundations of concrete dams.

It was deformation of the foundation on left abutment that caused the famous crash of Malpasset dam in France in 1959, in which more than 2,000 people lost their lives (see Chapter 13).

It is not within the scope of this book to present this theme, given the complexity and specificity of matter. But in the international bibliography there are several guidelines. In Brazil, there is a book titled *Instrumentation and Behavior of Foundations of Concrete Dams* (Silveira, 2003). This author presents in his book the results of several Brazilian works, collected and systematized over decades of experience in the subject. It has become the main reference on the subject for dam designers. It shows, in a clear and objective way, the importance of knowing the deformability of the foundation mass to subsidize the design of instrumentation of its discontinuities for measurement of settlements, displacements, etc. The theme was presented with the structure presented below:

- Deformability of rocks and foundations;
- Deformability of several types of rocky masses in the foundation of several Brazilian dams;

Geological and geotechnical studies   123

- Foundation overhangs during the construction period and deformability of the foundations during the reservoir filling period of several dams;
- Slow deformation of the foundation of so many dams;
- Instrumentation of the foundations for measurement of settlements, displacements, strains and suppressions;
- Methodology for the analysis of data of instrumentation in foundations of concrete dams;
- Anomalous behavior of dams associated with the deformability of the foundation, in which the author presents the cases of three dams, including recommendations.

6.7 CONSTRUCTION MATERIALS

In principle, those materials should be used in construction that are available in the borrow areas in the site, which decisively conditions the project.

The occurrences of materials with the requisite quality and quantity should be investigated on the basis of soils for use in earth works; sand for use on concretes and filters; gravel (rolled pebble) for use on concretes and slopes; and rock for use in rockfills, transitions and bulk aggregates (gravel) for concretes and filters.

The materials' quality should be classified according to international standards. The earthen materials for earthfills should be classified by tactile-visual analysis; characterized in terms of complete laboratory tests, granulometry, density, Atterberg limits, and compaction. The workability should be evaluated according to the soil clay content; and shear strength parameters should be determined in special laboratory tests, as presented in the reference Soil Testing for Engineers (Lambe, 1969), among others.

Likewise, granular materials, sands and gravels should be classified by tactile-visual analysis and full characterization tests in the laboratory, aiming to verify their suitability for use in filters, transitions and as an aggregate for concrete.

These materials should be totally clean and free of impurities, such as organic matter and fine materials (clay and silt). When contaminated, they should be washed and sifted before their application in the works of dam.

The aggregate, brittle or gravel, as well as rock, should have sufficient hardness to withstand the impact of hammer blows and not disaggregate when exposed to daily cycles of wetting and drying through time (cycling).

A complete program of aggregate testing should be prepared by a specialist, including testing the potential reactivity of the parent rock components to the cement alkalis.

Estimates of volumes of materials in the borrow areas, sand deposits, quarries, materials from the obligatory excavations, balancing of materials, etc. should be carried out using established methods.

It should be noted that the sand deposits in commercial exploitation of the region should be investigated. Its unit cost must be compared to that of exploration of the field. If there are no sand deposits in the region, the artificial sand alternatives obtained as a by-product of rock crushing should be considered.

The research of the stone material will always be conditioned to the quality and quantity of the rock surplus from the obligatory excavations.

Chapter 7

Dams

7.1 TYPES OF DAMS

The different types of dams are as follows:

- earth dams with homogeneous sections;
- rock-fill dams with impermeable core, asphalt core, with facing concrete, and with facing asphaltic concrete;
- concrete dam: conventional gravity type (weight) and arch type.

It should be noted that the types of layouts and their selection have already been presented in Chapter 3. Technically feasible solutions are diverse, and the choice of the section should be made using an iterative process following the criteria of safety and overall lower cost.

The projects are made around the world according to the criteria consolidated in the ICOLD Bulletin 61, "Dam Design Criteria" (1988), among others.

Figure 7.1a shows the technical terms on dams, standardized by ICOLD, which will be used preferably in this book.

7.2 EARTH DAMS

Earth dams have been used since the earliest days of civilization for retention and storage of drinking water, for irrigation, and for moving water wheels.

To this day, they continue to be the most common type because their construction involves the use of materials in their natural state with minimum processing.

These dams were designed on the basis of empirical methods. The bibliography records several cases of rupture of such dams. From 1930 onwards, the advancement of Soil Mechanics and construction equipment allowed for great advances in the design of earth dams, which began to be elaborated on the basis of more rational engineering procedures. These procedures included:

- detailed investigations to characterize the foundations and borrow areas of natural building materials;
- application of engineering techniques in the project; and
- detailed planning and control of construction methods.

DOI: 10.1201/9781003161325-7

126 Design of Hydroelectric Power Plants – Step by Step

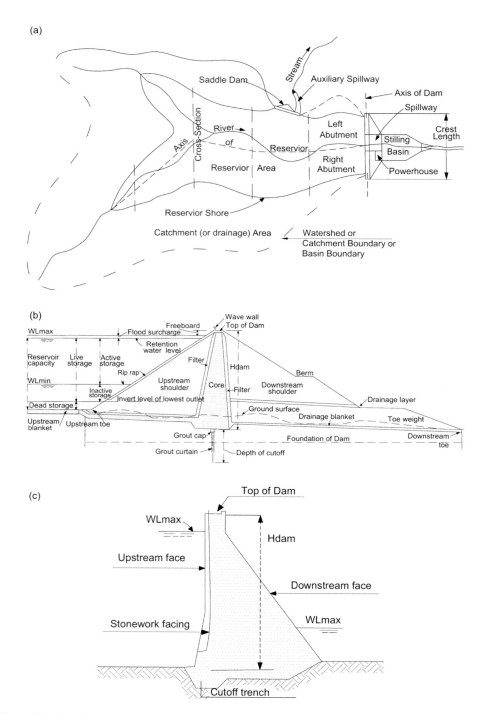

Figure 7.1 (a–c) (a) Dam terminology. Plant. Technical dictionary on dams. ICOLD (1978). (b) Dam terminology. (c) ICOLD, Technical dictionary on dams (1978).

7.2.1 Design criteria and section type

The art of designing a dam is linked to the art of controlling the flow of water through the dam-foundation. The stability must meet the basic safety requirements established according to the type of work and the various loading conditions. To meet these safety requirements, three basic principles of design must be obeyed: the principle of flow control, the principle of stability, and the principle of compatibility of deformations of the various materials.

7.2.1.1 Principle of flow control

With respect to the axis, from the upstream side, all the effort must be concentrated in order to seal the dam and its foundation to the maximum, introducing all the necessary sealing systems. From the downstream side, all efforts should be concentrated on facilitating the maximum output of the water by introducing all necessary drainage systems into the dam and foundation.

7.2.1.2 Principle of stability

The slopes of the dam must have resistance characteristics that guarantee stability. These characteristics must be compatible with the foundation materials to ensure the stability of the dam-foundation assembly for various loading conditions.

7.2.1.3 Principle of compatibility of deformations of the various materials

The compressibility of the materials of the various zones of the dam and its foundation must be made compatible by additional transition zones in order to reduce the differential and total settlements. Due to excessive total settlements, this may impair the performance of drainage and sealing systems, whether by the occurrence of cracks that render features of concentrated flow, or by inversion of flow gradients in the drainage systems.

The definition of the most appropriate typical section is made on the basis of the following aspects:

- characteristics, availability and workability of the existing materials in the borrow areas;
- site geology;
- seismic activity;
- regime regional hydrological regimen with regard to rainfall;
- shape and size of the valley (topographical aspects) and characteristics of the foundation;
- integration to the layout, construction schedule, and deviation scheme.

The design of an earth dam normally has the following characteristics:

- soil section compacted with materials sufficiently impermeable, with relatively smooth slopes, function of the stability criteria and with protections against superficial erosions;

- joints may be required depending on construction planning;
- the internal drainage system shall provide vertical filter and a drainage blanket under the downstream shoulder and a foot drain.

A homogeneous earth dam consists of a single type, or almost entirely of a single type, of impermeable material, excluding the protective material of the slopes – following Figure 7.2.

When the section is zoned, more impermeable materials should be placed in the central region and upstream side, and the more permeable materials should be placed in the downstream side. The projects of these dams are made according to the standardized norms. In summary, the following points stand out:

- the earthfill must be secured against a possible overtopping by the action of waves, during the occurrence of floods of the spillway project;
- the freeboard shall be estimated as shown in Chapter 4;
- the width of the crest is a function of the constructive process, and the minimum necessary for two-way traffic must be adopted with safety. In Brazil, CBDB/Eletrobras (2003) specifies a width of 10 m for earth or rockfill dams and 7 m for dykes;
- the declivity of the slopes should be defined considering the variation of the WLres and the characteristic resistance parameters of the materials of the embankment and of the foundation;
- earthfill slopes must be stable for all loading conditions (construction, operation, and rapid drawndown);
- the earthfill must be designed so as not to impose excessive stresses on the foundation;
- infiltration through the earthfill, foundation and abutments must be controlled in such a way that piping does not occur; the flow lost through percolation must be estimated and controlled so as not to interfere with the planned function of the project;
- the upstream slope must be protected against the action of waves; the crest and slope of downstream must be protected against the action of rains;
- if the dam is located in a region of seismic activity, the project must be done considering the occurrence of earthquake;
- the foundation in general must provide stable support for the dam, under all saturation and loading conditions, and provide sufficient percolation resistance to prevent excess water loss.

Hundreds of earth dams have been built in the last 70 years, with increasing heights without major accidents. However, ruptures of small dams continue to occur due to inadequate design and construction.

It is emphasized that the appropriate method of construction requires proper and rigorous treatment of the foundation, as well as the launching of the materials in the embankment of the dam in layers with pre-established thicknesses. These layers shall be compacted to the specified degree, observing the test and control methods throughout the construction process.

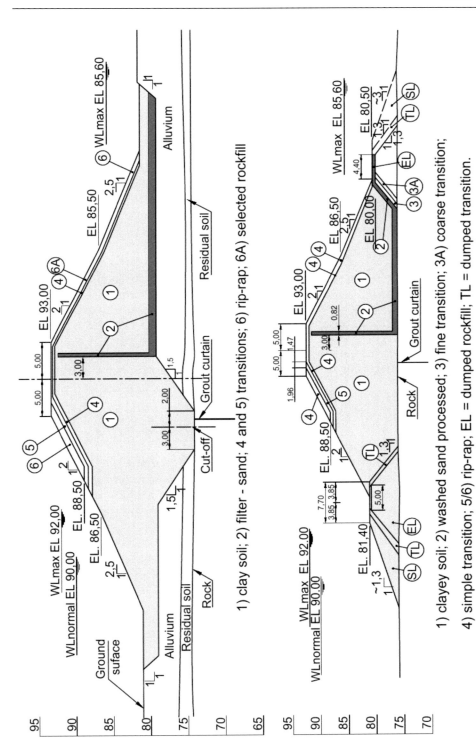

Figure 7.2 Earth dam – Aimorés HPP, Doce river, Minas Gerais. Typical sections (CBDB, 2009).

130 Design of Hydroelectric Power Plants – Step by Step

7.2.2 Percolation analysis

The laminar flow in a porous medium is governed by Darcy's Law (Equation 6.1):

$$Q = k\,i\,At \tag{7.1}$$

This equation is an adaptation of the Continuity Equation ($Q = V_d \cdot A$), in which the discharge velocity ($V_d = ki$) has been replaced by the coefficient of permeability (k), or Darcy's coefficient, which is percolation velocity V_s) through a unitary area under a unitary gradient; Q is the percolation flow in a cross section of area (A) normal to the direction of flow under hydraulic gradient (i), over a period of time (t). By rearranging the terms, the equation provides the basis for the experimental determination of permeability:

$$k = Q\,/\,i\,At \tag{7.2}$$

Darcy's percolation velocity multiplied by area (A), including voids (e) and solids (1), provides percolation flow (Q) under a given gradient ($i = h/L$).

It should be remembered, from Soil Mechanics, that the total volume of the sample is equal to the volume of solids plus the volume of voids ($1 + e$). The average percolation velocity (V_s) of a water mass flowing through the soil pores is equal to the discharge velocity ($V_d = k\,i$) multiplied by the inverse of the effective porosity ($1 + e)/e$.

Thus, the permeability is related to the rate of percolation by the following equation:

$$k = V_s n_e\,/\,i \tag{7.3}$$

where n_e is the effective porosity ($n_e = e/1 + e$).

For more details on the subject, one should consult Cedergren (1967). Fundamental considerations about the nature of the flow in a porous medium have led some researchers, as registered in Cruz (1996), to conclude that Darcy's law is a precise representation of the "law of that flow", because the velocities are low (laminar flow, low Reynolds number).

Although it is considered constant, the permeability coefficient varies according to the following factors: water viscosity; size and continuity of pores through which water flows; size and shape of soil particles; density of water; detailed arrangement of soil grains – the structure of the soil, the presence of discontinuities; and the great difference between horizontal (k_H) and vertical (k_V) permeability. The coefficient must be determined in the field tests: infiltration tests in the soil section in the drill holes; and tests of loss of water over pressure in the rock section in the rotary drilling holes, according to the methodology mentioned in Chapter 6.

The analysis should be based on a hydrogeotechnical model, using the finite element method (FEM), in order to estimate the pressure distribution and the value of infiltrated flows in the massif and foundations. They should provide subsidies for stability studies, drainage projects for the bulk and foundation, and the foundation's waterproofing project. The elaboration of the hydrogeotechnical model of the massifs should consist of:

- identification of materials with individualizable characteristics;
- definition of the parameters of each material, including its eventual anisotropy;
- spatial distribution of materials (two-dimensional and sometimes three-dimensional); the model should be as close as possible to reality, simplified, and with the ability to allow the application of mathematical analysis technique.

For the mass of the dam, it is emphasized that the model is a consequence of the section and the characteristics of each material. The establishment is more complicated for foundations, abutments, and occasional side saddles. The geotechnical investigations must be programmed to take care of the elaboration of this model. The selection of the hydrogeotechnical parameters of each material should be based on the statistical analysis of the results obtained in field and laboratory investigations, based on the experience with similar materials and conditions. The values of the permeability coefficient to be adopted should be adjusted to the level of effective stresses at the foundation and in the massif. Along with it, any anisotropy should also be considered.

The percolation analyses should be performed for the operating reservoir conditions, at their WLmax, and rapid drawdown, for the WLmax and WLmin. The definition of gradients, neutral pressures and percolation flows should be obtained from the flow nets. The flow in a porous medium can be represented through the Laplace equation. To develop this equation, it is assumed that the soil is homogeneous, the voids are completely filled with water, no consolidation or expansion of the soil occurs, soil and water are incompressible, the flow is laminar and Darcy's law is valid.

The Laplace equation can be represented by two families of curves (flow lines and equipotential lines) that intersect at right angles to form a pattern of square figures as a flow net. According to Cruz (1996), what occurs in many dams is that:

- the permeability of massif is variable from point to point;
- the permeability of the foundation has a dominant role in the flow, mainly in the lower part of the dam;
- at high altitudes, stress relief phenomena and incipient states of hydraulic rupture can cause a significant increase in horizontal permeability.

Establishing a flow net is a very complex task (see examples presented by Cruz, 100 Brazilian Dams). A network of all types can be drawn, with the aid, for example, of the MEF, manually adjusting the input and output conditions to meet the Laplace equation – flow lines and perpendicular equipotential and tangent conditions.

The flow net drawn for homogeneously isotropic or anisotropic media are inadequate for the masses of dams. The flow net that is established is a function of the permeability ratios kh/kv which, in turn, are related to the state of stresses.

As a practical recommendation for designs, Cruz (1996) recommends that the vertical drain lift be the same as the reservoir level and that the horizontal drain under the dam should be positioned in contact with the foundation.

Vertical drains (chimney type) are recommended for dams up to 30 m. For higher dams, the inclined drain provides a better distribution of stresses in the massif and avoids the inclusion in the embankment of a vertical wall of sand whose rigidity is always much higher than that of the adjacent massif, even in the case of rockfills.

132 Design of Hydroelectric Power Plants – Step by Step

7.2.2.1 Internal drainage system

In addition to toe drains, the internal drainage system shall consist of a vertical or inclined filter and a combined sub-horizontal drainage mat. The vertical or inclined filter shall have the top elevation at the reservoir WLmax max elevation. In designs of small height dam, of lesser responsibility, only one toe drain can be adopted. Where necessary, the foundation drainage systems shall consist of drainage trenches and regularly spaced downstream relief wells, along with sub-horizontal drain itself. It was to be used to control hydraulic outflow gradients downstream of the dam and reduce underpressure.

The thickness and distribution of the draining layers should be verified for the flow obtained in the percolation analyses, considering a safety coefficient of at least 10.

When designing the drainage devices, the WLmin-tw shall be considered. The minimum dimensions, for constructive reasons, are as follows:

- vertical or inclined filter: 0.60 m;
- sub-horizontal filter: thickness 0.25 m;
- foundation drainage trench: width 0.60 m;
- relief wells: diameter 0.10 m;
- transitions downstream of the core: width of 0.60 m;
- transitions in toe drains: width 0.40 m.

The design of the sub-horizontal filter should be done by applying Darcy's law or Dupuit's theory. The MEF can also be used, incorporating the filter itself in the percolation analysis.

7.2.2.2 Transitions

In the design and dimensioning of transitions, the conventional criteria for granulometric translocation of the following adjacent materials should be used, where "d" is the particle diameter of the materials to be protected (base) and "D" is the particle diameter of the filter materials.

In the definition of tracks, one can consider for "d" the average value and for "D" the upper limit of the range. The filter and transition materials, when used as draining elements, must have permeability compatible with their use and present percentage of non-cohesive fines passing in the sieve 200 of less than 5%. These materials should also meet the criteria: $D_{15}/d_{15} \geq 5$.

Depending on each case, the transition between clay soils (% that passes in the sieve 200% > 30%) and filters (sandy materials) should meet the following criteria:

- transition between the core and slopes and between granular soils: $D_{15}/d_{85} \leq 5$;
- transitions located in a region of lesser responsibility and/or in positions of low percolation gradients (between the core and the upstream backrest and between the protective rockfill and the upright backrest): $D_{15}/d_{85} \leq 9$.

The granulometric curves of the granular base and filter materials should be, if possible, approximately parallel. If it is not technically and/or economically feasible to obtain materials that meet the above requirements, laboratory studies and tests should be carried out with the aim of modifying or reducing these conditions.

7.2.2.3 Foundation waterproofing

Foundation waterproofing devices, intended to reduce the flow through foundation and high outlet gradients combined with the drainage system, may include waterproof trenches (cut-offs, diaphragm walls, mud trenches, etc.), injection curtains, and waterproof carpets. The design of these devices should be based on data from percolation analyses (flow nets).

7.2.3 Stability analyses

The stability analyses should be performed by effective stresses, considering the neutral pressure parameters for the following loading cases: reservoir drawdown, steady seepage condition, end of construction condition, and earth quakes. The downstream level indicating the most unfavorable combination of loading should be considered.

There are several methods of analysis, all considering the method of slices and lamellae, varying only the basic hypotheses about the efforts between lamellae. The choice of method to be applied must be a function of the shape of the rupture surface to be analyzed.

The most commonly used methods are: Simplified Bishop; Morgenstern and Price; Spencer; Generalized Janbu; Sarma; and Lowe and Karafiath.

These calculations are done using routine programs available in the market. As a general guideline, the safety factors in Frame 7.1 are recommended.

Frame 7.1 Safety Factors (CBDB, 2003)

Case	Safety Factor	Shear Strength	Observations
End of construction	1.3[a]	Q ou S[b]	Upstream and downstream slopes
Rapid drawdown	1.1–1.3[c]	R ou S	Minimum value for dilating soil Maximum value for soils that contract in shear Minimum value for dilating soil Maximum value for soils that contract in shear
Normal operation	1.5	R ou S	Downstream slope
Seismic analysis	1.0	R ou S	Upstream and downstream slopes

[a] For dams above 15 m height on relatively weak foundations use FSmín = 1.4;
[b] In areas where neutral pressures are not anticipated, use resistance form S tests;
[c] In cases where drawdown occurs frequently consider the coefficient 1.3.
Q, non-consolidated, non-drained tests (UU); S, consolidated and drained tests (CD); R, consolidated and non-drained tests (CU) (Lambe and Whitman, 1979).

7.2.4 Tension and strain analysis

The text on this subject that the author likes most is that presented by Cruz (1996), from which the points described below were extracted.

The proper evaluation of the safety of a dam requires an analysis of the state of stresses and displacements occurring within and within its foundations, as well as the knowledge of strain-strain characteristics of the materials involved.

Two dams with the same nominal safety factor of 1.55, for example, for downstream slope under permanent operation, may involve a real safety or a totally different risk of rupture, depending on the soils, rockfills and rocks that make up your body and your foundation.

Cruz (1996) cites examples of the Itaúba and Promissão dams. Itaúba is a rockfill dam with compacted clayey core, supported on basalt and without any unfavorable features in terms of resistance. This makes the stability of the downstream slope dependent solely on the rockfill itself. Promissão is a compacted earth dam, supported on collapsible porous soil (partially removed), followed by residual soil of sandstone and sandstone rock. These dams, constructed in the decades of 1960s and 1970s, respectively, were properly instrumented. They have been in operation for more than 40 years and are stable. According to this author, the correct evaluation of its stability could only be made if the states of tension and displacements of the massif and its foundations were known.

Hundreds of dams were designed and constructed without a correct evaluation of the states of tension and displacements that occur, and only a negligible portion suffered accidents that required treatments and partial reconstructions. A smaller number involved rupture, usually at the end of the construction period, with large material losses, but with few human losses. Cruz (1996) cites the accidents:

- Açu dams in Brazil (1982) and Carsington in England (1984), involving layers of low-resistance clay at the base of embankments;
- Waco dams in the USA (1961) and Gardiner in Canada (1964), which broke due to large displacements (metric) that occurred in discontinuous form in the foundation shale;
- from the very thin nuclei of the Balderhead rock dams in England (1962) and Hyttejuvet in Norway (1965), which occurred due to a transfer of core stresses to the transitions and slopes resulting from the large compressibility difference among the various materials used in construction.

Cruz (1996) points out that the professionals involved in these accidents were experienced and were all surprised. However, he records that in his professional experience, in more than 100 dams, he never had a reliable evaluation of the state of tensions and displacements, and the dams had a normal behavior, without any signs of instability. Cruz (1996) also registers that problems of differential pressures, deformability contrasts and stress transfer have been recognized and considered in the projects. The inclusion of thick layers of transitions between the clayey cores and the embankments of the slopes was carried out in the Capivara and Itaúba dams to try to reduce the difference in core-slope stresses. This was done in many other similar dams.

The inclined drain versus vertical option offers numerous advantages, and its use is recommended for any dam over 25 m in height.

Tension and displacement analyses were not systematically carried out in the projects because the calculation resources were limited. The advent of numerical methods in the 1980s altered this scenario for materials such as soils and rockfills, and behavior models based on laboratory tests and field observations were incorporated into the calculations. The modeling of soils and rockfills is very complex, since they can retract or expand, or deform by shearing, collapse and liquefy, as well as have a structure that includes suction, when not saturated, and neutral pressures of consolidation.

Add to this the limitations of laboratory tests on the rotation of stresses and the size of the assayable samples, especially in the case of saprolite soils and rockfills containing differentiated features of medium and large blocks. All these difficulties in modeling must be borne in mind.

According to Cruz (1996), mathematical modeling should be done for designs that present potential problems of concern. For example, he cites the designs of:

- rockfill dams with thin cores;
- a dam on sand that does not have a trench to seal the entire length of the foundation (e.g. Porto Primavera);
- a dam on weak rocks or with unfavorable features in the foundation.

7.2.4.1 Deformability and displacements

The materials of the foundation, as well as the materials that make up the compacted mass of a dam, suffer deformations in function of the tensions applied according to own laws.

Approaches to elasticity and plasticity theories and to rheological models are usually used to explain such behavior, but because they are approximations, they do not always lead to predictions very close to the deformations that occur in the prototype.

The mobilization of resistance, the possible generation of neutral pressures, the occurrence of cracks and the potential for formation of planes causing hydraulic rupture depend fundamentally on the volumetric variations that occur and, therefore, the interest to define permissible displacements is much more comprehensive than the conventional balance equilibrium stability calculations.

The presence of air and water in the voids brings additional complications and, depending on the permeability of the materials, the deformation phenomena are affected by the time factor. Add to this the breaking-up of the particles or blocks common in rockfill, and, to a lesser extent, in clay-bound particles and even in sand grains. Cementation components, present in lateritic soils and marine clays, and suction acting on unsaturated soils significantly affect soil deformability.

Figure 7.3 shows the main types of laboratory tests usually used to obtain soil compressibility parameters. The parameters necessary for the studies are defined in Tables 7.1 and 7.2 (Cruz, 1996). Details on the calculation methods can be found in Soil Mechanics books such as Cruz (1996).

The compressibility can be expressed by the classical deformability modulus (Ed = Young's modulus) or by the compressibility coefficient $Cc = (\Delta V/V)\ 1/\Delta\sigma$, as shown in Table 7.1.

136 Design of Hydroelectric Power Plants – Step by Step

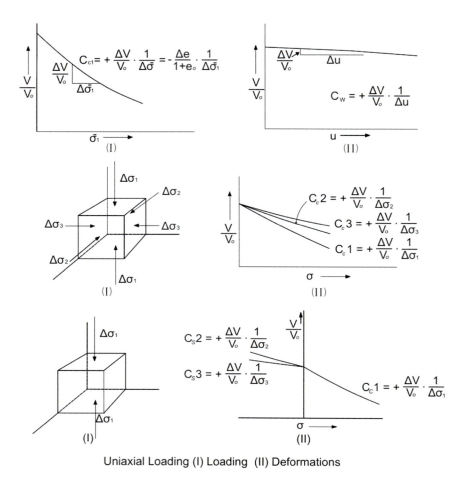

Uniaxial Loading (I) Loading (II) Deformations

Figure 7.3 Main laboratory tests for obtaining compressibility parameters (Cruz, 1996).

Table 7.1 Modulus of Deformability and Volumetric Variation (Cruz, 1996)

Type of Loading	Modulus of Deformability	Volumetric Variation
Uniaxial compression	$E = \dfrac{\sigma_z}{\varepsilon_z}$ (Mod. Young)	$\dfrac{\Delta V}{V} = \dfrac{\sigma_z}{E}(1-2\mu)$
Isotropic compression	$E = \dfrac{\sigma_z}{3ex}$	$\dfrac{\Delta V}{V} = \dfrac{3\sigma_0}{E}(1-2\mu)$
Confined compression	$E = \dfrac{\sigma_z}{\varepsilon_z}$	$\dfrac{\Delta V}{V} = \dfrac{\sigma_z - [(1+\mu)(1-2\mu)]}{E(1-\mu)}$
Triaxial compression	$E = \dfrac{\sigma_1 - \sigma_3}{\varepsilon_z}$	$\dfrac{\Delta V}{V} = \dfrac{1}{E}(1-2\mu)(\sigma_x + \sigma_y + \sigma_z)$

Dams   137

Table 7.2 Values for Deformability Modules and Compressibility Coefficients (Cruz, 1996)

Material	E_d (kg/ $cm^2 \times 10^3$)	C_c (l/kg/ $cm^2 \times 10^{-6}$)	Tension Level (kg/cm²)	References
Water		48	1 (20°C)	Bishop et al. (1977)
Quartz		2.66	100–600	Bishop et al. (1977)
Calcite		1.34	100–600	
Marble Vermont		1.42	100–600	
Concrete		16.8	1	Bishop (1976)
Rock	232	10.2	1	Ruiz (1976)
–Sandstone	700	2.9	1	Cruz (1996)
–Basalt	670	2.3	1	
–Granite	600	3.0	1	
–Gnaiss				
Água Vermelha	1.70	350	0–2	Silveira (1983)
Residual soil	0.80	750	0–4	
Basalt saprolite soil	0.60	100	0–6	
Jaguari	0.50	1,200	0–2	
Biotite residual soil	0.40	1,500	0–4	
Gneiss	0.35	1,700	0–6	
Dense sand		1,800	1	Skempton (1960)
Loose sand		9,000	1	
London clay		7,500	1	
Gosport clay		60,000	1	
Rockfill	0.800	875	0–4	Signer (1982)
Basalt of Capivara HPP	0.400	1,750	0–8	
Gneiss	0.67	1,050	0–4	Materon (1983)
Crushed stone	0.73	820	0–4	Signer (1982)
Artificial sand	0.28	2,100	0–4	

The stress and strain-strain analyze should be performed with the following objectives:

- to verify the compatibility of deformations between the various materials that constitute the dam, its foundations and adjacent structures;
- to evaluate the potential for progressive rupture of the dam and foundation;
- optimize the position of the core, in the case of zoned section dam;
- analyze the risks of cracking of the core caused by traction zones or by hydraulic fracturing;
- to subsidize the instrumentation design, identifying the critical points to be monitored;
- optimize the design of the excavations, in order to keep differential settlements within permissible levels;
- determine the over-elevation of the ridge to compensate for post-constructive settlements.

These analyses shall be carried out in the case of loading during construction, filling of the reservoir with establishment of transient flow and permanent operating regime with established flow net. Analyses are done using computer programs such as the Sigma of the University of Alberta – Canada.

138 Design of Hydroelectric Power Plants – Step by Step

The compressibility and deformability parameters of the various dam and foundation materials determined in the laboratory should be appropriately adjusted based on the experience with similar materials and loading conditions from the results of field geological-geotechnical investigations (CBDB, 2003).

In the design, special attention should be given to the research and identification of the occurrence of collapsible or expansive materials in the foundation.

7.2.5 Slopes protection

The region of the upstream slope around the WL is exposed to the action of waves generated by the action of the wind on the surface of the reservoir. The methodology for estimating these waves is presented in Chapter 4.

The region of the upstream slope subject to the erosive effect of these waves shall be protected as follows:

- protection with rip-rap: the blocks (diameter and grain size) should be dimensioned according to the Taylor criterion (ICOLD, 1973);
- non-rocky protection: where it is not possible to economically obtain blocks, other materials must be used; the alternative soil-cement must be analyzed;
- the submerged part of the slope is protected by granular material available on site.

The downstream slope should be protected against the action of rainfall with grass or other plant material. If there is excess of granular material, it can also be used in protection. The submerged part of the slope should be protected by granular material available on site.

7.3 ROCKFILL DAMS

A rockfill dam is composed of the dam body formed by compacted stone shoulders with an impermeable membrane to prevent the passage of water that freely percolates through the rockfill. The impermeable membrane may be positioned in the center, or core, of the dam, in the upright or inclined position, or may also be positioned on the upstream slope. The choice between one type and another depends on the technical and economic analyses of each project.

The origins of rockfill dams date back to the 1870s, during the gold rush in California. Until the 1930s, several rocky dams were built in the United States by the USBR (DSD, 1973).

Between 1930 and 1960, there was a reduction in the use of this type of structure, due to the increase in the costs of obtaining and placing large quantities of rocky materials.

The construction of this type of dam began again in the 1960s, a fact that was attributed to the more economical methods of quarrying, construction and placement of materials, the use of excavated materials in random areas and also to the optimization of project details.

The factors that determine the choice of a rockfill dam are as follows:

- local topographical and geological-geotechnical conditions, as well as the availability of abundant rock in the site, which can be easily explored, or which will come from the excavations required for the foundations of the structures provided for in the layouts, and as good foundation conditions for the dam, with sufficient strength to guarantee the stability of steeper slopes;
- the scarcity of earthy materials, difficult to obtain in the area, or requiring intensive processing to be usable;
- a short period of time available for the construction of the dam and also for the lifting of the dam body that may pass through the critical period of the project's schedule; and
- finally, the existence of excessively humid and rainy weather conditions in the region which may limit the release of large quantities of earthy materials.

Other factors may favor the use of rockfill dams, such as the possibility of placing the rockfill during the winter in cold countries and the possibility of treating the foundations through injections with the simultaneous placement of the embankment.

It should also be noted that during the rainy season the rockfill can withstand the pressures of any overtopping without problems. Design solutions with rockfill cofferdams foreseeing the overtopping were successfully used in Brazil – see cases of Serra da Mesa HPPs and Corumbá, in the publication Diversion of Large Brazilian Rivers (CBDB, 2009).

The rockfill dams are classified into three groups on the basis of impermeable membrane: with vertical central core, with inclined central core, and with membrane positioned on the upstream slope.

Each group has its advantages and disadvantages, which vary according to the type of membrane, materials available on the site, and the conditions of the foundation (DSD, 1973). Technical and economical analyses should be performed with the purpose of determining the type to be chosen, either with internal or external membrane. In this book, only the three main types are dealt with in a simplified way:

- rockfill dam with core (of clay and asphalt concrete);
- rubble dam with concrete face; and
- rockfill dam with asphalt concrete face.

According to Marsal and Nuñez (1975), rockfills are materials that, when subjected to a variation of stress, undergo structural transformations due to displacement, rotation and breaking of particles. In order to take into account these variations and their influence on the characteristics of deformation and resistance, it is necessary to study the distribution of contact forces and fundamentals of particle breaking. This subject, however, goes beyond the scope of this book. It is recommended that you consult this reference, as well as Signer (1973), apud Cruz (1996), which cover the following topics: basic concepts: structural void index; distribution of contact forces; particle breakage; factors that affect grain breakage; crushing, oedometer and triaxial compression tests; friction angle and resistance envelope; direct shear tests on rock samples; effect of particle disaggregation; fine rockfills; compressibility of compacted rockfills: Figure 7.4 shows data observed in the following dams: Capivara, Salto Osório, Itaúba, Paço Real (all in Brazil), Akosombo (Ghana), Muddyrun (USA), and El Infernillo (Mexico) (Signer, 1973; apud Cruz, 1996).

140 Design of Hydroelectric Power Plants – Step by Step

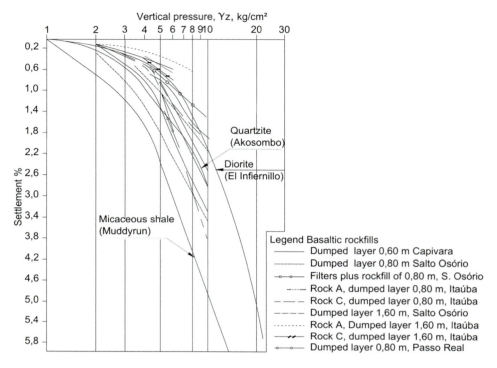

Figure 7.4 Compressibility of compacted rockfills (Signer, 1973, apud Cruz, 1996).

7.3.1 Rockfill dam with clay core

As depicted in Figures 7.5 and 7.6, the impermeable membrane (clay core) can be designed either vertically or in the inclined position in clay-core rockfill dams,. The point to note is that the vertical membrane rather than the inclined one is preferred because the contact pressure with the foundation will be maximal and the construction control may be less rigorous. The central membrane has the following advantages: it presents a smaller area exposed to contact with water, has shorter lengths of injection curtains, and has greater protection against weathering and external damage. On the other hand, it has the following disadvantages: one has to put the material of the core and the filters simultaneously to the placing of the rockfill, and the inaccessibility of the membrane for inspection and correction of any damages.

In the clay core rockfill dam, the arrangement of the materials should be made so as to provide a better compatibility of deformations between the core, the transitions, and the rocking of the slopes.

The crest width should be at least 10 m in order to meet the safety standards and to ensure proper constructive processes involving transportation and placement of various materials: impermeable material for the core, granular materials for filters, transitions, and rockfill.

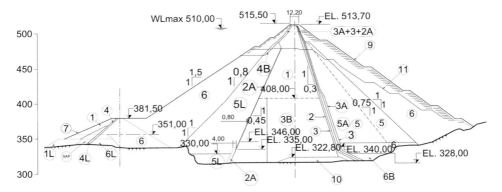

1) Clay; 2) Natural sand filter; 2A) Processed sand filter; 3) Fine transition;
3A) Medium transition; 3B) Clay gravel; 4) Coarse transition; 4B) Fine rockfill;
5) Medium rockfill; 5 L) Random rockfill; 6) Rockfill; 9) Protection rockfill; 10) Concrete.

Figure 7.5 Rockfill dam − vertical central core. Irapé dam (H = 208 m) (CBDB, 2009).

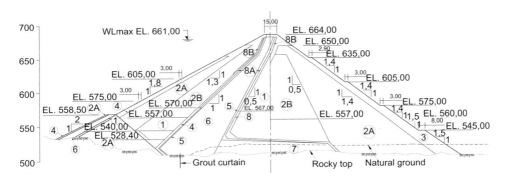

1) Compacted rockfill - 0,6 m; 2A) Compacted rockfill - 1,2 m; 2B) Compacted rockfill - 0,9 m;
3) Rock blocks; 4) Upstream transition; 4A) Fine transition; 4B) Coarse transition;
5) Sand filter; 6) Clay; 7) Random impermeable material;
8) Downstream Transition; 8A) Fine Transition; 8B) Coarse Transition.

Figure 7.6 Rockfill dam − inclined central core. Emborcação dam (H = 158 m) (CBDB, 2000).

The minimum core width at the base should be 0.3 H, where H is the height of the hydraulic reservoir load. At the top, the width should be 3 m, depending on the constructive aspects.

The internal drainage system shall consist of filtering layers (filter and transition), both on the upstream slope and on the downstream slope, and the filters shall meet the criterion defined in Section 7.2.2. In this regard, the particle size ranges of sand and gravel for clays suggested by Sherard (1984) are shown in Figure 7.7 (from Cruz, 1996).

It is recommended to read the aspects related to joint action of filtration and drainage presented by Cruz (1996). According to him, the filtering requirements suggest the adoption of smaller particle size filters and emphasize the importance of "fines" in blocking the particles or particle flakes of the base material. At the same time, Cruz (1996)

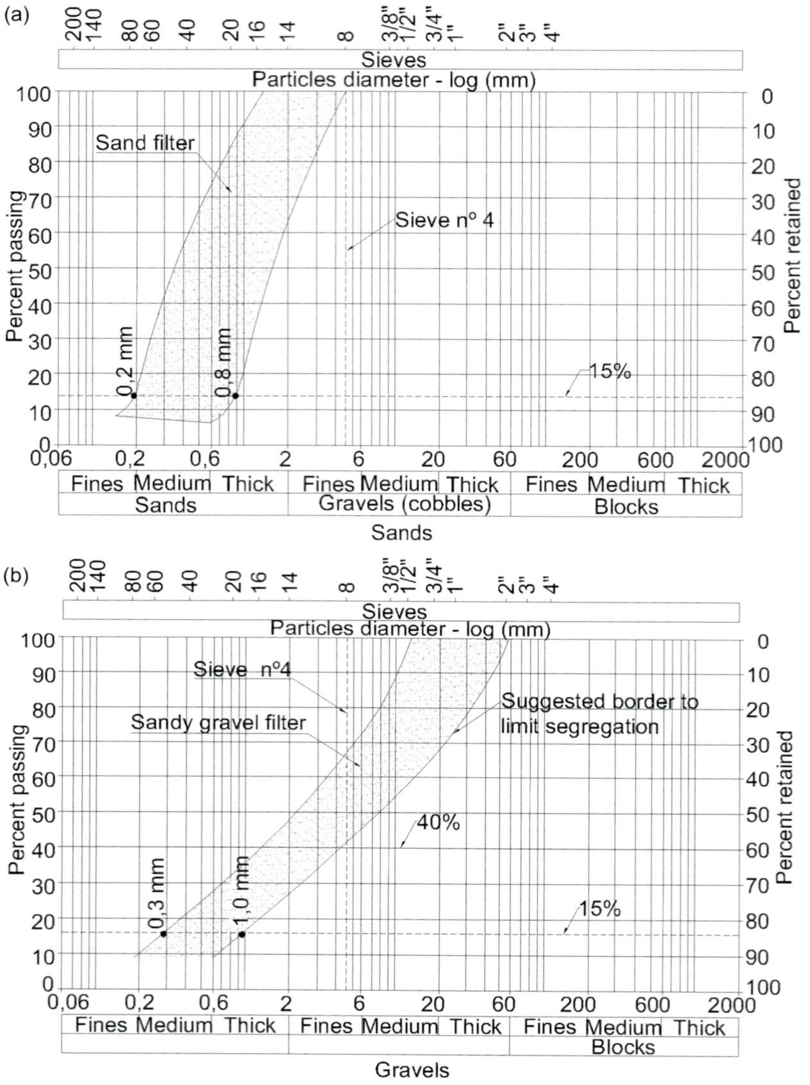

142   Design of Hydroelectric Power Plants – Step by Step

Figure 7.7 (a–b) Particle size distribution of sands and gravels advisable for fine clays (Sherard, 1984, apud Cruz, 1996).

calls attention to the fact that filters with cohesive fines are subject to cracks that totally invalidate the filtering function.

These requirements show that fine sand, or medium and coarse sand with a fraction of fine sand, would be the ideal filter for cohesive soils. However, the permeability of sand is directly associated with its granulometry. Fine sands have permeability as $k \sim 10^{-3}$ at 5×10^{-3} cm/s. In many cases, they are inadequate to flow to the foundations flow of the dams.

For this reason, Cruz (1996) recommends using medium and coarse sand, when available, which are adequate for the filtration function of the colluvial, residual and saprolithic soils, normally used in Brazilian dams. The permeability of these sands varies from 10^{-2} to 10^{-4} cm/s, and in many cases is sufficient to have outflow from internal drainage systems.

In Brazil, all geotechnical projects, excavations, treatment and sealing of foundations, rockfills, filtrations and drainages should be done only after taking criterions of CBDB/Eletrobras (2003) into consideration.

7.3.2 Concrete face rockfill dams

The following is a summary of concrete face rockfill dams. For details on this topic, it is recommended to consult the book "Concrete Face Rockfill Dam – CFRD" (Cruz et al., 2009). Besides this, it is also recommended to consult Cooke and Sherard (1985) and Cooke (1997, 1999).

The concrete face rockfill dam is the natural choice for places where there is no suitable soil for the core. It is worth mentioning that in many projects, this solution was chosen because it presented lower cost, even in those places where there were soil deposits at the core.

The lower cost of the concrete face rockfill dam compared to the conventional rockfill with earth core can be explained by several factors:

- the volume of rockfill is practically the same, but the unit cost is lower in the concrete face rockfill dam due to the allowed freedom following the construction of the various parts of the rockfill, with the easier use of internal access lanes and transport;
- the cost of the earth core and filters has been greater than the cost of concrete slab;
- the cost of the foundation treatment is significantly lower because the work can be performed in the upstream region, plinth area, regardless of the construction of the massif, and because the area of the surface to be treated is smaller; for rainy weather regions the advantage is widened (see Figures 7.8a and b of the Barra Grande dam).

In addition to the cost advantage of the structure itself, the smaller base width resulting from the steep slopes of the concrete faced dam often leads to shorter tunnels, power conduits and spillways, which implies additional reductions in total cost.

Adding the substantial advantage from a schedule point of view, in some cases, the completion of the work was anticipated within a year, the concrete face rockfill dam alternative became almost unbeatable.

After the end of the construction of the Foz do Areia dam in the Iguaçu river, several other dams with concrete face were successfully designed and built in Brazil (Frame 7.2). All of them were executed with basaltic and gneiss rocks.

The membrane upstream of this type of dam has the following advantages:

- are always available for inspection and repair, with the lowering of the reservoir;
- the membrane can be constructed after the construction of the rockfill section;
- the treatment of foundations can be carried out simultaneously with the rockfill;

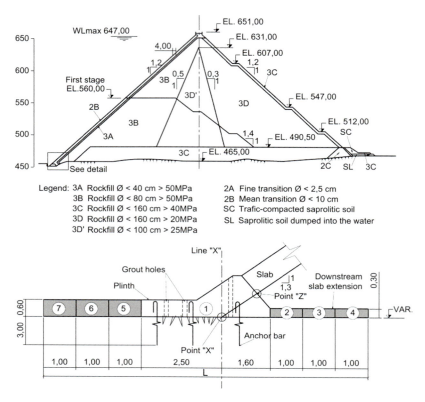

Figure 7.8 CFRD Barra Grande HPP. Height = 185 m (CBDB, 2009).

Frame 7.2 Dams with Concrete Face in Brazil (Falcão, 2007)

Name	River	Year	Upstream Slope	Height (m)	References
Foz do Areia	Iguaçu	1980	1.4 V;1.0 H	160	CBDB (1983)
Salto Segredo	Iguaçu	1994	1.3 V;1.0 H	145	CBDB (2000)
Xingó	San Francisco	1994	1.4 V:1.0 H	140	CBDB (2000)
Itá	Uruguay	2000	1.3V;1.0 H	127	CBDB (2009)
Machadinho	Pelotas	2002	1.3V;1.0 H	120	CBDB (2009)
Quebra Queixo	Chapecó	2003	1.25V;1.0 H	75	Xavier (2003)
Itapebi	Jequitinhonha	2005	1.25 V;1.0 H	110	CBDB (2009)
Barra Grande	Pelotas	2005	1.3V;1.0H	186	CBDB (2009)
Campos Novos	Canoas	2006	1.3V;1.0 H	200	CBDB (2009)

- the base of the dam will be available for stability against sliding;
- the membrane can be used as protection of the slope;
- it will be possible to raise the dam in the future, if necessary.

The recommendations summarized below should be considered in the design (CBDB 2003):

Dams   145

- upstream and downstream slopes may vary from:
  - 1.0 (V):1.3 (H) for basaltic and gneiss rockfills;
  - 1.0 (V):1.4 – 1.6 (H) for sedimentary rockfills.
- the characteristics of the rockfills should be determined in tests of simple compression, punctate compression, water absorption, Los Angeles abrasion, accelerated cycling and time cycling; large block tests, up to 1.0 m in diameter, may be required to obtain the deformability and shear strength parameters;
- experimental fills are widely used to prove the effectiveness of compaction equipment, number of passes and layer thickness;
- upstream of the dam axis, the materials shall be arranged in cross-section and compacted with wetting in layers up to 1.0 m thick; downstream of the axis are accepted lighter and compacted rockfills in layers up to 1.6 m; within a minimum range of 6–8 m of better-quality rockfill is left, next to the downstream facing for protection against bad weather;
- the width of the crest of the dam may vary from 7 to 10 m;
- concrete face thickness: it must be variable according to the hydraulic load to which it will be subjected, in order to guarantee a hydraulic gradient equal to 200. For this, the empirical formula used is:
  - up to 100 m height, $E = 0.3 + 0.002\ H$ (m), where H = vertical distance from the crest to the section considered;
  - height > 100 m, $E = 0.005\ H$ (m).
- concrete face reinforcement for crack prevention due to temperature variation, retraction, and other tensile stresses on the slab:
  - should be made of a mesh with an iron section corresponding to 0.4% of the theoretical section of concrete in the vertical direction and 0.3% of the theoretical section of concrete in the horizontal direction, about 50–65 kg/m^3;
  - in the region comprised by a range of 0.2 H, close to the encounters, 0.5% of the theoretical section of concrete is used in both directions. In the vicinity of the joints, it is convenient to establish a reinforcement of the structure in order to prevent any rupture of the corners due to compression on the neighboring slabs;
- joints: vertical joints and the perimetral joints are distinguished in the encounter with the plinth:
  - the spacing between vertical joints varies from 12 to 18 m;
  - vertical joints should be terminated in the normal position to the perimetral joint, so as to avoid a very sharp angle on the concrete slab, especially when the slope of the encounters is highly pronounced;
    - on softer slopes, keep the vertical joints straight until the encounter with the bottom slab;
    - horizontal joints occur only in the case of construction joints (dry joints) with continuous fittings, when concrete is interrupted (as juntas horizontais ocorrem somente no caso de juntas de construção (juntas secas) com ferragens contínuas, quando é necessária a interrupção da concretagem);
    - in the vertical joints of the central region of the face with a tendency to close, a metal plate made of copper or stainless steel;
    - at the bottom of the slab, the surface of the joint should be painted with asphalt; on the upper surface of the slab has recently been used a profile of neoprene.

146   Design of Hydroelectric Power Plants – Step by Step

- plinth: supporting structure of the concrete face which is located at the foot of the slope of the dam; this structure extends to the rocky massif, increasing the percolation path and therefore improving the sealing of the foundation mass; from the plinth, the injection curtain is implanted; the plinth should be anchored to the rock using 1 1/4" bars, every 1.20 m; the extent of the contact zone of the slab with the rock, in the flow direction (percolation), should be 1/20 to 1/10 of the pressure height, depending on the foundation rock;
- transition: a transition band between the rockfill and the concrete slab with a minimum thickness of 5.0 m shall be provided; on the outer face of the transition, primer should be provided with a rapid curing asphalt emulsion in an amount of approximately 4.0 l/m^2; the transition material should have $D_{max} = 15$–25 mm, well graduated, compacted horizontally in thin layers (40–50 cm) and according to the slope with upward vibrating roller (4–6 passes); the transition should be protected with profile in concrete with low cement content (55–75 kg/m^3); this procedure facilitates the construction, reduces losses of transition material, and increases the speed of dam elevation, while giving a better finish to the face of concrete;
- sealing curtains: in principle, it must be provided for the purpose of reducing the permeability of the foundation mass in the region of the plinth; three injection lines with spaced primary holes of a maximum of 10 m and spaced secondary holes of a maximum of 3 m should be provided.
- crest wall: a crest wall has been used 3–5 m high, with a curved upright face, to prevent the waves from reaching the crest of the dam.

7.3.3   Asphalt concrete face rockfill dams

The text of this item was made based on the following references, which should be consulted for a detailed examination of the subject: Sherard et al. (1963), USBR/DSD (1974), Grishin (1982), ICOLD, *Bulletin* 114 (1999), Falcão (2007), and Wilson (2013).

Asphalt concrete is the second most used type of coating for the face of the rockfill dam, with important advantages, such as: lower cost, greater flexibility when compared to concrete slabs, being able to withstand larger differential pressures without cracking and, moreover, it can be built faster by reducing the project's schedule. The first example of the modern era of dam construction is the Genkel dam in Germany (1950). Frame 7.3 presents several examples, but did not record the Montgomery dam in the United States (1957), which is presented in detail in DSD (1974). There is no doubt that it is an advantageous and safe alternative when properly constructed.

It is recommended that the upstream slope be 1.0 (V):1.7 (H), or softer (DSD, 1974). The hot bituminous concrete membrane is placed "in-situ" on the slope, as shown in Figure 7.9. Thickness of each asphalt concrete membrane layer is specified in Figure 7.10.

The transition layer between face and rockfill should be similar to that previously specified for the concrete face. Each layer, built upwards only, is placed in strips 3–4 m wide at right angles to the axis of the dam. A simple paving machine can lay 25–35 tons of asphalt concrete per hour. Plain rollers should be used for compaction. The last layer should be covered with asphalt mastic.

Frame 7.3 Asphalt Concrete Face Rockfill Dams (Falcão, 2007)

Name	Country	Year	Upstream Slope	Height (m)	Reference
Iron Mountain	USA	1397	1 V:2.00 H	48	ICOLD (1999)
BouHavina	Algéria	1398	1 V:0.80 H	55	ICOLD (1999)
Genkel	Alemanha	1950	1 V:2.25 H	43	Strabag (1994)
Croix	Suiça	1956	1 V:1.50H	15	ICOLD (1999)
Wahnbacht	Alemanha	1956	IV:1.60H	48	Strabag (1994)
Ohra	Alemanha	1966	IV:2.00H	59	ICOLD (1999)
La Preza	Venezuela	1967	IV:1.70H	60	ICOLD (1999)
Guarajaz	Espanha	1972	IV:1.75H	48	ICOLD (1999)
Futaba	Japão	1977	IV:1.85H	59.8	ICOLD (1999)
Oskenica	Alemanha	1978	IV:1.50H	46	ICOLD (1999)
Markersbach	Alemanha	1981	IV:1.75H	55	Walo (1997)
MerausZirimiliz	Itália	1987	IV:2.00H	53	Icold (1999)
Huesna	Espanha	1989	IV:1.60H	70	Strabag (1994)
Yashio	Japão	1992	IV:2.00H	90.5	Strabag (1994)
Alento	Itália	1993	IV:2.00H	43	Walo (1997)
El Agrem	Algéria	1994	IV:1.70H	63	ICOLD (1999)
Chiauci	Itália	1997	IV:1.60H	78	ICOLD (1999)
Goldisthal	Alemanha	1999	IV:1.60H	67	Walo (1997)
Midlands	Mauritius	2002	IV:1.60H	30	Walo (1997)

Figure 7.9 Asphalt concrete face. Enlargement of the Tierfehd reservoir. Glarus, Switzerland (2013).

It is necessary that the control of the construction is made from the sampling of random locations, along with carrying out tests of asphalt content, density, stability and permeability.

Joints between adjacent strips of the coating should be juxtaposed to seal the membrane. Transverse joints should be minimized and should be hot joints.

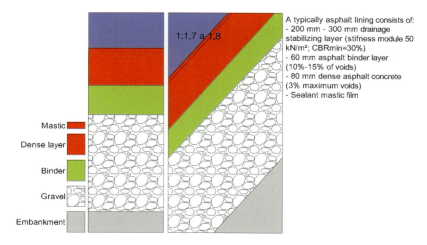

Figure 7.10 Concrete asphalt face – typical layers (2013).

Cold joints, between parallel or transverse bands in a single band, shall be treated as follows:

- applying an adherent coating of asphalt cement of the same type used in the design mix;
- place the asphalt concrete with an overlap of 7.5–15 cm;
- reheat the joint with an infrared heater to avoid open joints;
- compact the joint immediately after reheating.

The plinth should promote the easy release of the concrete layer in contact with the foundation.

The concrete asphalt face must be constructed in such a way that it is durable, flexible, impermeable, unbreakable and resistant to weathering. Materials of varying gradations of silt-to-graded gravel have been used to construct a suitable face (Figure 7.11).

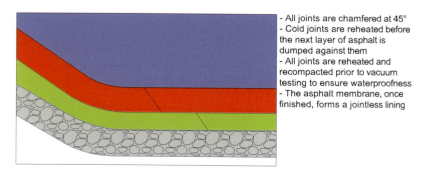

Figure 7.11 Concrete asphalt face – joints (2013).

Table 7.3 Asphalt Mixture. Grishin (1982).

Asphalt Mixture Type	Materials (%)				Weight (t/m³)
	Bitumen	Fines[a]	Sand[b]	Gravel[b]	
Mastic	20–40	50–80	-	-	1.60–2.00
Mortar	9–20	15–35	30–75	-	1.80–2.40
Bituminous Concrete	8–10	10–25	20–50	30–50	2.20–2.40

[a] Fines may be from rock, lime, brick, cement or other material with $0.05 < d < 0.50$ mm.
[b] Aggregate very well graduated, with a maximum diameter of 2.54 cm (1 in.) to fine sands.

It is not permissible to use clayey materials in the mixture because they tend to sway during the drying process and break when the mixture is compacted, leaving the dry material exposed to the water of the reservoir. The compositions of three mixtures are shown in Table 7.3.

In the last 15 years, the use of this type of membrane has increased greatly due to its low cost and reliability, which is greater than that of the concrete membranes, as well as the adaptability to the support prism deformations in the base (support prisms). It is believed that the use of asphalt (bituminous) concrete membranes will continue to increase, but it is necessary to keep in mind that selection of the composition of the mixture should take into account the climate and the conditions of service, and will require more detailed experimental studies. A detailed specification of composite materials of the asphalt concrete membrane is included in DSD (1974), including the asphalt tests.

All tests performed by USBR to determine the appropriate type of asphalt cement and the correct percentage to be used in the Montgomery dam led to a design mix incorporating 8.5% asphalt cement, which gives the percentage recommended by Grishin (1982).

7.3.4 Asphalt core rockfill dams

The use of rockfill dams with asphalt concrete core (Frame 7.4) is also increasing for the same reasons as for dams facing asphalt concrete. In the dams cited by Grishin (1982), the maximum thickness adopted for the core was H/20 (H is the hydraulic head on the dam). The asphalt concrete mix is placed in layers of 20–30 cm. The coarse aggregate, with a diameter of up to 30 cm, is mixed using a heavy vibrator. Asphalt consumption will be minimal if the mass of the outer layers of the core is composed of aggregates of uniform grains. The extensive use of bituminous concrete diaphragms is explained by their elasticity when compared to concrete cores, and by their lower cost.

In Brazil, the only example of this type is the Foz do Chapecó dam, in the Uruguay river, completed in 2004 (Figure 7.12). Figure 7.13 shows some photos of the Foz do Chapecó dam in August 2014, which illustrate the construction process. The photos show the mechanical method used by the paver machine specially developed for this service. Both the bituminous concrete core and the transition layers are applied simultaneously. This method is described in the Falcão thesis (2007).

Frame 7.4 Dams with Bituminous Concrete Core (Falcão, 2007)

Name	Country	Year	Core Thickness Top/Base (m)	H (m)	References
Henne	Germany	1954	1.0	58	Freiner et al. (1976)
Bigge	Germany	1961	1.0	52	Strabag (1990)
Dhunn	Germany	1962	0.5/0.7	13	Steffen (1976)
Mauthaus	Germany	1969	0.4	16	Steffen (1976)
Legardadi	Ethiopia	1969	1.0	26	Freiner et al (1976)
Ponza Honda	Ecuador	1969	0.6	28	Strabag (1990)
Wiehi	Germany	1971	0.4/0.6	54	Steffen (1976)
High Island	China	1973	1.4/1.0	95	Strabag (1990)
Vestredalstjern	Norway	1978	0.5	32	Hoeg (1993)
Finstertal	Austria	1978	0.5/0.7	100	Strabag (1990)
Strovant	Norway	1981	0.5/0.8	90	Hoeg (1993)
Megget	Scotland	1981	0.9	56	Penman and Charles (1985)
Pla de Soulcern	France	1981	0.6	29	Strabag (1990)
KleineZinzig	Germany	1981	0.5/0.7	70	Strabag (1990)
Shichigashuku	Japan	1985	0.5	37	Strabag (1990)
Hintermuhr	Austria	1989	0.5/0.7	40	Strabag (1990)
Muscat Flood	Oman	1992	0.4	26	Strabag (1990)
Storglomvant	Norway	1993	0.5/0.9	125	Hoeg (1993)
Ceres	South Africa	1997	0.5	60	Jones et al. (1999)
Goldisthal	Germany	1998	0.4	26	Strabag (1990)

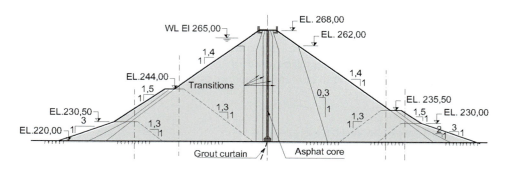

Figure 7.12 Rockfill dam with asphalt core. Foz do Chapecó dam (H = 47 m) (CBDB, 2009).

7.4 CONCRETE GRAVITY DAM

7.4.1 Gravity dam – conventional concrete

The name gravity dam is derived from the word gravitas, which means weight. This dam is a structure designed in such a way that its own weight resists the main horizontal forces, that is to say, the water pressure that acts in order to make the dam slide along the foundation, and to the vertical forces, such as the sub-pressure forces.

If the foundation is adequate and the dam is designed and constructed properly, the structure will be solid, and permanent and with little maintenance.

Masonry gravity dams were built thousands of years before Christ, with a base width of four times the height, as the ruins show. Over time, various types of mortar

Dams 151

Figure 7.13 (a) Foz do Chapecó rockfill dam (CBDB, 2009). (b) Equipment in operation – overview. (c) Detail of the compaction next to the concrete wall. (d) Half-height dam – overview.

152   Design of Hydroelectric Power Plants – Step by Step

Frame 7.5 Concrete Dams – Gravity Section. Ranked by Their Height

Dam	River	Height (m)
Grande Dixence - Switzerland	Dixence	284
Toktogul – Russia	Naryn	214
Krasnoyarsk – Russia	Yenisey	128
Bratsk – Russia	Angara	125

have been used to bind masonry, increasing stability and tightness along with realizing possibility of steeper slopes.

Concrete and cement mortar were used in the construction of cyclic concrete dams, which were precursors of the concrete dams (CCV), with gravity section, of concrete mass. Frame 7.5 presents some gravity concrete dams in the world.

The classical section is triangular, having the vertex opposite to the smaller leg at the height of the WLmax, plus a rectangular superstructure, which forms the crest of the dam, whose elevation is determined by the height of the wave that acts against the dam plus free-board.

External loads (sediments – Wsed and seismic – Wsism, Figure 7.17a, and W waves, Figure 7.17c) may imply changes in the classical section, as shown in Figure 7.17b, causing increase of the dam base (Δb) or slope of the upstream slope. The action of waves increases the horizontal force and can cause an increase in dam height (Δh) plus a free-board ($\Delta h1$), which adds a weight increase of the dam - Gs.

The conventional method of constructing concrete dams, gravity-like section, is based on a series of monoliths divided by contraction joints. The method has the advantage of preventing cracks caused by temperature, but the equipment needed for concrete cooling and construction joints makes it less economical than the conventional method of building earth dams. Another disadvantage is the limitation of the use of equipment due to small work place.

In the design of the structure, all the internationally recognized criteria, procedures, and standards summarized in the Project Criteria manual (2003) must be observed.

With regard to concrete, the materials to be used must comply with the provisions of Chapter 4 of this manual: Property of Materials.

Stability analyses should be performed for all current load cases described below:

- CCN, Normal Loading Condition: corresponds to all combinations of actions that present a high probability of occurrence during the useful life of the structure during normal operation or routine maintenance under normal hydrological conditions.
- CCE, Exceptional Loading Condition: corresponds to a situation of combination of actions with a low probability of occurrence over the useful life of the structure. In general, these combinations consider the occurrence of only one exceptional action, such as exceptional hydrological conditions, drainage system defects, exceptional maneuvers, seismic effects, etc. with the actions corresponding to the normal loading condition.

Figure 7.14 Great Dixence dam. 284 m high. Dixence river, Switzerland.

- CCL, Load Condition Limit: corresponds to a situation of combination of actions with very low probability of occurrence over the life of the structure. In general, such combinations consider the occurrence of only more than one exceptional action, such as exceptional hydrological conditions, drainage system defects, exceptional maneuvers, seismic effects, etc. with the actions corresponding to the normal loading condition (Figures 7.14-7.16).
- CCC, Construction Loading Condition: corresponds to all combinations that are likely to occur over the life of the structure. They may be due to shipments of construction equipment, partly completed structures, abnormal loading during the transport of permanent equipment, and any other similar conditions; occur for short periods in relation to their useful life.

In stability analyses, all loading cases described above must be considered, along with the peculiarities of each work. In these analyses, the concrete and foundation materials

Figure 7.15 Bratsk dam. Angara river, Central Siberia, Russia. 4,500 MW. 125 m high.

Figure 7.16 Krasnoyarsk dam. Yenisey river, Russia. 6,000 MW. 128 m high.

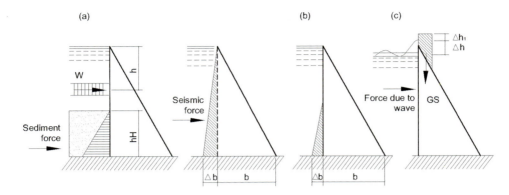

Figure 7.17 Change of dam section as a function of internal effort and conditions of operation (Grishin, 1982).

Table 7.4 Safety Factors (FSs) for Tipping and Floating

Safety Factors	Loading Cases			
	CCN	CCE	CCL	CCC
Tipping-FST	1.5	1.2	1.1	1.3
Floating-FSF	1.3	1.1	1.1	1.2

are considered as rigid bodies. In the calculations, spreadsheets are usually used. The concrete volumes, position of the centers of gravity and area of the foundation base are extracted from the computational models in 3D. The safety factors usually required for gravity structures are given in Table 7.4.

For structural and strain × deformations analysis, structural modeling is used by the FEM, which takes into account the elasticity and rigidity of the materials. The most used computational program in these analysis is SAP2000 (CSi).

7.4.2 Gravity dam – roller compacted concrete (RCC)

The need for a faster and more economical method of concrete dam construction was first addressed at the Asilomar Conference in California, USA (1972). The same issue was discussed a year later at the 11º ICOLD, Madrid (1973). Previous work has paved the way for suggesting alternatives to a faster and less costly method of building concrete dams. The idea of combining the advantages of continuous mixing of loose materials, transport and launch into the work place, with the advantages of concrete as a building material is from the 1960s.

Roller compacted concrete was first placed on a dam in 1960–1961. It was applied to the 65 m – high cofferdam core of the Shihmen Dam, Taiwan. The same aggregate continuously produced for use in conventional concrete with a maximum diameter of 76 mm was used. The RCC was made in the same plant used for the production of conventional concrete. The mixture used 107 kg/m^3 of cement. The material was transported by dump trucks and spread by tractors in layers of 30 cm. The compaction of the material in the optimum moisture (Modified Proctor) was done by the traffic of trucks and tractors D-8.

At the same time, between 1961 and 1964, the Alpe Gera dam was built in Italy, with metal plate face. The technology of the extended layer was applied, consisting of loose materials, and thin layer of concrete with 70 cm of thickness, placed from the outside to the outside (extended layer), avoiding construction joints. The compaction was done by vibrator batteries mounted on tractors. Cross joints were defined by cutting each layer. This same methodology was applied in the construction of the Quaira Della Miniera dam, also in Italy.

In 1965, Hydro Quebec designed two 18 m high gravity walls at the Manicouagan I dam in Quebec. In this project, the lean concrete was placed in the core of the dam and vibrated. In the upright face of the walls, a mixture of rich concrete was used, using forms. The joints and seals were spaced at intervals of 15 m. In the upstream side, precast concrete blocks were used. Hydro-Quebec estimated that the project saved 20% in cost and 66% of the time compared to that required to build the walls using conventional concrete.

Another important milestone in this constructive process occurred in the Tarbela dam in Pakistan, where 2.5 million cubic meters of CCR were launched between 1974 and 1982. The RCC was initially used to repair a rockfill protection wall destroyed during the collapse of the tunnel when filling the reservoir. The repair of the stilling basin and the cofferdam followed. A river aggregate with a maximum diameter of 150 mm and 133.5 kg/m^3 of cement was used. Trucks and tractors were used to transport, and vibration rollers were used for compacting. In the recovery of the spillway, a well-graded aggregate was used, with a maximum diameter of 150 mm and 10% of fines (passing through the 200 ASTM sieve). The aggregates were of two sizes, separated in the 19 mm sieve, continuously mixed with 110 kg/m^3 of Portland cement and water.

In Japan, surveys began in 1974 with two projects: the 89 m high Shimajigawa dam completed in 1980; and the foundation slab of Ohkawa dam (started in 1979). The RCC dosage included 130 kg/m^3 of Portland cement mixed with 30% fly ash and aggregate with a maximum diameter of 80 mm.

In 1974, USACE developed an RCC gravity dam as an alternative to the Canyon Zintel Reservoir in Washington, replacing an earth dam. This dam was not built, but the concepts served as the basis for the Willow Creek, Oregon dam.

In 1979, the Winbleball dam was concluded in England, which was the first with RCC with high ash content, with laser control of the forms.

The Willow Creek dam in Oregon (Figure 7.18), 52 m high, for flood control and recreation, was started in November 1981 and completed in October 1983, entirely in RCC. The upstream face is vertical and the downstream face is sloped 0.8 (H)/1.0 (V).

The total volume of 300,000 m^3 was completed in 5 months. The RCC was designed with a crushed aggregate with a maximum diameter of 76 mm and 4%– 10% of fines passing in the sieve 200. The cement dosage varied according to the part of

Figure 7.18 Willow Creek dam (1983).

the structure. For the inner part, 47 kg/m³ of Portland cement mixed with 19 kg/m³ of fly ash were applied. The thickness of the layer ranged from 24 to 34 cm (laser controlled).

The Upper Stillwater dam, Utah (Figure 7.19), 88.4 m high, for water supply and irrigation, was started in 1985 and completed in 1988. The upstream facing is vertical and the downstream slope is inclined with 0.6 (H)/1.0 (V). The total volume was 1,147,000 m³ RCC. The RCC mixture was designed with clean sand (borrow area 13 km from the dam), well-graded aggregate with a maximum diameter of 38 mm and 247 kg/m³ of binder material, including a high fly-ash fraction.

RCC was used for the first time in the Itaipu HPP in 1976. By the end of 1980, only two RCC dams were completed. From that year, its use increased progressively. By the end of 1986, there were 15 completed dams. Ten years later, in 1996, there were more than 150 dams in RCC.

The evolution of the concept of dams followed different paths:

- dams built with lean mixtures, with a cement paste content of 70 to 100 kg/m³ and with the laying of mortar between layers. This alternative was adopted by USACE and other researchers; the first major project was the Willow Creek Dam (1983);
- dams with high binder content from 150 to 270 kg/m³, with a high fly-ash fraction. The example is the Upper Stillwater dam project (1988), with 1,147,000 m³ of concrete with 247 kg/m³ of binder material;

Figure 7.19 Upper Stillwater dam (1988). USBR.

- dams with average pulp content, with mixtures ranging from 130 kg/m^3 (Les Olivettes – France) and 170 kg/m^3 (Craigbourne – Australia) of binder material;
- the dams were designed with a cement content of 70 kg/m^3 at 100 kg/m^3, with a high content of fines (8–12% less than 0.075 mm in diameter), with the placement of mortar between the layers and a conventional concrete-mass layer in the upstream facing part of the dam. The fine material may be silt without pozzolanic activity, as used in the Saco Nova Olinda dam on the Gravata river (Ceará), or may be rock dust with low pozzolanic activity, as used in the Salto Caxias dam, Iguaçu river, and the Jordan river dam (Paraná), Figure 7.20, whose features are shown in Table 7.5.

Table 7.5 Average Mix Design Used in the Jordão River Dam (Santi, 2000)

Materials	RCC		CCV (face)	
	Unit Quantity by Average Mix Design (kg/m^3)	Proportional Mix Design (kg)	Unit Quantity by Average Mix Design (kg/m^3)	Proportional Mix Design (kg)
Cement (CPIV-32)	74.90	1	185.30	1
Water	92.30	1.23	138.30	0.75
Artificial sand	1288.60	17.20	1039.40	5.61
Limestone powder	21.10	0.28		
Coarse aggregate 25 mm	714.70	9.54	583.40	3.15
Coarse aggregate 50 mm	512.40	6.,84	583.40	3.15

Figure 7.20 Jordan river dam. Brazil (95 m high).

The wide approval of the RCC dams is explained by the great advantage of this technology when compared to conventional concrete dams is due to higher concreting rate – can reach 2–2.5 m/week; various conventional equipment (dump trucks, tractors and rollers) can be used; as a result, cost reduction; and smaller impacts on the environment.

Most RCC dams are gravity-like. In some countries, such as South Africa and China, there are examples of arch RCC dams. Table 7.6 shows data from some RCC works carried out in Brazil.

Table 7.6 Use of RCC in Brazil (Andriolo, 1998)

Dam	Year	Height m	Volume m^3	Dosing (kg/m^3) Cement	Fines	Observation
Itaipu	1978	196-	26,000	91	26 (fly ash)	Warehouse slab
S. Simão	1978	95	40,000	-	-	Various struct.
Tucuruí	1982	77	12,000	65	38 (calcined clay)	Navigation lock
Saco N Olinda	1986	56	138,000	70	-	2,500 m^3/day
Xingó	1994	150	44,155	100	30 (fly ash)	Protection
S. Mesa	1989	22/13	28,600	60	133 (blast furnace slag)	400 m^3/day
Caraíbas	90-95	26	17,800	66	-	310 m^3/day
Gamelereira	90-95	29	27,000	70	-	
Juba I	90-95	21	17,000	70	-	Spillway steps
Jordão	1998	95	547,000	70–100	-	-
S. Caxias	1999	66	912,000	-	-	-

160   Design of Hydroelectric Power Plants – Step by Step

A dam in RCC can rise about 60 cm/day, depending only on the speed of production of the material. With this production, a 50 m high dam would take about 3 months to complete (for example, Saco de Nova Olinda).

RCC dams follow the same design procedures as conventional CCV concrete dams. Special attention should be given to thermal stresses and impermeability of the massif.

The diagrams of underpressure and the use of drainage galleries have been much discussed. In general, there is an internal line of drains to combat percolation and underpressure in the dams.

Thermal stress analyses have been routinely developed since the 1960s. However, due to the low cement content in the RCC mix, these analyses have shown that cracks are not problems in Brazilian dams.

The use of low cement content has the following advantages: it minimizes the risk of thermal cracking, reduces material for alkali-aggregate reactions and the cost of mixing.

When it is considered necessary to use pozzolanic material, its quantity should be as small as possible because of its high cost. Silt has also been used in some blends.

The major breakthrough was the use of crushed stone powder as a thin material in the mixture. The first experiment was that of the Itaipu HPP, in which it was proved that certain types of rock, when crushed, produce fines with pozzolanic properties. For this reason, there is a tendency to include a quaternary crusher at the crushing plant to produce this material.

In the construction, simple equipment for the production of RCCs has been used: common concrete plants or pugmills, small trucks (4–6 m^3), tractors (D4-D6), and rollers commonly available. The thickness of the layers increased from 25 cm at the beginning to the current 40 cm.

In the construction of the upstream face, conventional forms are used. On the downstream side, preformed elements may be used. In the execution of the construction joints and the seals, conventional procedures are also used.

It should be noted that the RCC was used for many years in sub-bases of highways and airports, where they were referred to as "lean concrete" or "dry lean concrete", in layers of 150–250 mm, mainly under the bituminous layer. Two photos of the RCC dam of the João Leite river, 53.50 m high, built for the water supply system of Goiânia, are presented in Figure 7.21. One can observe the layout of the dam, with a stepped spillway, as well as a detail of the embrace of the left wall by the clay core.

7.5   CONCRETE ARCH DAM

The arched concrete dam is curved (circular) in plan, works as a vault or bark, and resists horizontal loads, mainly due to its support in the abutments of the valley (throat). In Brazil, due to the topographic and geomorphological characteristics of the basins, there is practically no arc dam, except for the Funil HPP, with 85 m height, in the Paraíba do Sul river, Rio de Janeiro (Figures 7.22 and 7.23).

Figure 7.24 shows examples of types of arch dam (Grishin, 1982). The horizontal cross-sections (in plan) of the arched dams are usually circular in shape with the

Figure 7.21 (a–b) João Leite river dam. Height = 53.50 m. Water supply in Goiânia (CBDB, 2009). Side view. (b) Clay core embrace detail.

normal arches on the banks of the river at the abutments. The vertical cross-sections (cantilever) have different shapes and, in some cases, have curvilinear shapes. Stability is guaranteed by the frame supports on the abutments. Therefore, the dams should be designed with slender arches whose thickness is determined by the strength of the material of the structure.

The profiles of the arch dams are more compressed when compared to gravity dams and are characterized by the so-called β-form factor defined by the relation: $\beta = ef/H$, where ef = thickness of the dam at the foundation and H = height.

For slender arch dams, $\beta < 0.2$ (for gravity dams $\beta = 0.6$–0.8). Among the existing dams, the best ones are Tolla ($\beta = 0.023$–0.048) and Vaiont ($\beta = 0.084$), among others.

162 Design of Hydroelectric Power Plants – Step by Step

Figure 7.22 Funil HPP. Paraíba do Sul river (Rio de Janeiro, Brazil).

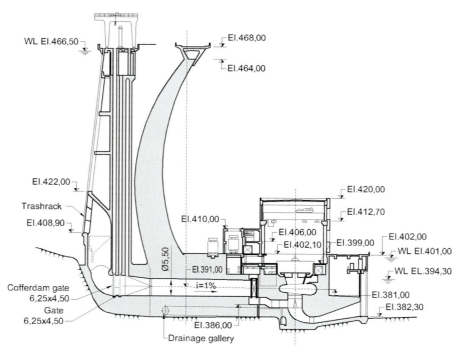

Figure 7.23 Funil HPP. Arch dam section. $H = 85$ m. "Topmost Dams of Brazil" (CBDB, 1978).

Dams 163

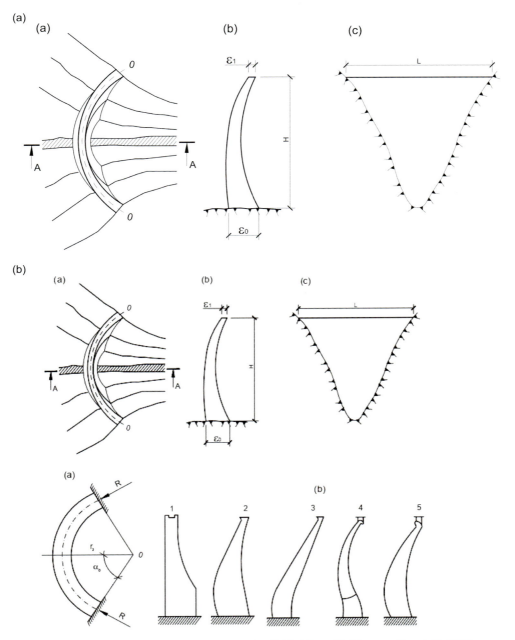

1) Tin Dam (H=180 m; e₀=44,5 m; L/H=1,63); (2) Mori Dam (H=65 m; e₀=18 m; L/H=2,86); (3) Anshane Dam (H=75 m; e₀=11 m; L/H=3,07); (4) Val-Galina Dam (H=92 m; e₀=11,2 m; L/H=2,48); (5) Oziletta Dam (H=77 m; e₀=10,8 m; L/H=2,91).

Figure 7.24 (a) Plant. (b) Arch dam sections. (c) Longitudinal section (Grishin, 1982).

Based on this parameter, arch dams are classified into three types:

- slender arch dams $\beta < 0.20$;
- dams in thick arch $\beta < 0.20–0.35$;
- arch-gravity dams $\beta > 0.35$.

Depending on the slenderness, the volume of concrete per meter of length of an arched dam is 2–4 times smaller; there are cases of 6–8 times lesser or more than the volume of a gravity dam. According to Grishin (1982), despite the higher cost of structural concrete, the cost per cubic meter does not exceed 10%–15%. The increased use of arch dams is explained by economy and reliability. In the late 1970s, more than 300 dams of heights greater than 30 m were under construction.

The first arched dam, Ponte Alto, was built in Italy in 1661, with a height of 5 m and a radius of 15 m. This dam was raised several times, and in 1883, it reached 38 m in height. In the 19th century, several dams were built in Europe and the United States, before the development peaked in the 20th century (Table 7.7).

Based on height, arch dams are classified into three categories: low height < 40 m, medium height 40–100 m, and high height > 100 m.

Based on their shape, arch dams are divided into the following types: the surface is curved only in the horizontal plane; the surface is curved in both planes: horizontal and vertical; if the profile is much curved over the height, the arched dams are known as dome dams.

According to the contact at the foundation, the dams can be divided into: elastic dam resting on the abutments; dams with perimeter joints (see Figure 7.25); dams with joints, or joints – notches, at the foundation of the dam.

Table 7.7 Arch Dams in the World (Grishin, 1982)

Country	Dam	Year	H (m)	e_c (m)	E_f (m)	$\beta = e_f/H$	L/H
USA	Hoover	1935	221.4	13.7	201.2	0.91	1.71
	Gene Wash	1937	42.7	1.52	8.08	0.19	2.73
	Copper	1938	61	1.52	10.67	0.17	1.32
	Hungry Horse	1953	171.9	11.9	100.6	0.59	3.75
	Ross	1957	165	10	63	0.38	2.30
	Glen Canyon	1964	216.4	7.6	91.4	0.42	2.20
	Yellowtail	1966	160	6.7	44.2	0.28	2.82
	Mossy rock	1968	185	8.2	38	0.21	2.06
	Boundary	1967	105	2.4	9.7	0.09	2.15
	Morrow Point	1968	142.6	3.7	15.8	0.11	1.55
	New Bullard's Bar	1969	194	8	58.7	0.30	3.46
Russia	Ladzhanursk	1960	67	4.5	13	0.19	1.90
	Ingursky	1987	271.5	7.5	59.7	0.22	2.77
	Sayano-Shushensk	1985	245	24	124	0.53	4.35
	Chikreysk	1976	236	7.6	42.5	0.18	1.18
Italy	Vajont	1960	266	3.9	23	0.084	0.71
	Pieve di Cadore	1950	57/112	6	26.2	0.23	7.45
Switzerland	Mauvoisin	1958	237	14	53.5	0.23	2.26
France	Tolla	1961	88	1.5	2	0.023	1.36
	Malpasset	1954	66.5	1.55	6.65	0.10	3.35
Zimbawe	Kariba	1959	125	13	27.5	0.22	2.00

Dams 165

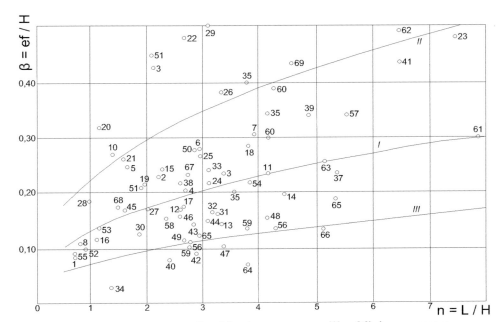

I - Average curve; II = Maximum curve; III = Minimum curve.

Figure 7.25 Graph between $n \times \beta$ (Grishin, 1982).

Figure 7.25 contains several points that refer to the existing dams. The following is the list of the best-known dams with their heights given to the right of their names: (**1**) Vajont, 262 m; (**3**) Glen Canyon, 216 m; (**19**) Kariba, 125 m; (**22**) Castelo do Bode, 115 m; (**23**) Piave di Cadore, 112 m; (**27**) Castillon, 100 m; (**28**) Picote, 95 m; (**29**) L'Aigle, 95 m; (**33**) Maréges, 90 m; (**40**) Caniçada, 82 m; (**41**) Piantelessio, 80 m; (**47**) Malpasset, 67 m; (**55**) Valle di Cadore, 61 m; (**57**) Clark, 61 m; (**61**) Schiffenen, 47 m; (**62**) Isola, 45 m; (**68**) Chirkeysk, 236 m; and (**69**) Sayano-Shushensk, 234 m.

Arched dams can also be classified according to less characteristic features, such as narrow valley or wide valley arc dams (L/H> 3–3.5), with L = length of dam along the crest; in symmetrical or asymmetric canyons, etc. It is desirable, whatever the classification, that an arched dam, from a technical-economic point of view, should be designed and constructed only in places that have adequate geological and topographic conditions.

In the present dams the values of the compression tensions exceed 200 kg/cm^2. To support such efforts, the abutments must be composed of sound, monolithic and low deformability rock masses. The solid must be waterproof, water resistant, and free from discontinuities, faults, fractures, cracks, and stratifications. It should be noted that geologically unfavorable sites require complex measures of foundations' treatment in order to consolidate the massif.

The topographic conditions of the valley have significant effects on the profile of an arch dam. The determining factor, in this case, is the coefficient of the site $n = L/H$, as defined above.

In narrow valleys, the arches of the dams can be made very slender, which lead to very economical structures. On the contrary, in broad valleys, the dam may not be economical.

Until recently, it was believed that arched dams could be constructed for $n < 3.0–3.5$ and slender arch dams for $n < 0.5–2.0$. Currently, several dams have been designed and built very economically for $n \sim 10$.

The range of applications of arch dams has been extended by the adoption of structural joints, which increase the effectiveness of arch effect. The effect of the coefficient n on the coefficient of form β for arc dam (its economy) can be seen in Figure 7.25.

For a same value and n, the effectiveness of an arch dam depends on the cross-sectional shape of the valley, which may be trapezoidal, rectangular, triangular, or a complicated non-symmetric section. The triangular shape is considered the best because, with it, the lower sections of the dam, subject to great hydrostatic pressures, not only have a shorter length and but are also slimmer.

Figure 7.26 shows schematics of arched dams for sites with complicated shapes. Valleys with composite sections are undesirable because a rapid change in dam height leads to stress concentrations. In such cases, the sections are modified in the excavation design to predict concrete gravity blocks on the abutments or concrete plugs in the foundation (see Figures 7.25a and 7.25c). The structure is separated from the foundation by a peripheral joint (Figure 7.25d), which ensures a smoother and more symmetrical contouring of the base, making it possible to obtain a more uniform stress state for the structure.

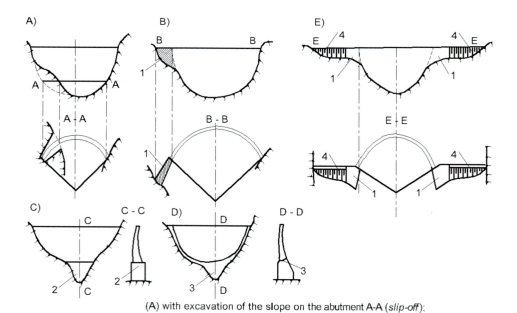

(A) with excavation of the slope on the abutment A-A (*slip-off*);
B) with support block on the abutment B-B (1 - coastal abutment);
C) with plug 2; D) with perimetral join 3; E) with wingwall (1); with closing dams (4).

Figure 7.26 Schematic arrangements for locations with complicated forms (Grishin, 1982).

The designs of the arched dams should consider all the aspects mentioned above with respect to the shape of the rings in plan, the selection of the rays, central angles, and profile of the dam. For the structural design of these dams, as well as for stability analyses of structures and their foundations, the specific publications listed in the bibliography at the end of the book should be consulted.

The Vajont dam is an example of an arched dam (Figure 7.27). Chapter 13 contains a summary of the accident with this dam in October 1963, with illustrative photos. The ground cover of the left slope (Monte Toc) slid into the reservoir, creating a 100 m high

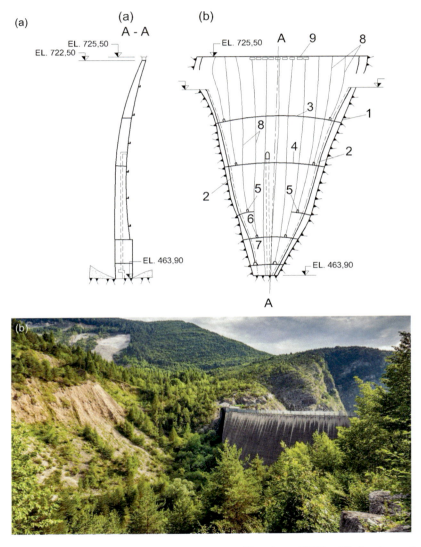

Figure 7.27 (a–b) Vajont dam (Itália). H = 266 m (Grishin, 1982). (1) Rock surface; (2) perimetral joint; (3–7) curvilinear joints; (8) construction temporary joints; and (9) spillway.

Figure 7.28 (a) Hoover HPP (1931–1936). USBR. $H = 221$ m; $L = 379$ m. Spillway – 2 side channel, $Q = 11{,}000$ m^3/s. Downstream bridge connects the states of Nevada – Arizona (2004--2009). The bridge is 600 m long and is 900 m high above the Colorado river. (b) Hoover HPP.

wave that reached the dam, which, to everyone's surprise, remained intact. The small town of Longeron, downstream, was razed. Two thousand people lost their lives.

Another example of this type is the Hoover Dam on the Colorado river (221.4 m high), USA, Figure 7.27, whose characteristics are presented in Table 7.7.

At that time, at 221 m, it was almost double the height of any existing dam. The installed capacity was 2078.8 MW: with 13 Francis machines of 130 MW; two of 127 MW; one of 68.5 MW; one of 61.5 MW; and two 2.4 MW Pelton machines. In the design, the First Trial Load Method was applied for the first time. In the power plant, new construction techniques were used with refrigeration tubing and stepped contraction joints. In addition, advanced concrete technology was used with low-heat hydration cement, as discussed in detail by Kollgaard (1988). It should be noted that the side spillways, for 11,000 m^3/s, were positioned on the abutments (Figure 7.28). The diversion tunnels, on both sides, were adapted for spillway tunnels (Figures 3.12 and 3.13).

To conclude this chapter, a cross section and a downstream photo of the 1000 MW Chirkeysk Power Plant on the Sulak river, Republic of Dagestan – Russia, constructed between 1955 and 1976 are presented in Figures 7.29 and 7.30. The dam has 236 m of height, crest thickness of 7.6 m, thickness at the foundation of 42.5 m and L/H = 1.18, as cited in Table 7.7.

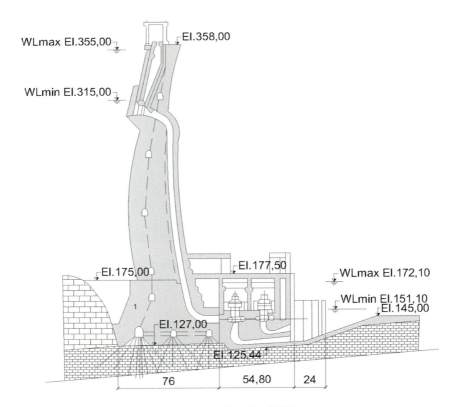

Figure 7.29 Chirkeysk HPP. Cross section. Grishin (1982).

Figure 7.30 (a–b) Chirkeysk HPP. Downstream view. Observe the tunnel spillway exit on the left bank.

Chapter 8

Spillways

8.1 TYPES OF SPILLWAYS AND SELECTION CRITERIA

Several types of spillways can be adopted, depending on the layouts as presented in Chapter 3. The topographical and geological aspects of the sites condition the definition of the layout, as well as the selection of type of spillway. Figure 8.1a shows the spillway of Tucuruí HPP in the Tocantins river, Pará.

The spillway comprises an approach channel, a control structure, a chute and an energy dissipator. In Tucuruí, the structure included 40 bottom outlets that were used in the third phase of diversion. After construction, they were plugged. But it should be noted that if some of them were definitive, they would be important for emptying the reservoir in case of need in the future.

At the beginning, the author would like to highlight the importance of having a spillway well positioned in the layout. This is done to avoid separation of the flow lines, and to achieve the best possible flow net for accommodating the approach flow. That is, the flow net must be the best possible. Figure 8.1b–h shows what was designed to accommodate the flow to the spillway in Tucuruí.

As seen in Chapter 3, spillways can be incorporated into the main dam body. They can be side-channel spillways or can be separate (isolated) dam structures positioned on the abutments. Some of the noteworthy examples are presented here:

- the side spillways of the Hoover dam in tunnel, Colorado river, Arizona/Nevada, USA (Figure 3.13);
- the spillways of the Serra da Mesa dam, Tocantins river (GO) (Figure 3.46) and Nova Ponte, Araguari river (Minas Gerais) (Figure 3.47), whose chutes have long sections not covered with concrete, that is, with the direct flow on the rocky massif;
- the spillway of the Karakaya dam, Euphrates river in Turkey (Figures 8.33 and 8.34), above the power house.

The spillways can be on the surface or at the bottom. This can be seen in various examples that will be presented:

- surface spillways, free or controlled by gates, allow the level of the reservoir to be lowered to its crest;
- bottom spillways, controlled by gates, allow the total or partial emptying of the reservoir.

DOI: 10.1201/9781003161325-8

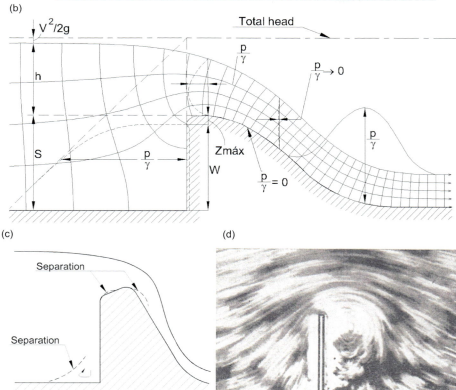

Figure 8.1 (a) Tucuruí HPP Spillway ($Q = 110,000 \, m^3/s$). (b) Typical flow net. (c) Typical separation of the flow. (d) Illustration of separation. (e) Illustration of the point of separation. (f) Tucuruí HPP. Part of the spillway and the right dam. Spigot detail designed to improve accommodation of the approach flow. (g) Spillway. Two-dimensional model. (h) Spillway. Two-dimensional model.

(Continued)

Figure 8.1 (Continued) (a) Tucuruí HPP Spillway ($Q = 110,000 \, m^3/s$). (b) Typical flow net. (c) Typical separation of the flow. (d) Illustration of separation. (e) Illustration of the point of separation. (f) Tucuruí HPP. Part of the spillway and the right dam. Spigot detail designed to improve accommodation of the approach flow. (g) Spillway. Two-dimensional model. (h) Spillway. Two-dimensional model.

The spillways have varied typical sections:

- sections with upright facing and sloping downstream facing Creager profile, over which water flows; after the Creager profile, one can have a stilling basin, a chute spillway with a ski jump at the end; this profile is often stepped spillways, without control of gates;
- sections of rounded crest with jet in free fall, or orifice with ski jump (or cross jets), in the cases of arch dams in narrow valleys;
- sections in shaft (morning glory spillway);
- sections in orifice (bottom), in conduit (under a dam) or in tunnel.

One should mention the siphon-type spillways, which are used when it is necessary to automatically discharge excess inflow to keep the WLres at a specified elevation. For this type, it is recommended to consult Grishin (1982) and USBR-DSD (1974), among other references (Figures 8.2–8.6).

Typical sections of some spillways, as well as some selected images, are presented in Figures 8.7–8.10.

174 Design of Hydroelectric Power Plants – Step by Step

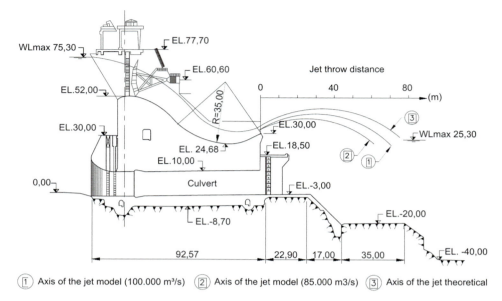

Figure 8.2 Tucuruí HPP – Ski jump spillway. Q = 110,000 m³/s (CBDB, 2002, Section 7.2.1).

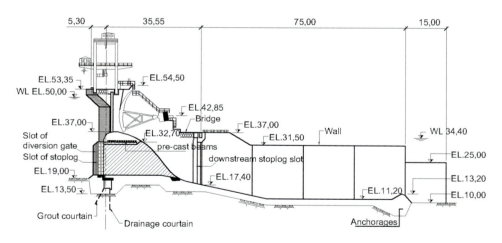

Figure 8.3 Balbina HPP – Spillway and stilling basin. Q = 5,840 m³/s (MBD, 2000).

Note: To settle the dam on the rock mass, andesite sound without any significant crack, the 60 m thick pre-glacial deposit has been removed. Thus, in the event of spills, the jets dive into a plunge pool of this depth. Underwater inspections carried out until 1978 didn't reveal any erosion problem (Figures 8.11–8.13).

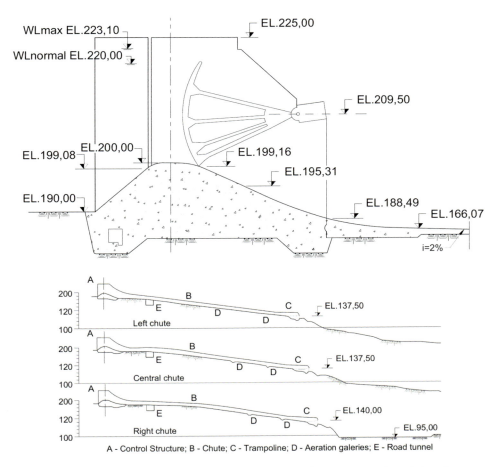

Figure 8.4 (a–b) Itaipu HPP – Chute spillway and ski jump in the right abutment. 72,000 m³/s (CBDB, 2002).

8.2 HYDRAULIC DESIGN

The following is a summary of the guide for the hydraulic design of the spillway. Initially, it is noteworthy that the flow for the design of the structure must meet the criteria specified in Section 3.4:

- for dams with heights greater than 30 m or whose collapse involves the risk of loss of life (permanent downstream housing), the design flow of the spillways will be the maximum probable flow (VMP);
- for dams less than 30 m in height or with a reservoir volume of less than $50 \times 10^6 \mathrm{m}^3$ and where there is no risk of loss of human life (no permanent downstream dwellings), the design flow will be defined by a risk analysis, noting that the minimum recurrence will be 1,000 years.

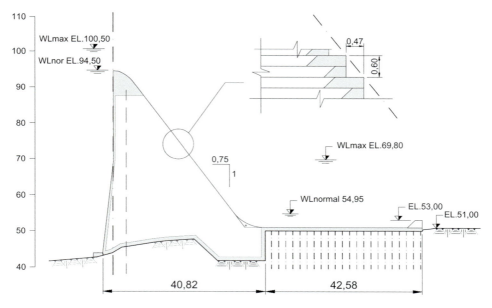

Figure 8.5 Dona Francisca HPP – stepped spillway and stilling basin. $Q = 10{,}600\,\text{m}^3/\text{s}$. (MDB, 2000).

Figure 8.6 Monjolinho HPP. Side spillway. $L = 210\,\text{m}$; $Q = 6{,}755\,\text{m}^3/\text{s}$.

Hydraulic sizing determines the width (L), or effective length of the crest of the structure, and the height of the hydraulic load on the crest (H).

As described in Chapter 4, it is noted that over 20% of dam accidents are due to insufficient spillway flow capacity (see ICOLD Bulletin, 99:1995). In these cases, it can be inferred that the estimated flow for the design of this structure has been exceeded.

Figure 8.7 Irapé HPP. Tunnel spillway. Q = 6,000 m³/s. Two gates 11 m × 20 m (MBB, 2009).

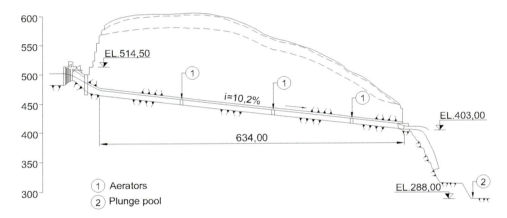

Figure 8.8 Irapé spillway – tunnel spillway profile (MDB, 2009).

This means that there may have been an error in determining the design flow and also in the hydraulic sizing of the structures.

The hydraulic dimensioning of an uncontrolled weir considers ideal crests with upstream and downstream quadrant geometries defined in the HDC (1977) charts 111-1 and 111-2/1. These profiles depend on the hydraulic load on the structure (H_o), the slope of the upstream face and the height of the crest on the spillway channel floor (P), which influences the approach flow velocity on the crest as defined in Figure 8.14.

The flow capacity of an uncontrolled sill is given by the equation:

$$Q = C_o L H^{3/2} \tag{8.1}$$

where:
Q = total flow (m³/s);
C_o = discharge coefficient;
L = effective crest length (m);
H = hydraulic head (m).

Figure 8.9 Mossyrock HPP. Cowlitz river, Columbia river. Tacoma City. Washington (USA). Dam height = 185 m. Falling jet spillway. $Q = 7{,}800 \, m^3/s$. (Kollgaard, 1988).

That is, the spillway flow capacity depends on the energy head on the crest (ht), the effective crest length (L) and the flow coefficient, which are determined as shown below.

Ho design head is used to define the crest downstream profile as shown in Figure 8.14. Hydraulic head can be equal to, less than or greater than Ho. For $H > H_o$, the discharge coefficient increases and negative pressures occur on the spillway face.

In determining the net length of the crest (L'), the contraction effect of the K_p pillars and the abutment K_a as defined in the letters 111-3/1 and 111-3/2 of the HDC (1977) or in the DSD (1973, p. 373) should be considered:

$$L' = L + 2\left(NK_p + K_a\right)H_o \qquad (8.2)$$

where:
L = effective crest length (m);
L' = net crest length (m);
N = number of pillars;
K_p = coefficient of contraction of the pillars;
K_a = coefficient of contraction of the extreme spillway walls.

The flow coefficient is influenced by a number of factors such as:

Spillways 179

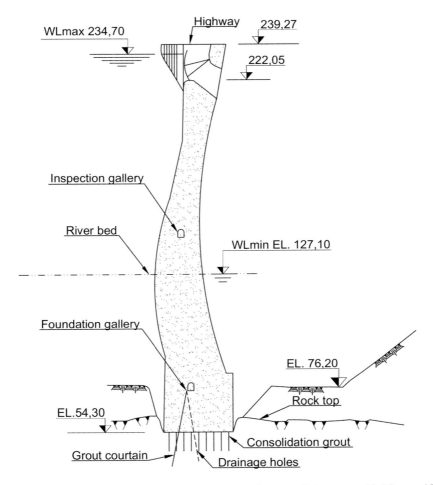

Figure 8.10 Mossyrock HPP. Falling jet spillway. four radial gates, 12.95 m × 15.24 m (Wengler, 1982).

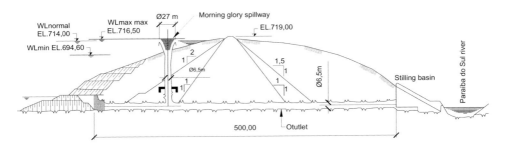

Figure 8.11 Paraitinga HPP. Morning glory spillway. $D = 27$ m at the crest. $Q = 5,400 \text{ m}^3/\text{s}$ (CBDB, 1982).

180 Design of Hydroelectric Power Plants – Step by Step

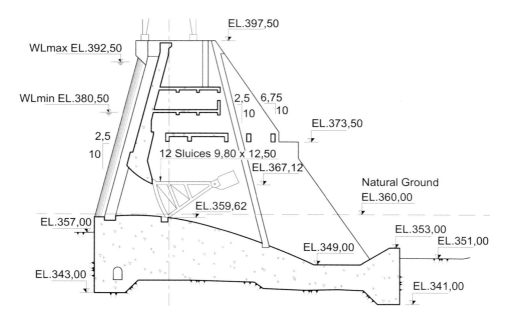

Figure 8.12 Sobradinho HPP Bottom spillway $Q = 16,000 \, m^3/s$ (CBDB, 1982).

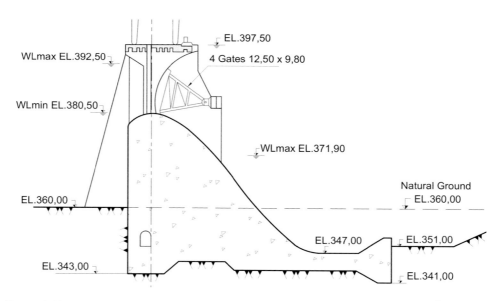

Figure 8.13 Sobradinho HPP. Surface spillway-short stilling basin. $Q = 6,855 \, m^3/s$ (CBDB, 1982).

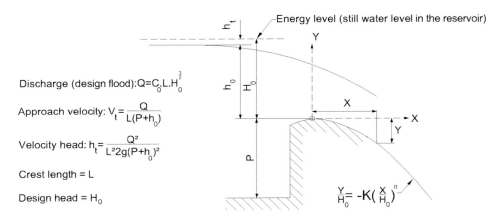

Figure 8.14 Standard ogee.

- the depth of the entrance channel (Figure 8.15);
- different heads of design head (Figure 8.16);
- upstream slope (Figure 8.17);
- downstream submergence interference (Figure 8.18).

In determining of the discharge coefficient, it is desirable, if possible, to rely on experimental studies of these structures in reduced models. For gate-controlled spillways, see Figure 8.15 and DSD (1973), Chapter 9, Chart 200 (Figure 8.19).

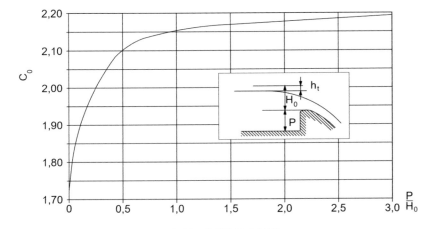

Figure 8.15 Discharge coefficients × P/Ho (NUST, 2003).

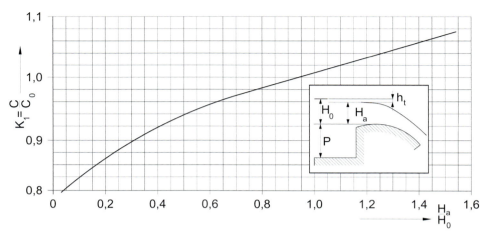

Figure 8.16 Discharge coefficient correction for other heads beyond the design head (Ho) (NUST, 2003).

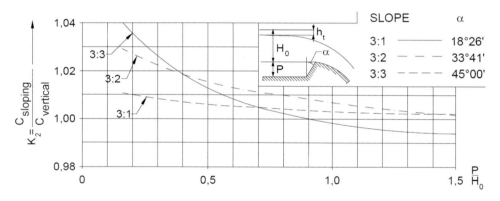

Figure 8.17 Coefficient of discharge for ogee with upstream slope (NUST, 2003).

For side-channel spillways, observe what is outlined in Chart 202. For morning glory spillways, observe what is outlined in Chart 212 and also in HDC Charts 140-1–140-1/8 (1977).

8.2.1 Design of the tucuruí spillway

Following are some "notes" about the Tucuruí HPP spillway, one of the largest in the world, a project in which the author worked for 5 years (1981–1986). Their participation extended until they completed the training of the plant's operation team on site (from 1984 to 1986). Table 8.1 shows the evolution of flood statistical studies throughout the project.

Spillways 183

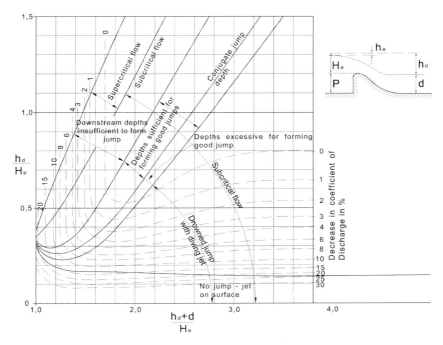

Figure 8.18 Discharge coefficient correction as a function of downstream effects (stilling basin elevation and submergence) (DSD, 1974).

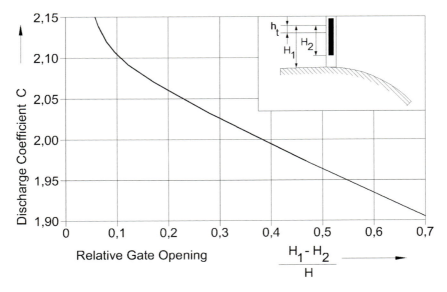

Figure 8.19 Discharge coefficients for controlled flow (NUST, 2003).

184 Design of Hydroelectric Power Plants – Step by Step

Table 8.1 Flood flow – Gumbel (Technical Memory, 1989)

TR (years)	Flows (m³/s)		
	Basic Design (1974)	Review (1979)	After the Flood (1980)
25	45,800	48,100	52,400
50	50,900	53,500	58,600
100	56,000	58,900	64,700
200	61,100	64,200	70,800
500	67,700	71,200	78,900
1,000	72,800	76,600	85,000
10,000	89,600	94,300	105,300

Studies were also developed for the definition of the Probable Maximum Precipitation (PMP), which leads to the determination of the Maximum Probable Flow (VMP), using the methodology recommended by the World Meteorological Organization (WMO), using the SSARR – Streamflow Synthesis and Reservoir Regulation.

The VMP was set at 105,800 m³/s. However, given the complexity of the alternative weather possibilities involved, the flow rate of 110,000 m³/s was adopted as the ultimate capacity.

The hydraulic design of the spillway was made for discharge $Q = 100,000$ m³/s with a design head $H_o = 22$ m. The normal WL was set at an elevation of 74.00 m. Therefore, the height of the crest at 52.00 m elevation was defined.

As the riverbed upstream of the spillway is at an elevation of 0,00 m, an upstream water height was characterized $P = 52.00$ m with respect to the spillway crest. So, $P/H_o = 2.36$. For this value in the graph in Figure 8.15, a flow coefficient $C_o = 2.18$ is given.

Using expression (7.1) gives the required total slope width $L \sim 445$ m. The project adopted a total width of 460 m, i.e., a spillway with 23 gates, 20 m wide.

Following the March 1980 flood of 68,400 m³/s, the flood studies were redone and the design discharge increased to $Q = 110,000$ m³/s (ultimate capacity). The spillway design was maintained. In the case of this event, the reservoir WL must reach the elevation 75.30 m – Wlmax at the max.

For normal reservoir elevation, 72.00 m, $Hd = 20$ m and the flow capacity 89,700 m³/s. The dimensions of the gates were fixed at 21.22 m high × 20 m wide. It should be noted that:

- the magnitude of the design discharge meant having to design a spillway with large gates;
- at that time, gates with heights around 20–22 m were, and still are, considered high;
- after the study of alternatives, a $H_o = 22$ m was adopted and the spillway width L was calculated; for a smaller H, the width L would be larger and, consequently, the volume of concrete would also be larger and the structure would be more expensive.

The study of gate width × height alternatives should define the most economical spillway, which results in the smallest volume of concrete.

8.2.2 Physical model studies

Due to the size of the project, the Tucuruí Hydroelectric structures and equipment were studied in several reduced models, which provided valuable subsidies for the project. Civil Works were tested in the Hidroesb hydraulic laboratory in Rio de Janeiro. Three models were built and operated:

- a three-dimensional model for studies of all works, at a scale of 1:150 (Photo 7.5);
- a two-dimensional model of the spillway, 1:50 scale;
- a two-dimensional model of intake, 1:40 scale (Figure 8.20).

A specific model was also built to study the Locks, at 1:25 scale, also by Hidroesb, at the facilities of the National Waterway Research Institute (INPH) in Rio de Janeiro.

In these models, extensive testing campaigns were conducted to test and optimize the general layout of the works, the spillway and intake structures, the forms of the dividing walls between these structures and the upstream hydraulic spike on the dam slope in the interface with the spillway as well as the detailed scheme of the river diversion phases. In summary, it is noted that:

- the approach flow conditions in relation to the structures have been verified;
- the flow capacity of the spillway and intake;
- the shapes of all structures, pillars and extreme walls;
- measures of mean pressures on the slope profile were taken;
- different mobile bottom tests were carried out on erosion trends in the plunge pool for several spillages (plunge pool);
- the three phases of river diversion (first and second phases in channel and the third phase by the culverts under the spillway) were investigated in detail;
- the spillway gate operation plan has been detailed (Figure 8.21).

In addition, it should be noted that the following models were used:

- the 1:20 scale turbines at Alstom Neyrpic in Grenoble (France);
- the intake gates, at 1:30 scale, at the FCTH-USP, in São Paulo;
- the gates of the culverts, in the 1:30 scale, at Badoni (BSI), in Sorocaba (São Paulo).

8.3 ENERGY DISSIPATION

The main types of power dissipators are as follows:

- the stilling basins in which part of the energy is dissipated in the "hydraulic jump";
- stepped chutes in which energy is dissipated on each step;
- ski jumps, free-falling jets and cross-jets in which part of the energy is dissipated in the trajectory of the jet in the air.

The criteria for defining dissipators are loosely defined and therefore depend on the experience of the designer. In this task, the following factors are involved:

Figure 8.20 (a) Reduced model of the Tucuruí HPP. Observe operating spillway and jet reach. Eletronorte News, Year III, No. 71, Feb. 2000. (b) Tucuruí HPP. Observe operating spillway and jet reach.

- topography, dam site geology and dam type;
- layout of the works: spillway incorporated into the dam or isolated;
- hydraulic parameters involved: head, discharge and flow rating curve; and
- economic comparisons between different types of dissipators.

Figure 8.21 Three-dimensional model with a scale of 1:150 For studies of all works, including Scour studies of the plunge pool, El. −40.00 m. $Q = 100{,}000 \, m^3/s$.

In addition to these, other important factors should be considered, namely:

- frequency of operation and maintenance facilities of the spillway;
- associated risks of damage and dam breaks; and
- the experience of the designer.

Given the complexity of the issue, even considering all these factors, erosions have been registered downstream of these devices. The important thing is to prevent erosion from threatening the stability of the dam and the spillway itself. It is noteworthy that the reduced models, used as an additional tool in the study of dissipators, have limitations, since they cannot reproduce some aspects that affect the prototypes, such as the flow aeration degree and the rock mass resistance.

It should be noted that the compact ski jumping solution provides great savings in civil works when compared to the dissipation basin solution, and this is a weighted argument in favor of this type of sink when hydraulic conditions exist for its adoption. In the case of Tucuruí, in 1988 it meant savings of US$ 100 million. It would not be technically and economically acceptable to consider adopting a stilling basin to protect the downstream rock mass that had no resistance to withstand the flow pressures free of erosion. This spillway has been in frequent operation since November 1984 without significant erosion.

8.3.1 Ski jump dissipators

In ski jumps, the kinetic energy of the runoff is harnessed to launch the jet from a distance, so that the dissipation takes place away from the spillway and other dam structures. The energy is dissipated basically in three phases:

- in the air phase of the jet, by air resistance, which decreases its range;
- if there is a draft downstream in the submerged phase of the jet diffusion process, it attenuates its erosive power; and
- in the rock mass after impact.

The trajectory changes with decreasing range, the cross-sectional area increases (relative to the theoretical value) and the jet energy decreases. Data regarding this evolution and the estimated effective range are presented by Martins (1977).

According to Hartung and Häusler (1973), the diffusion process of the jet into the draft can be accurately analyzed using the free jet turbulence theory. Energy dissipation occurs in the momentum exchange process: the jet enters the draft. In this case, the dispersion is almost completely linear. After these two phases, aerial and submerged, the residual energy of the jet is transmitted to the rock mass with the consequent development of the pit (Figure 8.22).

The development of the erosion pit was presented in detail by Spurr (1985). Research on erosion pit development is done in qualitative trials on small models. The impact of the jet exerts a combination of forces on the rock mass, which vary as the pit develops. Initially, when the pit (or pre-excavation) is shallow, the pressure gradient is high and the rock mass will deflect the jet.

The penetration of high pressures in the stratification planes, faults and fractures, causes the fracture of the rock mass and erosion develops. The pressure propagation

Figure 8.22 Diffusion process of the jet into the draft (downstream water). See also Figure 8.29.

rate is a function of the degree of fracture, fracture conditions (open, sealed, decomposed) and the orientation of the discontinuities in relation to the angle of incidence of the jet. These characteristics govern the evolution of erosion.

When the rock mass is homogeneous and strong enough to confine the currents in the jet's impact zone, the hydraulic action on the pit contours will be more intense. In this case, erosion is concentrated and deepens, as happened in Jaguara HPP in the Grande river between Minas Gerais and São Paulo states. This case was presented in detail by Magela and Brito (1991).

As this is a spectacular case, it was considered opportune to present some photos of erosion obtained during a trip to the plant in July/1989, showing details of the rock massif, a healthy but very fractured quartzite, without any decomposed zone (Figures 8.23–8.26).

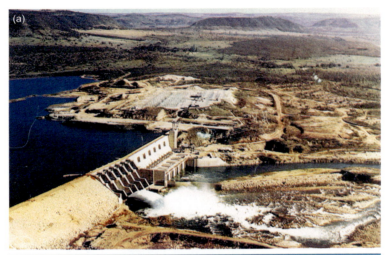

Figure 8.23 (a) Jaguara HPP. Spillway with dissipator in ski jump. $Q = 12,600 \text{ m}^3/\text{s}$. (Magela and Brito, 1991). (b) Jaguara HPP. Downstream view of the spillway. Aspects of the downstream rock massif: sound quartzite but very fractured (Magela and Brito, 1991).

Figure 8.24 (a–c) Jaguara spillway. Detail of fractures and joints of the downstream rock mass. In picture C: X = schistosity; JD = strain relief; JC = shear joint (Magela and Brito, 1991).

In heterogeneous or low-strength massifs, lateral erosion can develop rapidly, asymmetrically, and evolve towards walls or the foot of the dam. An example of this type of erosion is what happened downstream of the Tarbela HPP service spillway on the Indus river (Lowe, 1979) (Figure 8.27).

Return currents (or recirculation currents) usually affect the side areas of dam slopes, such as the Água Vermelha HPP, and upstream of the jet's impact zone, and may or may not erode them, depending on their intensity and of the strength of the massif. The Água Vermelha HPP spillway was designed for a discharge capacity of 20,000 m^3/s, with six segment gates of 15.0 m × 19.6 m. On January 20, 1980, for 5,750 m^3/s, the return currents caused an erosive process on the downstream slope of the left bank dam embrace (see details in Oliveira, 1985). In 2 days, 40,000 m^3 of basalt blocks were removed from the slope. In Photo 7.14, you can see the dam's foot protection wall on the left side of the spillway, built between May and December 1980 (see CBDB, 1982 for details) (Figure 8.28).

In the development of the pit itself, pressure fluctuations cause dislocation of the blocks in fractured massifs. These recirculating blocks within the pit intersect, fragment, and are then expelled from the crater, which may form bars (deposits) immediately at the exit or be transported downstream.

It is recommended to consult the work of Xavier (2003), "Unconventional Solution for the Itapebi HPP Spillway", which describes the unique solution adopted to adapt

Figure 8.25 Jaguara HPP. Downstream View: Spillway – span 01. Wall and rockfill dam. Observe detail of intact dam protection wall close to the right bank natural channel (Magela and Brito, 1991).

the project to unforeseen geological conditions, identified only during construction in July 2001 (Figure 8.29).

For free-fall or orifice jet dissipators used in narrow-valley arched dams, the same is true for ski jumping. In the case of free fall, due to shorter range of the jet, additional safety measures may be necessary, such as construction of submerged sills downstream to raise the water level and increase the protection of rock massif. This is the case of the Katse dam in Lesotho (Figure 8.30).

Figure 8.31 shows the evolution of erosive process downstream of the Kariba HPP (600 MW) on the Zambesi River, Zambia/Zimbawe, Africa. The dam, with a height of 128 m, was settled at 362.00 m. The spillway is made up of six orifices (9 m × 9 m), each with a flow capacity of 1,400 m³/s (Figure 8.32).

Figure 8.26 Jaguara HPP. Spillway – ski jump ($Q = 12,600 \text{ m}^3/\text{s}$). (a) Jaguara HPP (CBDB, 1982) (b) Jaguara HPP (CBDB, 2001). (Photo (a) shows the ancient erosion pit. Photo (b) shows the new channel built by CEMIG in 2001.)

Local geology is composed of sandstones, conglomerates, sand/gravel (from Triassic), clayey stratum (from Permean), quartzite (from Precambrian), gneisses, amphibolites and schists (from Archean). The riverbed and the abutments are formed by very healthy gneiss. On the right abutment, approximately 90 m above the riverbed, the gneiss is covered by a jointed quartzite with strong foliation (Anderson, 1960). In both abutments, the materials were quite weathered (Figure 8.33).

Spillways 193

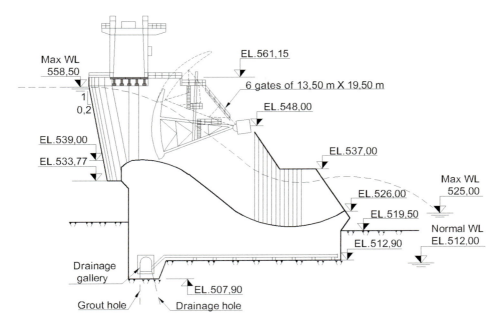

Figure 8.27 Jaguara spillway – ski jump (CBDB, 1982).

Figure 8.28 Água Vermelha HPP (Cesp). Fonte: CBDB.

Figure 8.29 Katse Dam (Lesotho). Spillway jet free fall. $Q = 6,250 \text{ m}^3/\text{s}$. Dam height = 180 m; $\Delta h = 140$ m.

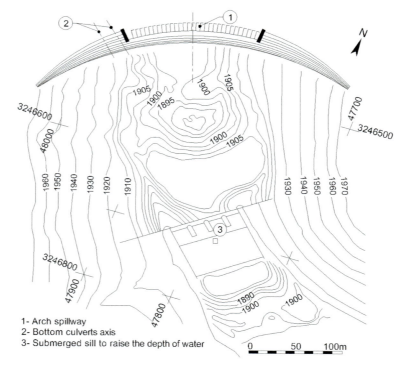

1- Arch spillway
2- Bottom culverts axis
3- Submerged sill to raise the depth of water

Figure 8.30 Katse spillway. Erosion pattern for $Q = 6,250 \text{ m}^3/\text{s}$ (Furstenburg, 1991).

Spillways

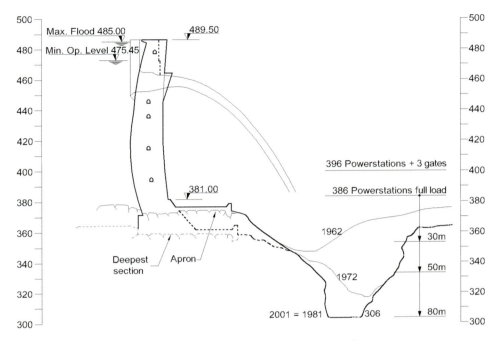

Figure 8.31 Kariba dam. Ski jump orifice spillway. $Q = 8,400 \, m^3/s$ (Hartung and Häusler, 1973; Whittaker, 1984; Mason, 1986; Magela and Brito 1996; CBDB, 2002).

Figure 8.32 Kariba dam. Downstream view.

196 Design of Hydroelectric Power Plants – Step by Step

Figure 8.33 Kariba dam. Spillway working.

The plant went into operation in 1960. The erosion process evolved rapidly, as shown in Figure 8.31. In 1982, the erosion crater was 125.5 m deep, of the same order as the dam height. The author searched the internet but did not find the latest information (Figure 8.34 and 8.35).

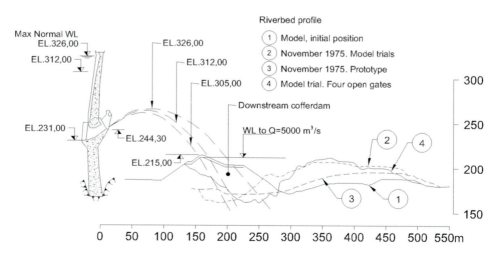

Figure 8.34 Cahora-Bassa HPP, Zambezi river, (Mozambique). Spillway – cross jets energy dissipator. $Q = 13,100 \, m^3/s$. Dam height = 170 m (Lemos, 1982).

Figure 8.35 (a–b) Cahora Bassa HPP. Mozambique (2,000 MW). (a) In construction. (b) Spillway – cross jets energy dissipator (Lemos, 1982).

8.3.2 Hydraulic jump energy dissipators – stilling basins

Hydraulic jump is defined as the sudden and turbulent passage from a low-energy stage below the critical depth to a high-energy stage above the critical depth during which the velocity changes from supercritical to subcritical as shown in Figure 8.36B. That is, the hydraulic jump is the process of dissipating the kinetic energy of supercritical flow into potential energy in subcritical flow in the dissipation basin. The jump consists of an abrupt rise in the water level in the impact region between fast and slow flows (Figure 8.36A).

A hydraulic jump will form when a small water depth supercritical flow attacks a body of water with a considerable depth and subcritical velocity. It should be noted that the jump is formed only if the depth is smaller than the critical depth. For the jump to form, the pressure along with the momentum after the jump must be equal to the pressure and the momentum before the jump. The requirements of pressure plus momentum remain valid regardless of channel shape or inclination.

The stilling basins are presented in detail by Elevatorski (1959), Rudavsky (1976) and Peterka (1983). The following is just a summary considering the schematic diagram of the hydraulic jump of Figure 8.36C. In the analysis, the following assumptions are made:

- the channel is rectangular with parallel sides and bottom slab is horizontal;
- all friction losses are neglected;
- the cam is assumed to happen instantaneously;
- there is an orderly streamline flow upstream and downstream of the jump.

P_1 is the hydrostatic pressure in Section cd, and P_2 is the hydrostatic pressure in Section ef. For simplicity, it is assumed that channel width is unitary.

Figure 8.36A Example of classic stilling basin – Garrison dam (model). Missouri river, Riverdale, North Dakota. USACE 1947–1953. Design flow = 23,400 m³/s (Elevatorski, 1959). Q = total discharge (m³/s); W = basin width (m); $Q/W = q =$ specific discharge (m/s/m); V_1, d_1 and E_1 = velocity (m/s), depth (m) and energy, entering jump (Section 1); V_2, d_2 and E_2 = velocity (m/s), depth (m) and energy at the end of jump (Section 2); $F_1 = V_1(gd_1)^{0.5}$; L = jump length (m).

Spillways 199

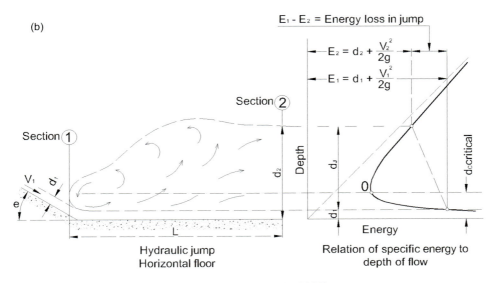

Figure 8.36B Definition of symbols (HDSBED-USBR, 1983).

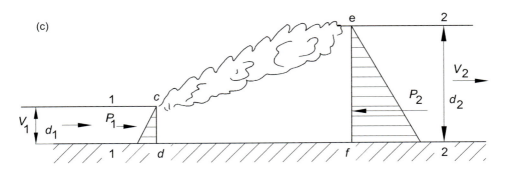

Figure 8.36C Schematic diagram of the hydraulic jump (Elevatorski, 1959).

The change in hydrostatic pressure ΔP from Section 1 to Section 2 for a unit width channel is:

$$\Delta P = P_2 - P_1 = \frac{1}{2}\gamma\left(d_2^2 - d_1^2\right) \tag{8.3}$$

The momentum variation is:

$$\Delta M = \frac{\gamma q(V_1 - V_2)}{g} \tag{8.4}$$

According to Newton's Second Law, the change in hydrostatic pressure is equal to the change in momentum. Equating $\Delta P = \Delta M$, we have:

$$\frac{\gamma q V_1}{g} + P_1 = \frac{\gamma q V_2}{g} + P_2 \tag{8.5}$$

For rectangular channels, the equation becomes:

$$d_2^2 - d_1^2 = \frac{2q}{g}(V_1 - V_2) \tag{8.6}$$

As reported by Elevatorski (1959), several researchers worked on the subject until the classic dimensionless equation presented in the graph of Figure 8.37 was reached.

$$\frac{d_2}{d_1} = \frac{1}{2}\left(\sqrt{8F^2 + 1} - 1\right) \tag{8.7}$$

The efficiency of a stilling basin is shown to be a function of the flow kinetic factor, Froude – F_1 number, at the basin entrance and the basin type. This dimensionless factor is given by Equation 8.8, where V_1 and d_1 are the velocity and the depth at the basin entrance, respectively, and g is the acceleration of gravity.

$$F_1 = \frac{V1}{\sqrt{gd_1}} \tag{8.8}$$

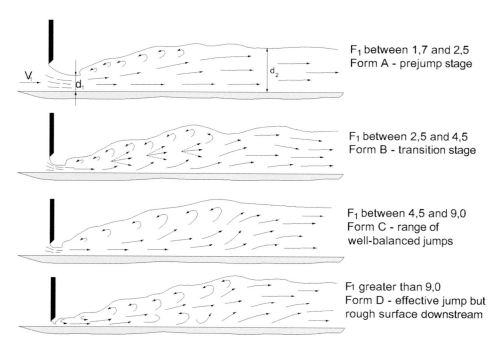

Figure 8.37 Jump forms (function of Froude Number) (HDSBED, 1983; NUST, 2003).

Spillways 201

As a function of F_1, there are four different forms of hydraulic jumps shown in Figure 8.20C (DSBED and NIT). All of these forms are found in practice.

Figure 8.38 shows a Type II basin and graphs used for sizing its dimensions as a function of Froude Number and the conjugate depths of the jump (d_1 and d_2). It should be noted that as F_1 increases, the efficiency of the basin also increases. Otherwise, the

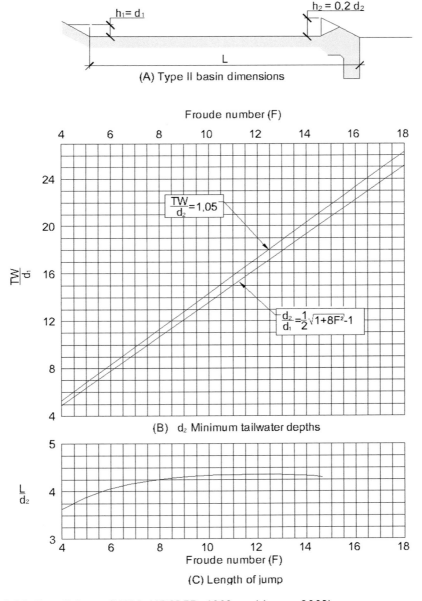

Figure 8.38 Type II Basin (USBR-HDSBED, 1983, and Lysne, 2003).

efficiency is higher as the height of the jump (d_2-d_1) increases, with d_2 being the depth at the exit of the basin.

The length of the basin (L_1) is a function of the height of the jump. When the rocky bed is fractured, Elevatorski (1959) recommends:

$$L_1 \cong 7(d_2 - d_1) \tag{8.9}$$

and 60% of this value when the bed is made of sound rock.

From design practice, it is known that basins with more than 30%–40% efficiency are not common. It turns out that 70%– 60% of the energy that enters the basin comes out and must be dissipated in the water mass and the downstream rock mass.

The downstream flow of the end sill impacts the rock mass with a much smaller incidence angle than those observed in ski jumping dissipators. Lower back currents are formed and often bring loose blocks to the basin, which erode the concrete by abrasion. This fact is almost always aggravated during asymmetric spillway gate operations. There are several examples of these occurrences. We highlight the case of Ilha Solteira, Figure 8.39, reported by Neidert (1980).

The erosion pattern determined in reduced model tests is shown in Figure 8.40. The areas of erosion and deposit/bar formation are clear. It should be noted that downstream erosion effects of this plant are larger than expected, resulting from its undersizing or inadequate evaluation of the resistance of the rock mass, due to the absence or insufficient amount of geological and geotechnical investigations in the downstream region of the spillway. Examples of such occurrences are the erosions occurring downstream of Coaracy Nunes, Moxotó spillways reported by Neidert (1980).

Figure 8.39 Ilha Solteira HPP (3,200 MW), Paraná river, São Paulo/Mato Grosso. Ltotal = 6.2 km; Powerhouse: L = 633 m; 20 Francis turbines of 160 MW; H_d = 46 m; Spillway: L = 351.50 m, 19 gates 15 × 15.46 m, Q = 40,000 m³/s; energy dissipation: baffle blocks in the descending branch of the Creager profile and a 71.50 m long stilling basin.

Figure 8.40 Spillway of Estreito HPP. Reduced model tests with moving bed ($Q = 28,700 \text{ m}^3/\text{s}$). Downstream erosion pattern of dissipation basins (Fudimori, 2013).

It is also worth mentioning the case of Sobradinho HPP (Figure 8.41), São Francisco river, Bahia, owned by CHESF. In Figure 8.42, you can see the cofferdam built to make repairs possible, the bottom outlet and the dam back wall. Figure 8.43 shows the scar of the large diameter block torn downstream from the bottom outlet. In Figures 8.44 and 8.45, one can notice the size of the blocks compared to those of the people. Two factors contribute to this type of occurrence:

Figure 8.41 Sobradinho HPP. substation, right bank dam, powerhouse, divisor wall, tailwater channel/surface spillway, divisor wall/bottom outlet, left sidewall/left bank dam.

204 Design of Hydroelectric Power Plants – Step by Step

Figure 8.42 Sobradinho HPP. Downstream view: note cofferdam built to enable repairs (Photo obtained during the visit to the plant August 1982).

Figure 8.43 Sobradinho HPP. View of downstream erosion of bottom outlet. Observe scar of torn large diameter block (Photo: August 1982).

Figure 8.44 Sobradinho HPP.

- incomplete knowledge of the flow rating curve, especially for high flow rates; and,
- the partial operation of the spillway, so that, for a given energy concentration, no downstream levels are available to counterbalance it and maintain the jump within the basin.

In operations with a reduced number of gates, it is very common to request the corresponding section of the structure and the rock mass under more adverse and more frequent conditions than those corresponding to the design flood itself.

Loose block view of granite-gneiss downstream of the bottom outlet. Note block size by comparing it to close people. The divider wall is in the background, with the surface spillway (Photo: August, 1982).

In the case of roller buckets, with similar operation, some problems of bucket abrasion by riverbed materials have been reported due to shorter structure, such as the Grand Coulee HPP (6,480 MW) on the Columbia river (Figure 8.46A), reported by Rudavsky (1976).

This plant was built between 1933 and 1942. The geology of the site is made up of basalts and granites. The reservoir, at level 393,20 m, has the following purposes: irrigation, flood control, municipal and industrial supply. The dam is 168 m high. The reference head is 105 m. The spillway, at the central stretch of the dam, is 503 m wide and has 11 gates 41.14 m wide and 8.53 m high. The discharge capacity is 28,315 m^3/s (Kollgaard, 1988). The downstream slope of the structure is 0.8 H:1.0 V. The spillway has 40 outlet works of 2.6 m in diameter, with a capacity of over 15,430 m^3/s. Spillway and the outlets discharge into a stilling basin. The roller bucket is at an elevation of 266,52 m.

Figure 8.45 Sobradinho HPP. View of downstream erosion of the bottom outlet near the dividing wall with the surface spillway (Author photo: August 1982).

It is worth showing the historical case of the spillway of the Wilson dam, where, in 1926, there was an erosion of 4 m at the toe of the structure (Figure 8.46B). The jump did not form on the apron because of a deficiency of tailwater. This difficulty was alleviated by the construction of a toe wall, which minimized the erosion by lifting the flow away from the riverbed (Elevatorski, 1959).

Spillways 207

Figure 8.46A Grand Coulee HPP. Columbia river, Coulee City, Washington, USA (USBR, 1986).

Figure 8.46B Erosion (4 m) at the toe of Wilson dam spillway. Tennessee river (Elevatorski, 1959).

208 Design of Hydroelectric Power Plants – Step by Step

8.3.3 Efforts downstream of dissipators

Hydrodynamic forces downstream of dissipators can be estimated in two ways: through measurements on reduced hydraulic models, using pressure transducers, or through hydraulic calculations. In dam design practice, these efforts have inexplicably remained unmeasured or uncalculated. There are often works on the subject "large or high pressures" on the massif, without specifying what they are in numerical terms.

Apart from the work done in Russia in the 1960s, until recently there have been few attempts to understand the dynamic interaction of hydraulics with geology in the designs of these heatsinks, an understanding that passes through the knowledge of these efforts. Regarding the geo-mechanical aspects, it is recommended to read Brito's work (1991).

The methodology for calculating downstream stresses of the dissipators was presented by Magela and Brito (1991) and Reinius (1986) for stilling basins. The methodology for effort measurement can be found in Pinto's work (1991).

8.3.4 Erosion pit dimensions assessment

The prediction of the extent of the fossa that will form downstream due to the incidence of jets is still quite imprecise, given the large number of factors that intervene in the phenomenon, namely:

- the shape of the incident jet;
- jet energy or, in other words, the existing head;
- the specific flow rate;
- the degree of aeration of the jet;
- the height of the downstream water;
- the rocky matrix of the riverbed and its degree of homogeneity;
- the degree of alteration and displacement of rock and the possible existence of geological faults;
- the frequency of operation of the spillway and the frequency of asymmetric gate operations.

The jet influences erosion through its range, which conditions the location of the excavation, and its area and shape on the impact on the bed, which conditions the geometry of erosion. These two aspects are linked with the speed of the jet.

The behavior of the jets depends on their initial velocity and turbulence, their characteristic outlet dimension, diameter or height, shape factors and air travel. Due to air resistance and the consequent aeration, the effective range of the jet is lower than theoretically obtained. Emulsifying water is beneficial for reducing its erosive power, as previously explained. Currently, this influence is already considered in the erosion estimate.

The higher the depth of the downstream water, the lower the residual energy with which the jet will focus on the rock mass and, consequently, the lower the erosion.

The increase in specific flow, driven by advances in gate drive and operation technology and ever larger floodgates, tends to aggravate the risk of erosion.

The evolution of the erosion process is closely linked to the frequency of spillway operation, and how fast it is on the basis of magnitude of the flow rates. The asymmetric

Spillways 209

gate operation provides the intensification of the return currents, accentuating the erosive power of the flow.

Two paths have been used to evaluate erosive effects downstream of spillways: the reduced models, in moving bottom tests and with a certain degree of cohesion, and the calculations through formulas deduced from the test data and prototypes.

The erosive process takes place as previously described. Whatever the erosion prediction method, the difficulty of a more accurate calculation runs into the difficulty of assessing the strength of the rock mass.

The knowledge of prototype erosions, their association with the operating conditions that caused them and the detailed geological survey of the area susceptible to the erosive process, constitute valuable information to better guide the model studies.

Concerning the use of cohesive materials, cases of Grand Rapids (Feldman, 1970) and Picote (Lencastre, 1966) are cited, where, after the prototype erosion survey, the model material was progressively adjusted to reproduce the same degree of alteration of the river bed as observed on the site.

As for the theoretical estimate of erosion depth, several formulas were developed. Mason (1986) makes a detailed critical analysis of 31 formulas developed for this purpose. In summary, the currently used formula has the following expression:

$$D \cong K \frac{q^x H^y h^w}{g^v d^z} \tag{8.10}$$

where:

$x = 0.60 - H/300$; $y = 0.05 - H/200$; $w = 0.15$; $v = 0.30$; $z = 0.10$; D = erosion depth; $K = 6.42 - 3.10\, H^{0.10}$; q = specific discharge;
H = head: difference between upstream and downstream levels;
h = the height of the downstream water;
g = gravity acceleration;
d = characteristic block diameter.

Figure 8.47 shows the erosion pits measured at various spillways compared with the pits estimated by formulae of various authors. Note that Veronese equation is the envelope.

Based on this figure, we sought to define the coefficient K values as a function of the rock characteristics in the jet's diving basin, as described by Magela and Brito (1996).

$K = 0.525$: for very hardy rocks (Foz do Areia HPP, São Simão HPP)

$K = 1.2$: for medium rocks

$K = 1.4$: for less hardy rocks (Tarbela), alluviums (Colbun), or hard rock with long exposure to large flows (Itaipu HPP)

$K = 1.9$: limit value, Veronese envelop: very fractured and potentially erodible rocks (Jaguara HPP).

It is also worth mentioning the erosion depth estimation method developed by Yuditskii (1963), based, after years of laboratory research, on more than 200 small model tests and which was probably the first work developed for rocky beds. This work is a complete study on the subject and should be consulted.

To illustrate the theme, the layout of the Tarbela dam (6,300 MW) on the Indus river, Pakistan, is shown in Figure 8.47A, in which the two spillways stand out.

210 Design of Hydroelectric Power Plants – Step by Step

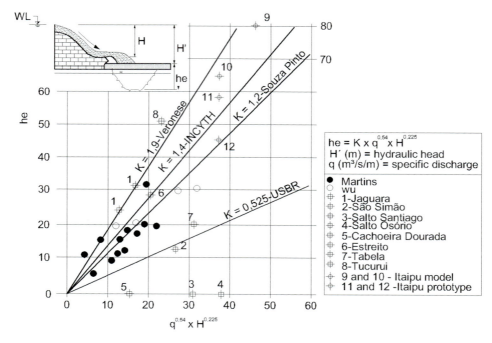

Figure 8.47 Erosion measured downstream of spillways × estimated erosion using formulas (CBDB, 2002).

Figure 8.47A Tarbela dam spillways view ($H = 143$ m). Indus river. Pakistan. Main spillway: 18,400 m³/s, seven gates of 15.2 × 18.6 m. Service spillway: 24,070 m³/s, nine gates of 15.2 × 18.6 m.

8.4 CAVITATION

8.4.1 Conceptualization and characteristic parameters

Cavitation is a dynamic phenomenon consisting of the formation and subsequent collapse of cavities (hence the name of the phenomenon), or vapor bubbles, in a flowing liquid (Abecasis, 1961). Bubble formation occurs in regions where, under any circumstances, the local pressure drops below the water vapor pressure. The collapse happens because bubbles are dragged by the flow to areas immediately downstream where the pressure is greater than the vapor pressure of the water. The sudden collapse of the bubbles creates very high localized pressures resulting in pressure fluctuations, vibrations and noise.

When this phenomenon occurs repeatedly close to the solid contour that defines the flow boundary, the structure suffers high-intensity shock actions. When the forces resulting from the impact exceed the internal cohesive forces of the solid surface material, it ruptures. This action is called cavitation erosion. This erosion may also be due to a prolonged cavitation flow action causing fatigue requests resulting from the repeated action of the phenomenon. It is generally found in cases of flow separation.

In the case of concrete walls or floors, the destructive action is felt mainly on the less resistant constituent, i.e., the binder (cement-binder). Erosion around aggregate particles increases wall roughness, and cavitation conditions may become more critical. The particles eventually come loose, and the erosive phenomenon tends to progress downstream. Finding favorable situations can reach very important proportions and cause complete destruction of the coating.

Experience has shown that cavitation problems in the outlets, considering the border-defining concrete surfaces with suitable finishes, occur at speeds greater than 30–35 m/s, depending on the value of pressure vicinity of the area where cavitation occurs. Problems with slower speeds are the result of imperfect surface finish or local turbulence caused by sluice gates, columns, spreading blocks, damping and falling blocks, etc., or the reduction of pressure caused by the profile of the works (siphons, divergent).

Cavitation initiation and intensity depend on the dynamic flow structure characterized by velocity distribution, boundary layer characteristics, along with mean pressure fields and pressure fluctuations. If the viscosity and compressibility forces can be neglected, the pressure varies with the velocity square and the characteristic cavitation parameter (σ) has the expression:

$$\sigma = \frac{p - p_v}{V^2 / 2g} \tag{8.11}$$

p = absolute pressure at a point in the system outside the cavitation zone (m);

p_v = vapor pressure within the bubbles, equal to the water vapor tension at water temperature (m);

V = flow velocity at a flow reference point (m/s);

g = gravity acceleration (m/s^2).

212 Design of Hydroelectric Power Plants – Step by Step

The critical parameter, σ_{CR} or σ_I, which characterizes the onset of cavitation in various hydraulic structures for different operating situations, is obtained experimentally in reduced models.

Practical results for design effect are presented next for abrupt and gradual irregularities. However, it is necessary to take into consideration the factors that may affect the pressure field and influence the transposition of the results to prototypes. For details, see Quintela and Ramos (1980).

8.4.2 Cavitation caused by irregularities

Ball (1976) classified the main irregularities observed on solid surfaces, indicating corresponding regions affected by cavitation erosion, which are shown in Figure 8.48. USBR (1963) undertook an extensive program of experimental investigations to define incipient cavitation conditions (σ_{CR}) for abrupt and gradual unevenness along the solid contour of hydraulic surfaces.

The recommended results to guide the finishing projects of these surfaces are presented in Figure 8.49 (Pinto, 1979). In this figure, the results (σ_{CR}) are observed for an abrupt jump and for an abrupt recess, positioned transversely to the direction of flow. You can also see the result (σ_{CR}) for a softened surface. The favorable effect of the mitigation of these irregularities is evident. A step of 0.70 cm in softened height, in the ratio of 20 (H): 1 (V), has a value of 10 times smaller (2.05 to 0.2).

It is noted that in these studies, no mention is made of the dynamic flow structure characterized by velocity distribution, boundary layer characteristics, and mean pressure and pressure fluctuation fields. It is known that the development of the boundary layer is the main cause of better performance of spillways when compared to bottom outlets.

In these structures, the steep curvature of the downstream flow of the gate eliminates the boundary layer and any millimeter irregularity is subjected to very high speeds. In spillways, the increase in flow velocity is concomitant with the development of the boundary layer. Minor irregularities of the order of centimeter can be tolerated.

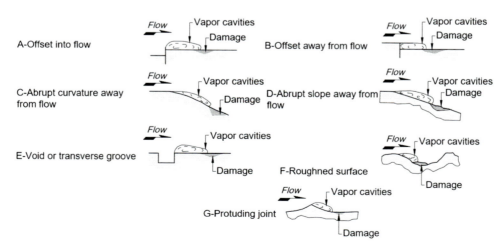

Figure 8.48 Flow action and cavitation damage at flow surface irregularities (Ball, 1976).

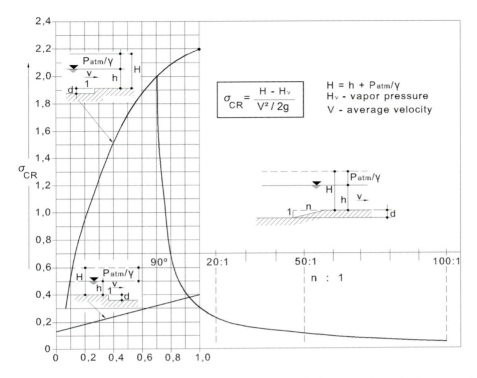

Figure 8.49 Incipient cavitation values σ_{CR} for offset d abrupt and gradual contouring surface (Ball, 1976, Pinto, 1979).

For a deeper and more detailed analysis of these aspects, it is recommended to consult all the references cited.

On the same subject, we present Figure 8.50, which shows Incipient Cavitation Number for small surface irregularities (NUST, 2003).

8.4.3 Protective measures specifications

The most frequent measure for the protection of concrete surfaces of discharge organs subjected to high velocity free surface, flows from cavitation erosion caused by irregularities has been the adoption of stricter specifications for the finishing of these surfaces.

Abrupt irregularities on these surfaces are not tolerated. The specifications require them to slow down to inclined surfaces by beveling as a function of the flow velocity in the region.

In addition, for the most vulnerable sections of surfaces, it has been recommended to use stronger materials – special concretes, in place of steel shielding, until recently considered by some experts to be the ideal but cost effective and impractical building solution in larger surface areas.

In sections of the discharge organs with pressure flow, cavitation erosion can be prevented in many cases by fixing appropriate cross section to the duct.

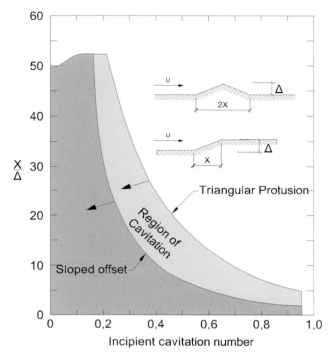

Figure 8.50 Incipient cavitation values σ_{CR} for small gradual surface irregularities (NUST, 2003).

Increasing the section of the duct reduces the speed and consequently increases the pressure. These aspects are combined in the favorable sense of the non-appearance of cavitation.

USBR surface finish specifications state that:

- offsets (steps) or recesses (depressions) are limited to 3.2 and 6.4 mm, depending on whether the surface is shaped or not and whether the irregularity is transverse or in the direction of flow; when the irregularity is downstream of a gate, it must be completely removed;
- Unacceptable irregularities are removed or reduced to an appropriate size by beveling. Slopes shall be less than 20 (H):1 (V), 50 (H):1 (V) and 100 (H):1 (V) for speeds from 12 to 27 m/s, 27 to 36 m/s and greater than 36 m/s, respectively. Slopes 50 and 100 (H): 1 (V) are impractical. Under these conditions, the use of aeration should be recommended.

The finishing conditions imposed by any of the above values are extremely restrictive and therefore difficult and costly to implement since the irregularities are treated by grinding the surfaces.

Analyzing Ball's curves in detail, it turns out that for conditions where speeds are of the order of 35 m/s and absolute pressure of the order of 5 m water column, the surface irregularities should be ground at the ratio 100 (H):1 (V).

Several experts consider that this restriction is materially impossible. In Figure 8.49, it can be seen that for the milder slopes of 50 (H): 1 (V) to 100 (H): 1 (V), no significant reduction of σ_{CR}, and, therefore such a decision would not be technically and economically justified (Figure 8.51).

Under these conditions and considering the executive difficulties of special concrete in large areas, the most logical alternative is liquid vein aeration. For an analysis of this aspect, it is recommended to consult the references already cited, including Neidert (1980) (Figure 8.52).

Figure 8.51 Shahid Abbaspour HPP. $H = 200$ m, Karun river, Iran (1976)

Figure 8.52 Shahid Abbaspour HPP. Spillway working.

Figure 8.53 Shahid Abbaspour HPP. Cavitation in chute slab (Pinto, 1979).

For a detailed analysis of cavitation erosion in grooves, spreader blocks and stilling basin blocks, and for details on the use of more resistant materials (shields, steel fiber concretes, epoxy mortars, use of resins and impregnation), it is recommended to consult Quintela and Ramos (1980) (Figure 8.53).

These authors include the experience of several Russian authors, who tolerate greater heights for irregularities, as shown in Table 8.2.

Table 8.3 shows the values of σ_{CR} for transverse and longitudinal irregularities as a function of slope.

Table 8.2 Chamfer Tilt to Prevent Cavitation Erosion (Oskolov and Semenkov, 1979)

Head (m)	Offset Height (mm)	Chamfer Tilt/offset		
		Transversal		Long.
40–50	5	-	-	-
	5–10	1:4	1:8	1:2
	10–20	1:8	1:10	1:3
	20–40	1:12	1:14	1:3
60–70	2.5	-	-	-
	2.5–5	1:7	1:11	1:2
	5–10	1:14	1:18	1:3
	10–20	1:16	1:20	1:3
	20–40	1:20	1:24	1:3
80–100	10–20	1:32	1:38	1:4
	20–40	1:36	1:42	1:4

Table 8.3 Cavitation Erosion Parameters – σ_{CR} (Oskolov and Semenkov, 1979)

Slope d/a	Chamfer Tilt/Offset		Longitudinal
	Transversal		
1:2	0.73	0.77	0.40
1:3	0.70	0.74	0.30
1:4	0.67	0.71	0.23
1:6	0.61	0.66	-
1:10	0.52	0.57	-
1:14	0.45	0.49	-
1:18	0.39	0.43	-
1:22	0.34	0.38	-
1:26	0.31	0.34	-
1:30	0.28	0.31	-
1:34	0.25	0.28	-
1:38	0.23	0.26	-
1:40	-	0.24	-

8.4.4 Cavitation cases

The following are some examples of spillways where cavitation problems attributed to irregularities in the concrete surface occurred. At the Shahid Abbaspour HPP (old Karun 1), 2,000 MW, erosion occurred in a section subjected to a 100 m load/speed of 43 m/s. Cavitation eliminated the entire chute slab.

The spillway of Guri HPP, Figure 8.54, has nine gates of 20 m × 15 m. It has been operated 9 months a year in the first ten years of the plant operation from 1968 with discharges from 10 to 14,000 m^3/s. At the launching edge the jet ($\Delta h = 120$ m), the velocity was 47 m/s and negative pressures of up to −5.0 m. Erosion is shown in Figure 8.55 (4–7 m depth). For details, see Chavarri et al. (1979). Repair work included changing the profile of terminal bucket. The studies were done in the laboratory of Saint Anthony Falls.

Another case of cavitation occurred at the Dworshak dam spillway in Idaho (Figures 8.56 and 8.57).

This dam has a crest length of 1,002 m, a height of 212 m, a spillway capacity of 4,000 m^3/s, two gates; three bottom outlets, 3.7 m × 5.2 m, 1,180 m^3/s under $H = 81$ m; $H_{max} = 208$ m between the WLres max and the stilling basin. The velocities at the basin entrance are of the order of 60 m/s. Both the basin and bottom spillway had serious cavitation problems.

The same problem occurred at the Libby dam spillway in Montana (Figure 8.58). This dam has a crest length of 684 m; height of 136 m, a spillway capacity of 4,100 m^3/s, two gates; three bottom outlets, 3.7 m × 5.2 m, 1,180 m^3/s under $H = 81$ m; $H_{max} = 118$ m between the WLres max and the stilling basin.

The velocities at the basin entrance are of the order of 48 m/s. Both the basin and bottom spillway had serious cavitation problems (Figure 8.59).

Another example is the Keban HPP (1,240 MW), Euphrates river, Turkey (Figures 8.60 and 8.61). The dam is 158 m high. Construction was completed in 1973.

Figure 8.54 Guri HPP, Caroni river, Venezuela. $H = 162$ m, $Q = 27,000$ m^3/s (1978).

Spillways 219

Figure 8.55 Guri HPP. Cavitation detail of jet launching lip (Chavarri et al., 1979).

Figure 8.56 Dworshak HPP (400 MW). North Fork Clearwater river, Idaho. US Corps of Engineers, Walla Walla District (1988).

Downstream has the Karakaya and Atatürk dams. The spillway operation (1,700 m³/s; 6 gates of 24×16 m) began in 1979 after the reservoir was filled. $H = 106.57$ m. Therefore, the flow velocity is of the order of 44 m/s. Soon after the beginning of the operation, cavitation occurred in the chute due to irregularities in the concrete surface (offsets) in the region between the joints between the concrete slabs (Aksoy et al., 1979).

Figure 8.57 Abrasion in Dworshak spillway chute (Pinto, 1979; Neidert, 1980).

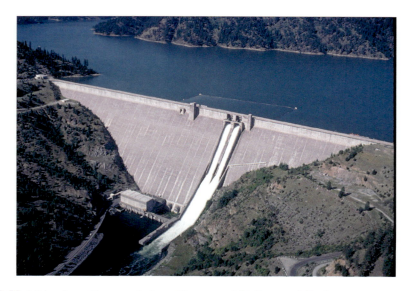

Figure 8.58 Libby dam. Kootenai river, Montana. US Corps of Engineers.

Figure 8.59 Abrasion at Libby Dam spillway. US Corps of Engineers (Neidert, 1979).

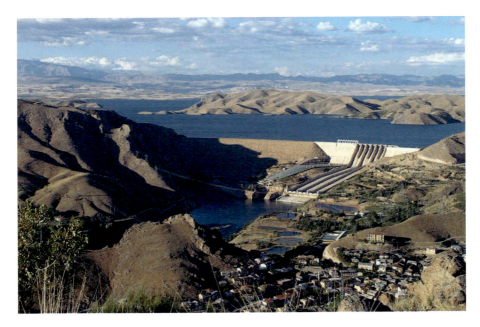

Figure 8.60 Keban HPP.

Figure 8.62 shows Tarbela HPP in Pakistan. According to Lowe (1979), tunnel basins and spillways were used many months a year at full capacity. Basin speeds reached 44 m/s. Figure 8.63 shows the cavitation in the left trough of the spillway four of the service spillways.

Figure 8.61 Keban Dam. Cavitation in Spillway Chute. Aksoy (1979 apud Demiröz, 1991).

Figure 8.62 Tarbela HPP. Service spillway.

Refer to Pinto (1979) for some examples of deep spillways where cavitation accidents occurred: Palisades, Hoover and Yellowtail Dams (USA), Serre-Ponçon and Sarrans Dams (França), San Esteban and Aldeadávilla (Spain), and Infiernillo (Mexico).

It is recommended to consult the work of Oliveira et al. (1985), which presents cases of erosion and cavitation of surfaces of concrete structures of spillways and power houses of CESP plants, as well as a description of the repairs made.

Spillways 223

Figure 8.63 Cavitation. Tarbela service spillway basin four left chute (Lowe, 1979).

8.5 AERATION

Aeration is used to minimize the risk of cavitation occurring at high velocity flows. To compose this book, the following is a summary of the theme extracted from Pinto's works (1979, 1989). According to this author, a concrete surface, even when performed to strict finishing specifications, is not free from cavitation damage when the flow velocities reach or exceed 40 m/s, as well as for high specific discharges.

Still according to Pinto (1989), for high specific discharges that are common in current projects, the air entrainment of the water surface due to development of the boundary layer does not always reach the bottom region of the channel. In the absence of the protection provided by the water-air emulsion any surface irregularity capable

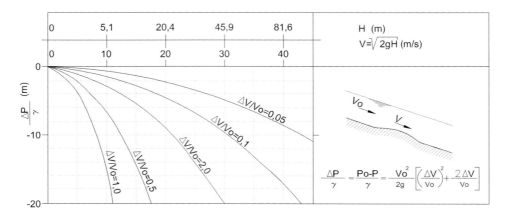

Figure 8.64 Local pressure reduction as a function of speed variation (Pinto, 1989).

of reducing the pressure locally to water vaporization level (vapor tension) becomes important. The higher the flow velocities, the greater the possibilities of cavitation. For speeds of 40 m/s or greater, the pressure field is particularly sensitive as shown in Figure 8.64. A speed increase of 5%– 10% is shown to be sufficient to cause pressure reductions of 10 m water column.

Experience shows that it is very difficult, or practically impossible, to build concrete surfaces that are smooth enough to prevent cavitation at these speeds.

In these cases, steps or ramps can be used to promote air intake (artificial aeration) under the lower stream bed, directly on the contact surface, as a logical alternative to the adoption of very smooth and regular gowns whose execution is difficult, expensive and not always successful.

Several dams were built providing aeration devices, namely: Bratsk, Ust-Ilim, Nurek and Toktogul-Narim in Russia, Yacambu in Venezuela, Itaipu and Foz do Areia in the Iguaçu river, Paraná, and Emborcação in the Paranaíba river, Minas Gerais. According to Pinto (1989), designing an aeration system requires to answer three main questions:

a. At what speed should the first aerator be provided?
b. How much air is drawn into the aerator?
c. What is the space between aerators to maintain a given level of protection?

The answer to the first question relates to the concept of aeration as a cavitation protection system. Care taken during construction certainly contributes to reducing irregularities and consequently the risk of cavitation. Table 8.4 shows some incipient cavitation index (σ_i) values, which indicate the effect of surface quality. Quality improvement is naturally related to cost increases due to difficulty of executing surfaces with tighter tolerance levels.

One of the basic ideas that underlies aeration systems is cost reduction, resulting from less rigorous specifications for finishing concrete surfaces. A proper assessment of the incipient cavitation index (σi) for expected site irregularities and economic considerations should support the decision of where to place the first aerator.

Table 8.4 Incipient Cavitation Index Values (σ_i)

l: n	σ_i [a]
1: 5	0.62
1:10	0.50
1:20	0.35
1:40	0.25

[a] See Figure 8.49: n = slope of the ramp; Equation 7.12: $\sigma_i = (H - H_v) V^2/2g$.

Available information on the effects of aeration (Pinto, 1989) indicates that an air concentration of 5–10% ($C = [\text{Var} / \text{Var} + V_w]$) near the surface to be protected, virtually eliminates the risk of cavitation. However, proper design of an aeration system depends on correct estimate of the amount of air to be drawn by the aerator (second question) and the development of an air concentration near the surface to be protected (third question). Additional aerators should be provided in sections where the concentration heads are below the required minimum level.

Ramps, steps, recesses, grooves, etc. inserted into the spillway chute are the devices used to cause natural flow aeration near the concrete surface. The sudden discontinuity in bottom alignment creates an air-water interface in which high velocity of water drags air through an intense mixing process. Defining aerator geometry and its influence on aeration performance is not an easy task.

For details on the subject, specifically on the topics air drag mechanism, dimensional analysis, prototype results, and model×prototype conformance, it is recommended to consult the works of Pinto (1979 and 1989). The following is a summary, considering the study by Pinto (1989) based on measurements of the Foz do Areia, Emborcação, Amaluza, Colbun and Tarbela prototypes. The aerator geometry is defined in Figure 8.65.

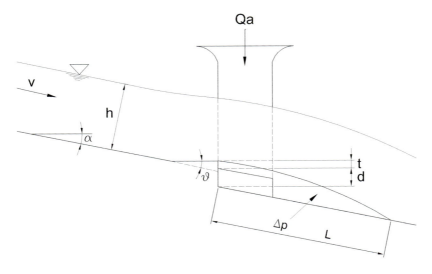

Figure 8.65 Aerator geometry (Pinto, 1989).

226 Design of Hydroelectric Power Plants – Step by Step

where:

α chute inclination angle;

t ramp height (m) and φ, ramp angle with respect to the bottom of the chute;

B chute width, corresponding to each aerator (m);

A duct area under the water lamina (m^2);

Q total flow (m^3/s);

q specific flow per unit chute width (m^3/s/m);

Q_a total air flow (m^3/s);

q_a specific air flow per unit chute width (m^3/s/m);

h water height in aerator section (m);

V average flow velocity: $Q = q/h$ (m/s);

Fr Froude number: $Fr = V/(gh)^{0.5}$.

The air duct is considered as a short orifice or duct, defined by its cross section and the consequent loss of air flow pressure. D is the effective duct area per unit chute width and is calculated by $D = CA/B$ (m^2/m). C is the discharge coefficient for the duct air discharge formula by means of ducts:

$$q_a B = CA\sqrt{2\frac{\Delta p}{\rho_a}} \tag{8.12}$$

where:

Δp is the difference between atmospheric pressure and mean pressure under the jet as measured along the vertical face of the step or ramp (N/m^2); and

ρ_a is the air density (kg/m^3).

Aerator performance is measured by the ratio of air to water flow:

$$\beta = \frac{Qa}{Q} = q_a/q \tag{8.13}$$

In general, as quoted by Pinto (1983) in "Modeling Aerator Devices – Dimensional Considerations", XXII Congress IAHR, Lausanne, Switzerland, the performance of an aerator, as far as the air drag effect is concerned, can be represented by the function:

$$\beta = f\left(Fr, t/h, \, D/h\right) \tag{8.14}$$

where:

Fr, Froude number, reflects water runoff conditions;

t/h is a measure of the relative height of the step or ramp;

D/h characterizes the effect of the aerator.

When the dimensionless parameters Fr, t/h and D/h are uniquely dependent on water flow and β is a unique function of water flow, as shown in Figure 8.66, we have:

$$\beta = f(q) \tag{8.15}$$

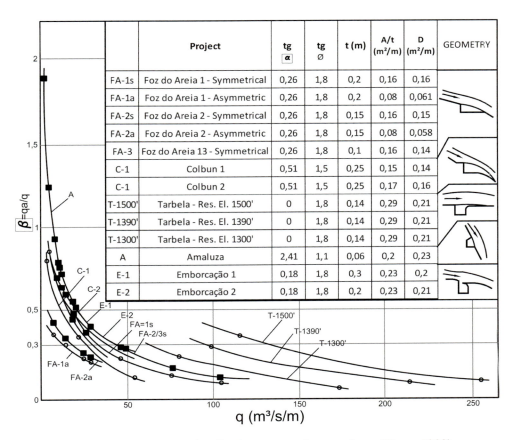

Figure 8.66 Performance curves ($q \times \beta$) of aerators of some plants (Pinto, 1989).

A more detailed analysis of the Foz do Areia data (measurements spanning a large range of q – ranging from 7 to 100 m³/s/m and closing an aeration tower) provided a unique opportunity to study the effect of strangulation on variation of the independent parameter D/h. The Foz do Areia data were plotted on a $\beta \times Fr$ graph (Figure 8.67). Curves A and B were identified to illustrate the $\beta + f(Fr)$ function for two different strangulation conditions of aerator 1 for symmetrical and asymmetrical conditions. Similar curves can be identified for aerators 2 and 3. Note the high rate of change of β when the number of Fr increases to an apparent limit.

To represent the prototype data analytically, disregarding the geometry differences of the aerators, a simple function of the type

$$\beta = a(Fr - K)^b \, x(t/h)^c \, X(D/h)^d \tag{8.16}$$

was adjusted by the minimum square method.

The influence of t/h, because it was small, was neglected and the following equation was found with a correlation factor of 99%:

228 Design of Hydroelectric Power Plants – Step by Step

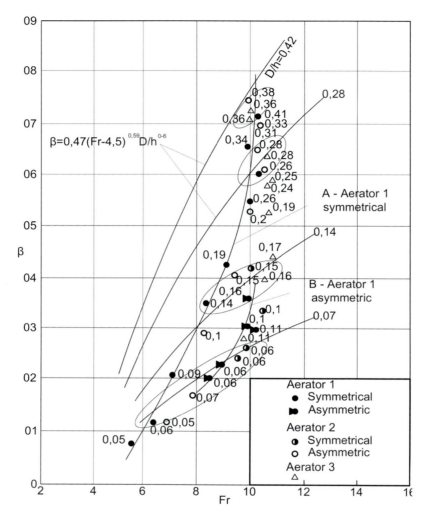

Figure 8.67 Curves $Fr \times \beta$ of the Foz do Areia aerator n. 01 (Pinto, 1989).

$$\beta = 0.47(Fr - 4.5)^{0.59} \times (D/h)^{0.60} \tag{8.17}$$

This equation is presented in Figure 8.67 for some D/h values. The indicated D/h values of each measured prototype data clearly confirmed the expression.

Following the good results of the analysis, an expression for general data was attempted, despite the diversity of aerator geometry and the lack of a theoretical basis for simple monomial function. It was concluded by the following formula, which fit well with the measured results with a correlation factor of 97.62%:

$$\beta = 0.29(Fr - 1)^{0.62} (D/h)^{0.59} \tag{8.18}$$

Pinto (1989) suggests that this expression can be used to estimate the amount of air entrained in a project's initial phase. As mentioned earlier, a 5%–10% air concentration near the surface to be protected virtually eliminates the risk of cavitation.

On this subject, it is worth mentioning and recommending reference to more recent works such as Falvey (1990), Kramer (2004) and Brito (2011). The first presents a procedure for aerator localization; the other two mention the difficulties in determining the minimum amount of air for cavitation protection and critical cavitation index, which remain open topics.

According to these authors, the need for further experimental studies on the evolution of air concentration decreasing along the flow remains evident. As there is not yet a defined criterion for estimating this evolution and effective protection, based on the experience of Bratsk, Nurek, Foz do Areia, Emborcação and Guri, Pinto (1989), it seems reasonable to consider that a well-designed project of an aerator should secure a 50–100 m chute stretch.

The Foz do Areia HPP no Iguaçu river (1,676 MW, 4 units), a 160 m high concrete face rockfill dam is depicted in Figure 8.68.

The spillway on the left abutment, with a total capacity of 11,000 m^3/s, consists of a classic ogee with four sector gates of 14.5 × 18.5 m, followed by a 70.6 m wide and 400 m long, with a slope of 25.84% to the flip bucket deflector, at an elevation of 118.50 m below the reservoir WLmax. The velocity in the bucket is on the order of 47 m/s and the specific discharge is 155.80 m^3/s/m.

As can be seen in Figure 8.69, the aerators were positioned 145.5 m from the spillway crest and 92.50 m from the jet launch edge. The three aerators were spaced 72 and 90 m apart. According to Pinto, in 1989, the operation was adequate, but it was not possible to conclude if two aerators would have been sufficient. The main details of the aeration system are shown in Figure 8.70.

Figure 8.68 Foz do Areia HPP (CBDB, 2002).

230 Design of Hydroelectric Power Plants – Step by Step

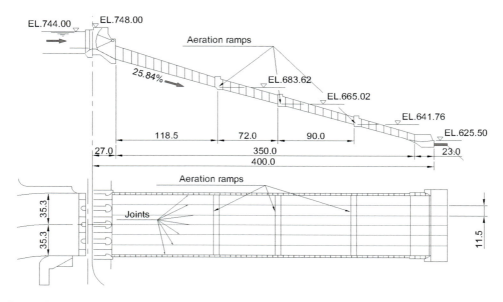

Figure 8.69 Foz do Areia spillway (Pinto 1989).

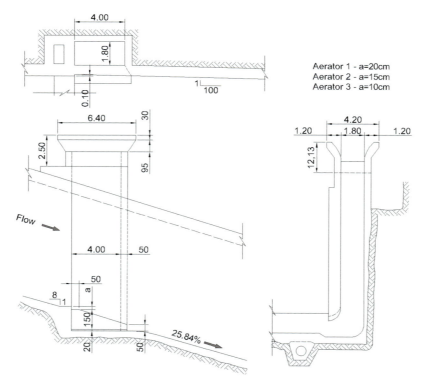

Figure 8.70 Foz do Areia HPP. Aeration device (Pinto 1989).

The Emborcação HPP (1,192 MW) has a 158 m high core rockfill dam. The spillway on the left abutment (Figures 8.71 and 8.72), 7,600 m³/s of discharge capacity, consists of a classic ogee with four sector gates of 12.0×18.77 m.

The chute has the following characteristics: 58.5 m wide and 330 m long and 18% inclined to flip bucket deflector at 83.5 m below the WLmax; two aerators spaced 103 m apart were provided, the second being 65 m upstream of the jet launch edge approximately. The velocity in the bucket is of the order of 40 m/s and the specific discharge is 129.91 m³/s/m.

Special mention should be made of the 3,000 MW Xingó HPP spillway design which began operating in December 1994 (see the following figures – CBDB, 2002). It has a capacity of 33,000 m³/s, with 12 gates of 14.8 m×20.7 m, in two chutes: the right one lined with concrete, is the service chute and has two aerators; the left chute, with a

Figure 8.71 Emborcação HPP (CBDB, 2002).

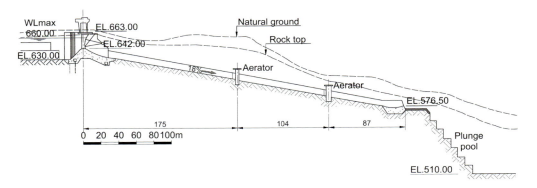

Figure 8.72 Emborcação spillway profile (CBDB, 2000).

long unlined stretch, only operates when the flow exceeds 17,500 m³/s ($TR = 200$ years) (Figures 8.73 and 8.74).

The chutes are 251 m long and 109 m wide each. The flow velocity in the bucket is in the order of 44 m/s and the specific discharge in the right chute is 160.55 m³/s/m. The structure is founded on a gneiss rocky massif sound of excellent quality (Figures 8.75 and 8.76).

Although not directly linked to aeration, it should be noted that in December 1995, several erosion tests were performed on the unlined chute of the Xingó spillway, aiming to confirm the suitability of the adopted solution. In the first test, on 12/11/1995, the flow of 1,200 m³/s was passed for 5 hours.

In the second test, on 12/12/1995, erosions were observed. The flow rate was 2,400 m³/s for 7 hours. Considerable erosion was observed near the fault crossing the spillway at a 45° angle at the top of the chute. The erosion was along the foot of the central wall that divides the chutes.

Downstream, at the end of the chute, the behavior was good. In the third test, on 12/14/1995, the flow rate of 4,000 m³/s was passed for 30 minutes. This test lasted 2 hours. All unstable blocks, as well as dental concrete, were removed and a major depression was formed between the two main discontinuities (see Figure 8.77). There was no regressive erosion near the concrete lining. The erosion near the central wall remained as in the previous test. At some points, it reached 1.0 m deep.

Chute behavior was considered adequate. Additional treatment of reinforcement and dental concrete has been programmed in a few places along the central wall to minimize future maintenance work. The spillway operation was fully released, with the recommendation to continue monitoring erosion in case of future discharges with higher flow rates. As previously mentioned, it is noteworthy that this chute will only be

Figure 8.73 Xingó HPP view (CBDB, 2002).

Spillways 233

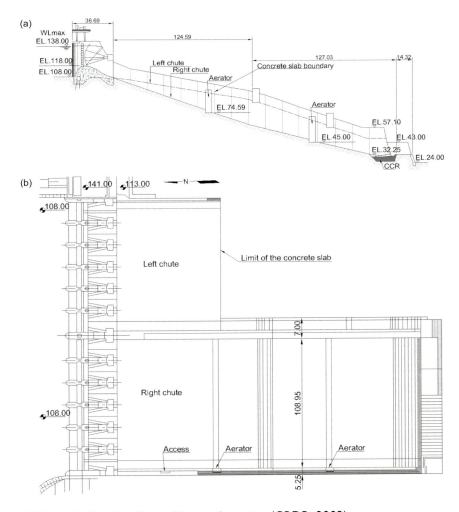

Figure 8.74 (a–b) Xingó spillway. Plant and section (CBDB, 2002).

Figure 8.75 Xingó HPP spillway operating (CBDB, 2002).

Figure 8.76 Xingó spillway – test of the left chute: $Q = 4,000 \, m^3/s$ (CBDB, 2002). This chute will only operate for $Q > 17,500 \, m^3/s$ ($TR = 200$ years).

Figure 8.77 Xingó spillway. Picture taken after the third test – $Q = 4,000 \, m^3/s$ (CBDB, 2002).

operated if flow rates greater $17,500 \, m^3/s$ occur, which have a recurrence time greater than 200 years. Large Brazilian Spillways, published in 2002, reports that the spillway had not operated since 1998.

8.6 OPERATING ASPECTS IN SPILLWAY MONITORING

The spillways are operated considering the Gate Operation Plan (GOP), prepared on the basis of flows to be spilled over their useful life. The various possible operating configurations are tested on reduce models. These plans must be strictly followed in view of the safety of the works. Prior to the start-up of the plants, specific training of the operator team is recommended for this purpose.

The rules should be clear and well defined for the various levels of operational decision and should cover all normal, exceptional and emergency operating situations. In these plans, the following conditions must be observed:

- it is important to define the ranges of flow variations to be spilled, in order to minimize the rapid variation of the downstream water level, which can cause downstream accidents (either with the riverine populations, whether with fishing boats, commercial boats, etc.); In emergency cases of sudden variation in flows, the population that may be affected should be warned to some extent;
- minimize asymmetrical gate maneuvers wherever possible;
- consider the possibility of gates jamming and define procedures for such situations;
- provide for the possibility of maintenance on the spillway and downstream of the river, taking into account the time period, depending on the hydrological regime, available for these activities; the alternative of diverting the river to enable the repair of the structure should be considered;
- consider all assumptions aimed at increasing the operational flexibility of the structure.

Consulting the list of "questions" at ICOLD's congresses since 1933, it can be seen that the topics related to the safety of structures, safe operation, learning lessons from design and construction errors in order not to repeat them are always present and it could not be different. Experience in the operation of these structures records several important facts. We will mention in this book just a few.

The Val Grosina dam was implanted between 1940 and 1950 on the Adda river basin, on Rhaetian Alps, in Italy, close to Swiss border, between the Grosio and Eita cities. It was known that the basin had a high solid discharge, with a predominance of gravel. Then a project was conceived with several devices for passing this sediment to guarantee the useful life of the reservoir. A concrete buttress dam was designed with 77 m high and 450 m long, with side spillway, bottom outlet discharge, and a diversion tunnel that shunts the whole reservoir. This imposed revision of dissipation structure with the purpose to guarantee the expulsion of the material notwithstanding the progressive heightening of and without prejudice of stilling basin runoff. The structures were tested on reduced model carried out at the Laboratory of Works Department of A.E.M. of Milan (Figure 8.78).

During 1947, extensive erosion damage of the shale foundation underlying the apron of the Waco Dam, Brazos river, Texas, caused not only destruction of the entire apron but carried away large quantities of the foundation. Seven meters of shale below the apron was removed, and a total of $40,000\,m^3$ of concrete and shale was lost in the erosive process. The damaged section was quickly refilled to preclude any downstream movement of the dam (Figure 8.79).

The Pit river is tributary of the Sacramento river, and the two developments, Pit 6 and Pit 7, are located upstream of the USBR Shasta dam. They were completed in 1965. In 1969, after four years of operation, it was found that the energy-dissipating teeth of both spillways were well damaged by cavitation. A frequent problem in spillways where this type of dissipating tooth has been used. The teeth were repaired with stronger concrete but for operational reasons repair work took a long time (Figure 8.80).

Figure 8.78 (a–c) Val Grosina dam and dissipation structure. Meaning of letters: (a) Tunnel that discharges into reservoir, the waters returned by the upper. Premadio power-station and those of residue gathering ground (67 m³/s). (b) Power tunnel (under pressure). (c) Diversion tunnel.

The Bonneville dam, 1,242 MW – 20 units, 60 m high, was implanted by the USACE between 1934 and 1938 to improve navigation in Columbia river and provide hydropower. The second powerhouse was implanted between 1974 and 1981. The dam is 60 km east of Portland, Oregon. In 1993, it was completed with a larger navigation lock. The spillway, 442 m long, has 18 gates, 15 m wide × 16 m high, and a flow capacity of 57,000 m³/s approximately. Figure 8.81 shows the cavitation damage in the baffle piers in the stilling basin.

At Jaguara HPP, the first gate of spillway to operate in 1971, at the left side, should be the last according to the plan (see Section 8.3.1 and Magela and Brito, 1991). This was because the gate was the first to be ready, which demonstrates that for some reason the assembly planning was not in line with the GOP. As previously seen, the operation of this extreme span immediately caused an intense erosion near the wall between the spillway and the tailwater channel that luckily did not fail.

In Kariba, the minimum flows corresponded to extremely low downstream water levels and this situation was more critical than that for design discharge (Hartung,

Figure 8.79 Waco dam. Brazos river. Texas. Damage due to scour and uplift pressure.

Figure 8.80 (a–b) Pit 6 dam. Spillway energy dissipator problems. Typical failure pattern of floor blocks. Strassburger, A. G. ICOLD, Q. 41, R. 16 (1973).

1973). Downstream erosion in 1982 (see also Section 8.3.1) was already 130 m deep from the downstream WLmax at elevation 400.00 m, as reported by CBDB (2002). This case was also reported by Anderson et al. (1960), Hartung (1973) and Whittaker and Schleiss (1984).

In Tucuruí, during the drought period of 1986, a special spillway operation criterion was adopted, though not foreseen in the GOP. It aimed to improve the downstream water quality standards consumed by the riverside population along 10 km. It was found that this aspect could be improved and change the spillway operating rule during this period. The special assembled scheme sought to maintain the turbine discharge/total discharge ratio around 0.55. This was intended to cause adequate mixing

Figure 8.81 Bonneville dam. Stilling basin damaged baffle piers (Elevatorski, 1959).

of the flow from the spillway, consisting of a high oxygen flow, and the flow from the intake, a low oxygen flow, to improve water quality while minimizing critical environmental effects downstream as well as in the reservoir.

In Grand Coulee, it was found that the spills increased the nitrogen content in downstream water affecting the ichthyofauna (Jabara and Legas, 1986).

These last two examples do not seem to have any interference with the issue of downstream spillway erosion, but special operation schemes, not provided for in the GOP and designed to address environmental issues, should adequately consider such interference to eliminate unwanted surprises. The environmental impacts mentioned draw attention to the fact that this variable must be considered in the preparation of the GOP. In specific cases, other effects of spills, such as wave action on the shore or impacts on downstream navigation may be associated with operative aspects and should be considered in the GOP.

After commissioning, the performance of the structures should be systematically monitored aiming to verify, among other aspects, the development of erosion processes. This monitoring is of fundamental importance not only to guide the eventual maintenance work, but also for obtaining data for retro analysis of the quality of studies and design carried out.

Chapter 9

Hydraulic conveyance design

9.1 INTRODUCTION

The hydraulic conveyance of hydroelectric plant includes the following structures: power or headrace canal, intake, surge tank if necessary, penstock (or tunnel), powerhouse (scroll case, turbine/generator, draft tube) and tailrace, as illustrated in Figure 9.1.

Depending on the layout, the solution of the hydraulic conveyance may be in a tunnel or in a penstock. If the layout is in a tunnel, or underpressure pipe, generally depending on its length, a surge tank may be required, as will be seen in this chapter.

9.2 POWER CANAL

Many plants have a power canal (or tunnel) for conducting flow to the intake, whose layout must be carefully studied. The approach flow should be subcritical, quiet, well accommodated to the contour surfaces and without detachment, to minimize vortex formation and head losses.

Upstream of the intake, the final stretch of the power canal, a forebay (or head pond), a sand trap and often a side spillway may be required. The design criteria are summarized below.

With the reservoir at its minimum operating water level, the power canal must be designed for maximum turbine flow. The freeboard should be designed to the maximum level.

The maximum flow velocity is defined by economic criteria and is generally less than 2.5 m/s for dug or concrete-lined canals.

For dug canals, speeds should not exceed 1.0 m/s to prevent erosion. The slope may be equal to the waterline profile for the reservoir water level corresponding to the maximum flow and the velocity adopted.

The head loss along the canal should be estimated using the Manning formula:

$$V = \frac{1}{n} R^{2/3} I^{1/2} \text{ or } h_f = L \left(\frac{nV}{R^{2/3}} \right)^{1/2} \tag{9.1}$$

where:
V = average velocity in section (m/s);
I = slope (m/m);

DOI: 10.1201/9781003161325-9

240 Design of Hydroelectric Power Plants – Step by Step

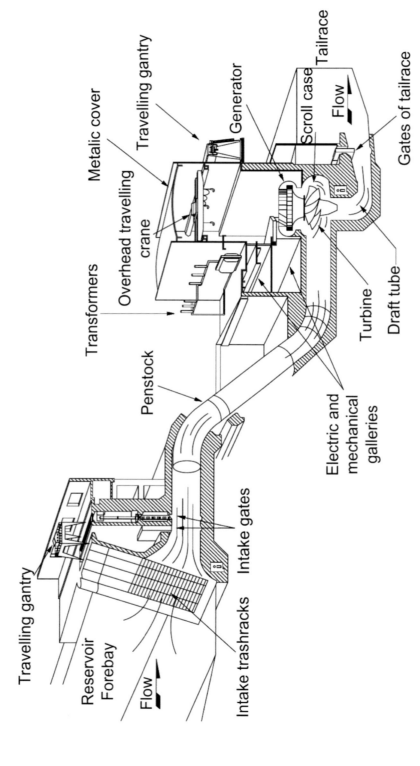

Figure 9.1 Schematic of a hydroelectric facility.

Hydraulic conveyance design 241

Table 9.1 Roughness Coefficient *n* or Manning Coefficient

TYPE	Min.	Medium	Max.
Canals excavated on land: clean	0.018	0.022	0.027
Canals excavated on land: with some vegetation	0.030	0.027	0.033
Canals excavated on land: lined with rockfill	0.028	0.035	0.045
Canals excavated on rock, or tunnels	0.030	0.033	0.038
Projected concrete canals, or tunnels	0.020	0.022	0.025
Concrete channels and ducts	0.013	0.014	0.017
River or canals with rocky bottom without vegetation on the banks	0.030	0.035	0.040
River or canals with rocky bottom with vegetation on the banks	0.040	0.045	0.050

R = hydraulic radius = A/P, (m);
A = wet section area (m^2);
P = wet section perimeter (m);
L = length (m);
n = roughness coefficient (Table 9.1);
hf = continuous head loss (m).

It should be noted that the transformed Manning formula allows checking the canal slope at a predetermined speed.

$$\frac{h_f}{L} = I = \left(\frac{nV}{R^{2/3}}\right)^2 \tag{9.2}$$

V ranging from 1.2 to 2.0 m/s and R from 4 to 2, for n = 0.015, we find that I ranges from 0.05‰ to 0.4‰, which are mild slopes – such as waterline profile in the canal.

9.3 INTAKE

9.3.1 Geometry

The intake is a transitional structure between a free flow in the reservoir and a constricted flow in the penstock in which the velocity must be slow to limit stresses and the head loss. In design, a structure with geometry should be sought that receives and accommodates flow evenly, without vorticities, and promotes a progressive and gradual acceleration of the flow without phenomena of separation and/or detachment, so that the best turbine efficiency is obtained. This is especially important for low head power plants that use horizontal machines, where the distance between the intake and the turbine is small.

Intakes are typical tower or gravity section structures, as shown in Figures 9.1 and 9.2. This type of structure around the world already incorporates a design standard to achieve a flow with minimal head loss. On the upstream face, there are trashracks, and upon entry, the gate grooves before the penstock.

The design of the intake is complex due to the following details: position of the structure in plan and in depth; interfaces with approach flow to neighboring structures

242 Design of Hydroelectric Power Plants – Step by Step

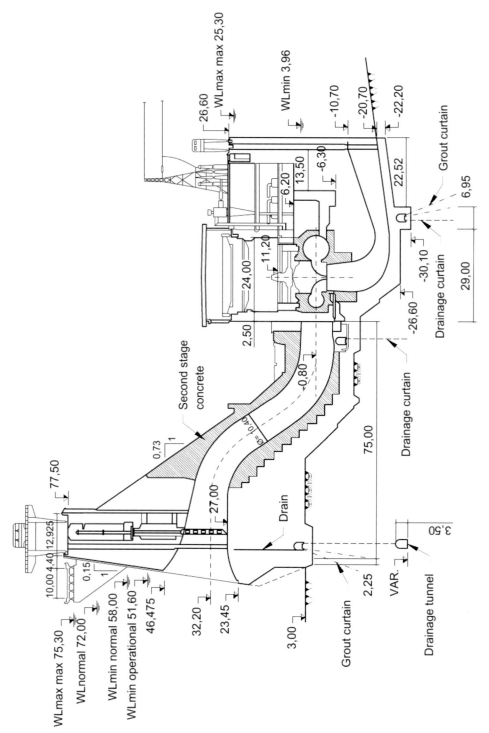

Figure 9.2 Typical structure of intake and powerhouse of Tucuruí HPP. First phase: 12 turbines of 330 MW; $Q \sim 600$ m^3/s; $H_d = 60.8$ m. Technical Memory (1989).

Hydraulic conveyance design 243

(e.g., dam, spillway, fish ladders, ditches); the approach angle of the flow; as well as the constriction of the flow itself. In large projects to check the sizing and performance of the structure, a reduced model is used.

The CBDB/Eletrobras (2003) document sets the following limits for trashracks flow velocity:

- with a load less than 30 m, the speed in the section of the trashracks should be 1.0–1.5 m/s;
- with a load greater than 30 m, the speed in the section of the trashracks should be 1.5–2.5 m/s;
- in the gates section, the maximum speed should not exceed 6.0 m/s.

The dimensions of the hydraulic passageway, width × height, are defined in each case according to the maximum flow to be turbined.

9.3.2 Minimum submergence

To avoid vortexes near the structure, the submergence of the upper edge of the inlet mouth from the minimum operating water level that should be verified using Formula 9.3 (Gordon, 1970).

$$S = CV d^{0.5} \qquad (9.3)$$

where:
  $C = 0.7245$ and 0.5434 for asymmetric and symmetric approximation flow, respectively (Figure 9.3a);
  $V =$ flow velocity in the gate region (m/s);
  $d =$ duct height (m).

The maximum elevation of the upper duct generator, in the section where dimension d was taken, is defined by the quota of the minimum normal reservoir level subtracted from the value of S (Figure 9.3b).

In addition to this condition, the upper edge of the water intake inlet portal shall be at least 2.0 m below the minimum operating water level of the reservoir (Section 3.7.2, CBDB/Eletrobras, 2003).

The formation of unfavorable vortices is greatly influenced by the flow circulation in the approach canal. Therefore, the submergence criterion should only be considered as a preliminary estimate for the outlet design, which should then be verified in a reduced model.

9.3.3 Ventilation duct

The ventilation duct should be sized for the maximum turbine flow, with a maximum speed of 60 m/s. If the intake is fitted with an upstream sealed wagon gate, the free space in the gate niche may replace the ventilation duct. The over-rise in water level caused by load rejection should not be at risk of operation in the ventilation duct and in the gate niches.

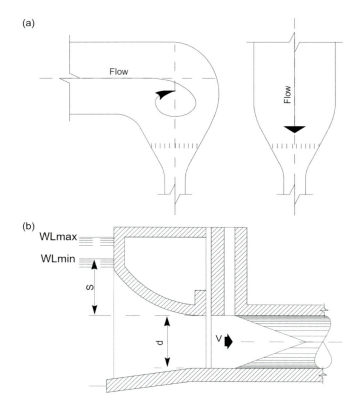

Figure 9.3 (a–b) Intake (Gordon, 1970). (a) Approximation of the flow in the intake. (b) Intake submergence.

9.3.4 Vibration in the trashracks

The flow, when passing through the trashracks, induces vibrations of the bars, whose frequency should be kept at less than 1/3 of the natural frequency of the bars, in order to avoid resonance phenomena that may cause damage or destruction of the bars themselves.

9.3.5 Head losses

Losses of head in the intake incorporate the loss in the trashracks and the loss in the concrete structure until the penstock inlet, including the loss in the gate grooves. These losses should be estimated using Formula 9.4 (CBDB/Eletrobras, 2003):

$$h_{ta} = K_g \frac{V_g^2}{2g} + K_e \frac{V^2}{2g} \qquad (9.4)$$

where:
h_{ta} = total head loss at intake (m);
K_g = coefficient of head loss on trashracks;

K_e = head loss coefficient between the inlet port and the penstock section, including the head loss in the gate grooves;

V_g = speed in the trashracks section (m/s);

V = flow velocity downstream of the penstock or tunnel inlet (m/s);

g = gravity acceleration (m/s^2).

For the calculation of the head loss coefficient on the trashracks, Levin and Berezinski's Formula 9.5 (DSD-1973) can be used.

$$K_g = 1.45 - 0.45\frac{a_l}{a_b} - \left(\frac{a_l}{a_b}\right)^2 \tag{9.5}$$

where:

a_l = net area through the trashracks;

a_b = gross area of the trashracks.

The value of the K_e coefficient can be obtained from the tables presented in Graphs 221-1 to 221-1/3 (HDC, 1977), which are summarized in Table 9.2.

Table 9.2 Losses on Entry – Concrete Ducts

Outlet Ceiling Shape	Project	Duct			K_e
		A/D	R_e	h_v (ft)	
HDC 221-1		Intake – one mouth			
Circular	Pine flat	54	2.9–3.6 × 10^7 – P	65–81	0.16
		Intake – two mouths			
Elíptica	Denison	40	1.2 × 10^8 – P	66	0.19
		47	8.2 × 10^5 – M	61–82	0.12
	Ft Randall	39	0.7–1.5 × 10^8 – P	10–72	0.25
		39	0.9–1.0 × 10^6 – M	46–86	0.16
		Intake – three mouths			
Elíptica	Tionesta	98	1.5–4.1 × 10^5 – M	7–50	0.33
HDC 221-1/1	Tionesta	Three gates; L/D = 98 – M			0.53
Three gates		Gates 1 and 3; L/D = 98 – M			0.45
	Wappapello[a]	Three gates; L/D = 13 – M			0.50
		Gates 1 and 3; L/D = 13 – M			0.37
	Arkabutla[a]	Three gates; L/D = 20 – M			0.79
		Gates 1 and 3; L/D = 20 – M			0.68
HDC 221-1/2	East Blanch	Two gates; L/D = 125 – M			0.23
Two and four gates		One gate; L/D = 125 – M			0.45
	Ft Randall	Two gates; L/D = 39 – M			0.44
		One gate; L/D = 39 – M			0.78
	Sardis	Four gates; L/D = 31; M			0.32
		Three gates 1.2 and 3; L/D = 31; M			0.33
		Three gates 1.3 and 4; L/D = 31; M			0.37
HDC 221-1/3	Youghioghney	Three gates; L/D = 64 and 34; M			0.29
Mid-tunnel control		Two gates 1 and 3; L/D = 64 and 34 – M			0.41
structures	Abiquiu	Two gates; L/D = 25 and 25 – M			0.30
	Oahe	One gate; L/D = 60.8 and 77; M			0.44

A, distance from upstream face to gate; D, conduit diameter, or equivalent diameter for non-circular conduit; Re, Reynolds number; $h_v = V^2/2g$; V = speed after gate; K_e = head loss coefficient; P, prototype; M, model.

[a] no-circular conduit; For more details, see the HDC graphics (1977).

246 Design of Hydroelectric Power Plants – Step by Step

9.4 PENSTOCKS

Penstocks lead the water from the intake to the turbines. Pressure flow through the penstocks may be permanent or non-permanent, uniform or non-uniform. Permanent and uniform flow is one in which its characteristics do not vary over time. The flow in any section remains constant over time. Nonpermanent flow is one in which flow and pressure changes occur. The flow may still be laminar or turbulent, depending on the Reynolds number ($R_e = VD/\upsilon$), as briefly described below. For details, refer to the indicated references.

Penstock length is variable in each project. As a general guideline, it is sought to reduce its design length, bearing in mind that its cost is high because it is conduit under to the highest pressures. The maximum flow velocity in the penstocks results from economic considerations. In general, the maximum flow velocity in concrete-lined penstocks can reach up to 7.0 and 8.0 m/s in steel penstocks according to CBDB/Eletrobras (2003).

9.4.1 Head losses

The primary sources of penstock losses are friction with the walls, curves and bifurcations. These losses are estimated using the Darcy-Weisbach formula or the Manning-Strickler formula. The Darcy-Weisbach formula is given by the expression:

$$h_f = f \frac{L}{D} \frac{V^2}{2g} \tag{9.6}$$

where:

h_f penstock loss (m);
f universal coefficient of resistance from Darcy-Weisbach formula;
L penstock length (m);
D penstock diameter ($4 \times$ hydraulic radius in non-circular sections) (m);
V average flow velocity in the penstock (m/s);
g gravity acceleration (m/s^2).

The coefficient of resistance, as a function of wall roughness, penstock diameter and flow velocity, is given by the formula of Colebrook-White (1938/1939):

$$\frac{1}{\sqrt{f}} = -2\log\left[\frac{2\varepsilon}{D} + \frac{2.51}{Re\sqrt{f}}\right] \tag{9.7}$$

where:

f Darcy-Weisbach resistance coefficient, taken from the Moody graph (Figure 9.4). This coefficient can also be obtained from HDC graphs 224-1 to 224-1/5 (1973);
ε wall roughness (m);
ε/D relative roughness (or $\varepsilon/4R$, where R = hydraulic radius)
Re Reynolds number = VD/υ;
υ kinematic viscosity = 1.01×10^{-6} m^2/s (for water at 20°C).

It is worth making a brief history on this subject, extracted from Moody (1944). The coefficient of resistance is a dimensionless quantity, and at current velocities it is a function of two, and only two other dimensionless quantities, the relative roughness and the Reynolds number.

As stated by Moody (1944), Pigott (1933) had already published this graph based on an analysis of 10,000 experiments from various sources, without the benefit of the latest developments in the functional relationships of curves credited to Nikuradse (1933), von Kármán and Prandtl (1933), Bakhmeteff (1936) and Rouse (1938).

Colebrook and White (1938/1939) developed a function that gave practical shape, filling the transition zone gap left in the analysis by von Kármán and Prandtl (1933).

Based on this, Rouse in 1943 plotted a new graph of very complex use because it involved problematic interpolations.

Moody then, in 1944, incorporating the already accepted new functional relationships, published his new graph (Figure 9.4) in the conventional form used by Pigott (1933), that was more convenient to use.

Moody's diagram shows that laminar flow, in which fluid particles flow in linear paths parallel to the conduit wall and without mixing, is observed for Reynolds numbers up to 2,000, and follows the Prandtl-von equation. Kármán for flow in plain tubes. For Reynolds numbers around 2,000, a critical zone appears, extending to Re = 4,000, before the flow becomes turbulent. For rough pipes, a transition zone is found for Reynolds numbers above 4,000. The Reynolds number in the range between 4,000 and 10,000 characterizes the transition zone. The flow is completely turbulent for Reynolds numbers greater than 10,000. From that point on, f no longer varies and the curve becomes smooth (flat).

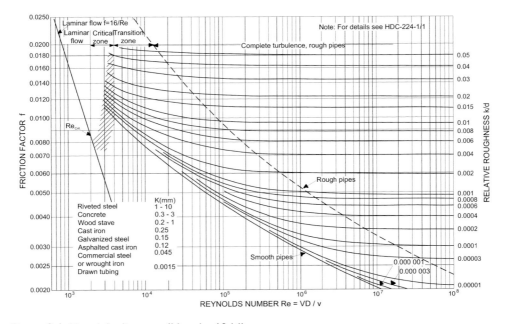

Figure 9.4 Moody's diagram (Moody, 1944).

248 Design of Hydroelectric Power Plants – Step by Step

The flow in penstocks and tunnels is usually turbulent and follows different trajectories in the Moody diagram depending on relative roughness. The zone for Reynolds numbers, greater than 4,000 is normally applied for the hydraulic design of most penstocks and tunnels. It should be borne in mind that f varies when calculating head losses for different flows in the same pipe.

Localized losses should be obtained through the equation:

$$h_f = \sum K \frac{V^2}{2g} \qquad (9.8)$$

where:

h_f sum of localized losses (m);
K head loss coefficient;
V average flow velocity in the penstock (m/s);
g gravity acceleration (m/s^2).

For curvature losses, the value of K can be defined using several references: for example, the HDC graph, 228-1 (1973) as well as the graphs presented in Figure 9.5a and b, taken from NUST, vol. 8, (2003). It is recommended that the radius of the bend be greater than two to three times the pipe diameter.

Losses in transitions, expansions or contractions can also be estimated by a coefficient applied to the velocity head. Figure 9.6 shows the loss coefficient for a conical gradual expansion.

Expansion loss can be estimated by the expression:

$$h_f = \left(1 - \frac{A_1}{A_2}\right)^2 \frac{V_1^2}{2g} \qquad (9.8a)$$

The contraction loss can be estimated by the expression:

$$h_f = 0.42 \left[1 - \left(\frac{D_2}{D_1}\right)^2 \frac{V_2^2}{2g} \right] \qquad (9.8b)$$

where D_1 and D_2 are the diameters at the input and output of the contraction, respectively.

For more details, see the HDC (1977), among other references already cited.

9.4.2 Economic diameter

The optimal penstock diameter is one that minimizes the sum of construction and maintenance costs plus the present value of unproduced energy due to head losses. To determine the optimal diameter, several solutions can be studied and budgeted by varying the diameter of the penstock. Complex computer calculations can be used to solve a system of linear equations, as proposed by Deppo and Datei (1984). It is possible to make an approximate calculation using empirical equation proposed by Sarkaria (1979), based on the historical relationships between penstocks costs and energy value.

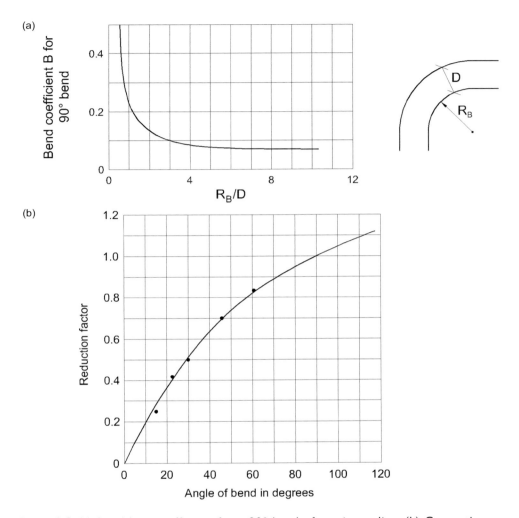

Figure 9.5 (a) Bend loss coefficient for a 90° bend of varying radius. (b) Correction coefficient for other angles bends, as compared with 90° bend of Figure 8.5a. Reduction factor.

The results would be ±10% accurate if compared to the actual diameters used in the 40 conduits used as the basis of the study (Gulliver and Arndt, 1991).

$$D = 0.71 \frac{P^{0.43}}{H_l^{0.65}} \tag{9.9}$$

where:
  P = installed power (kW);
  H_l = net head (m).

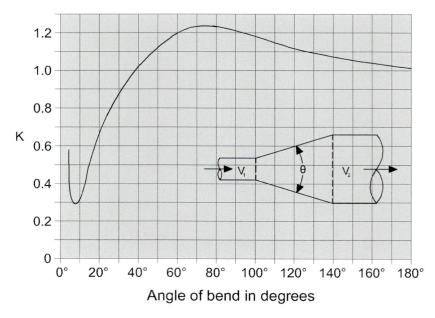

Figure 9.6 Typical values of loss coefficient for a conical gradual expansion.

9.4.2.1 Annex Support and anchor blocks

References:

- USBR: Welded Steel Penstocks, Engineering Monograph No. 3 (1977);
- IS-Indian Standards Code 5330-Criteria for Design Anchor Blocks for Penstocks with Expansion Joint (1984);
- SA Water, South Australian Water Corporation, Technical Guideline TG 96 – Guidelines for the Design of Anchors and Thrust Blocks on Buried Pipelines with Unrestrained Flexible Joints and for the Anchorage of Pipes on Steep Grades (May, 2007).

Two types of concrete blocks are used to support the penstock:

- support block, or saddles, where the duct simply rests, being allowed to slide over it;
- anchor block, which has the function of absorbing the stresses that develop in the duct, in long straight sections and at points of change of direction.

Alternatively, "structural steel rings" can be used, conveniently fixed to a concrete base.

- Support block or saddles

Figure 9.7 shows a schematic section of a duct section, with two support blocks.

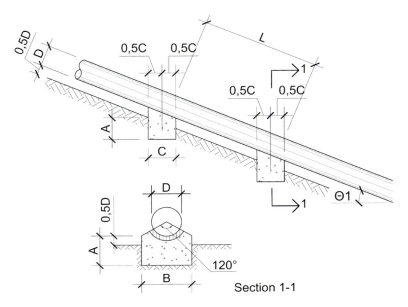

Figure 9.7 Support block.

Table 9.3 Support Block: Base Length C*

D (m)	Inclination of the Duct θ_1			
	0°	15°	30°	45°
0.20	0.35	0.35	0.45	0.65
0.40	0.65	0.65	0.65	1.00
0.60	1.00	1.00	1.00	1.20
0.80	1.30	1.30	1.30	1.40
1.00	1.60	1.60	1.60	1.60
1.20	2.00	2.00	2.00	2.00

(*) For $\sigma_{c\,adm}$ = 1.5 kgf/cm² (allowable compression stress).

Table 9.3 shows the value of the width of the base C of the blocks for the following physical conditions of the support system:

L = saddle spacing ≤ 6 D ≤ 5.0 m; A = block height = 1.2 D; B = width of the base = 1.6 D; C = length of the base of the block, tabulated according to the diameter and the angle of inclination (θ_1) of the duct, which meets the conditions specified below.

Active efforts
The distributed unit load (q) attached along the length of the duct is equal to:

$$Q = q_1 + q_a \tag{9.10}$$

where q_1 = duct unit weight (tf/m) and q_a = water unit weight (tf/m).

252 Design of Hydroelectric Power Plants – Step by Step

In the simplified dimensioning, presented below, the highest values of the main efforts were considered, disregarding the others.
> Normal force due to unit load (q)

$$F_n = QL\cos\left(tf\right) \tag{9.11}$$

> Tangential force due to temperature differences: as the duct is simply supported, this force is transmitted to the support corresponding to the maximum frictional force.

$$F_t = f_a F_n\left(tf\right) \tag{9.12}$$

f_a = friction coefficient between the conduit and the support block, adopted equal to 0.25 – corresponding to the friction between the conduit and a metallic support device in the block head, poorly lubricated.
> Support block weight

$$Gc = A \times B \times C\gamma_c \tag{9.13}$$

γc = concrete specific weight = 2.40 tf/m^3.
Block base length: C values are shown in the following table, as a function of D and θ_1.

They satisfy an admissible compression ratio of the foundation $\sigma_{c\,adm}$ = 1.5 kgf/cm^2, or work rate, corresponding to a compact coarse sand or a hard clay that is difficult to mold with the fingers, considering the stability conditions of the result of the efforts should pass through the central third of the block.

Safety Factor
$F_t / F_n > 2.0$, for blocks supported on rock;
$F_t / F_n > 2.5$, for blocks supported on the ground.
Effort transmitted to the foundation

$$\frac{\sum F_V}{A_B} < \sigma_{c\,adm} \tag{9.14}$$

$\sum F_V$ = sum of vertical forces (kgf/cm^2);

A_B = block area = $B \times C$ (cm^2).
Anchor block: is used in long straight sections of the conduit and in places of change of direction (Figure 9.8).
Active efforts
In addition to the efforts considered for the support block case, two other efforts should be considered:
> Tangential force due to distributed unit load (q)

$$F_t = q\,L\,sen\theta_1\left(tf\right) \tag{9.15}$$

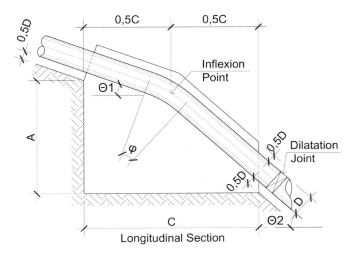

Figure 9.8 Anchor Block.

> Radial force, due to internal water pressure in the pipe curves:

$$F_R = \frac{\pi D^2}{2} P_t \operatorname{sen} \frac{\varphi}{2} \qquad (9.16)$$

P_t = total water pressure (m) in the duct, equal to the hydraulic load (H) between the reservoir and the block location plus the overpressure due to possible waterhammer. It is recommended to adopt $P_t = 1.35\ H$.

The influence of the centrifugal force on the curve, due to the flow speed, was not considered because of its small magnitude when compared to other forces.

Anchor block dimensions

The spacing between the blocks, the height and width of the base are fixed:

$L \leq 30$ m = maximum spacing (m); A = block height $\geq 2.0\ D$; B = width of the base = 3.0 or 4.0 D.

The values of length C of the base of the blocks are presented in Tables 9.4 and 9.5 for foundations in soil and rock, respectively, according to D, $\theta 1$, $\theta 2$ and H.

When sizing the block, account must be taken of:

$\sigma_{c\ adm} = 10$ kgf/cm² for rock foundations: altered rock resistant to disassembly by pick;
$\sigma_{c\ adm} = 1.5$ kgf/cm² for foundations on soil: compact coarse sand or hard clay that is difficult to mold with your fingers.

An expansion joint must be installed in the penstock downstream of the blocks. In the case of a block involving the duct, the duct must be attached to the block using a steel stirrup, as shown in the following Figure. The stirrups should be $\varnothing = \frac{3}{4}$ at least spaced every 20 cm and embedded in the base, as shown in Figure 9.9.

Table 9.4 Anchor Blocks on Earth Foundation - Base Length (m)

H = 5 m

D (m)	$\theta_2=0°$				$\theta_2=15°$				$\theta_2=30°$				$\theta_2=45°$			
	$\theta_1=0°$	15°	30°	45°	$\theta_1=0°$	15°	30°	45°	$\theta_1=0°$	15°	30°	45°	$\theta_1=0°$	15°	30°	45°
0.2	1.8	3.3	3.8	3.3	1.8	2.8	3.3	3.3	2.3	3.3	3.3	2.8	2.3	3.3	3.8	2.8
0.4	1.7	3.2	3.1	2.6	2.2	3.2	3.1	3.2	2.2	3.2	2.6	3.2	2.7	2.6	3.1	2.7
0.6	2.3	3.3	3.8	3.3	2.3	2.8	3.3	3.3	2.3	3.3	3.3	2.8	2.8	3.3	3.8	2.8
0.8	2.4	3.4	3.4	3.4	2.4	3.4	3.4	3.4	2.9	3.4	3.4	3.4	2.9	3.4	3.4	3.4
1.0	3.0	3.5	4.0	4.0	3.0	3.5	4.0	4.0	3.0	3.5	4.0	3.5	3.0	4.0	4.0	3.5
1.2	3.8	4.1	4.1	4.1	3.6	4.1	4.1	4.1	3.6	4.1	4.1	4.1	3.6	4.1	4.1	4.1

H = 10 m

D (m)	$\theta_2=0°$				$\theta_2=15°$				$\theta_2=30°$				$\theta_2=45°$			
	$\theta_1=0°$	15°	30°	45°	$\theta_1=0°$	15°	30°	45°	$\theta_1=0°$	15°	30°	45°	$\theta_1=0°$	15°	30°	45°
0.2	1.8	3.3	3.8	3.8	1.8	2.8	3.8	3.3	2.3	3.3	3.3	3.3	2.8	3.8	3.8	2.8
0.4	1.7	2.6	3.1	3.1	2.2	3.2	3.1	3.7	2.7	2.6	2.6	3.2	3.2	3.1	3.1	2.7
0.6	2.3	3.3	3.8	3.8	2.3	2.8	3.8	3.8	2.8	3.3	3.3	3.3	3.3	3.8	3.8	2.8
0.8	2.4	3.4	3.9	3.9	2.9	3.4	3.4	3.4	2.9	3.4	3.4	3.4	3.4	3.9	3.9	3.4
1.0	2.4	3.5	4.0	4.0	3.0	3.5	4.0	4.0	3.0	4.0	4.0	4.0	3.5	4.0	4.0	3.5
1.2	3.0	4.1	4.6	4.6	3.6	4.1	4.6	4.6	3.6	4.1	4.1	4.1	4.1	4.6	4.6	4.1

H = 15 m

D (m)	$\theta_2=0°$				$\theta_2=15°$				$\theta_2=30°$				$\theta_2=45°$			
	$\theta_1=0°$	15°	30°	45°	$\theta_1=0°$	15°	30°	45°	$\theta_1=0°$	15°	30°	45°	$\theta_1=0°$	15°	30°	45°
0.2	1.8	3.3	4.3	4.3	2.3	2.8	3.8	3.8	2.8	3.3	3.3	3.3	3.3	4.3	3.8	2.8
0.4	1.7	2.6	3.6	3.6	2.2	3.2	3.1	3.1	3.2	3.1	2.6	3.1	3.1	3.6	3.1	2.7
0.6	2.3	3.3	4.3	4.8	2.3	2.8	3.8	4.3	2.8	3.8	3.3	3.3	3.8	4.3	4.3	2.8
0.8	2.4	3.4	4.4	4.4	2.9	3.4	3.9	3.9	3.4	3.4	3.4	3.4	3.9	4.4	3.9	3.4
1.0	2.4	3.5	4.5	4.5	3.0	3.5	4.0	4.5	3.5	3.0	4.0	4.0	4.0	4.5	4.0	3.5
1.2	3.0	4.1	4.6	5.1	3.6	4.1	4.6	4.6	3.6	4.1	4.1	4.1	4.1	4.6	4.6	4.1

(Continued)

Table 9.4 (Continued) Anchor Blocks on Earth Foundation - Base Length (m)

	D (m)	$\theta_2 = 0°$				$\theta_2 = 15°$				$\theta_2 = 30°$				$\theta_2 = 45°$			
		$\theta_1 = 0°$	15°	30°	45°	$\theta_1 = 0°$	15°	30°	45°	$\theta_1 = 0°$	15°	30°	45°	$\theta_1 = 0°$	15°	30°	45°
H = 20 m	0.2	1.8	3.3	4.3	4.8	2.3	2.8	3.8	4.3	2.8	3.8	3.3	3.8	3.8	4.3	4.3	2.8
	0.4	1.7	3.1	3.6	4.1	2.7	3.2	3.6	3.6	3.2	3.1	2.6	3.1	3.6	4.1	3.6	2.7
	0.6	2.3	3.8	4.8	5.3	2.8	2.8	4.3	4.8	3.3	3.8	3.3	3.8	4.3	4.8	4.3	2.8
	0.8	2.4	3.4	4.9	5.4	2.9	3.4	3.9	4.4	3.4	3.9	3.4	3.9	4.4	4.9	4.4	3.4
	1.0	3.0	4.0	4.5	5.0	3.0	3.5	4.0	4.5	3.5	4.0	4.0	4.0	4.5	5.0	4.5	3.5
	1.2	3.6	4.1	5.1	5.1	3.6	4.1	4.6	5.1	4.1	4.6	4.1	4.6	4.6	5.1	4.6	4.1

	D (m)	$\theta_2 = 0°$				$\theta_2 = 15°$				$\theta_2 = 30°$				$\theta_2 = 45°$			
		$\theta_1 = 0°$	15°	30°	45°	$\theta_1 = 0°$	15°	30°	45°	$\theta_1 = 0°$	15°	30°	45°	$\theta_1 = 0°$	15°	30°	45°
H = 25 m	0.2	1.8	3.8	4.8	5.3	2.3	2.8	4.3	4.8	3.3	3.8	3.3	3.8	4.3	4.8	4.3	2.8
	0.4	1.7	3.1	4.1	4.6	2.7	3.2	3.6	4.1	3.1	3.1	2.6	3.1	4.1	4.1	3.6	2.7
	0.6	2.3	3.8	5.3	4.4	2.8	2.8	4.3	5.3	3.8	4.3	3.3	4.3	4.8	5.3	4.8	2.8
	0.8	2.4	3.4	4.9	5.9	2.9	3.4	4.4	4.9	3.4	3.9	3.4	3.9	4.9	5.4	4.4	3.4
	1.0	3.0	4.0	5.0	6.0	3.0	3.5	4.5	5.0	4.1	4.0	4.0	4.0	5.0	5.5	4.5	3.5
	1.2	3.6	4.1	5.1	5.6	3.6	4.1	4.6	5.1	4.1	4.6	4.1	4.6	5.1	5.1	4.6	4.1

Notes: Block height ≥ 2.0 D; Block base width $B = 3.0$ D. except in the marked region. where it should be 4.0 D.

Table 9.5 Anchor Blockson Rock Foundation – Base Length (m)

		$\theta_2 = 0°$			$\theta_2 = 15°$				$\theta_2 = 30°$				$\theta_2 = 45°$			
	D (m)	$\theta_1 = 0°$	15°	30°	45°	$\theta_1 = 0°$	15°	30°	45°	$\theta_1 = 0°$	15°	30°	45°	$\theta_1 = 0°$	15°	30°	45°
H = 5 m	0.2	1.6	2.3	2.3	1.8	1.6	1.8	2.3	1.8	1.3	2.3	2.3	1.8	1.8	2.3	2.3	1.3
	0.4	1.7	2.7	2.7	2.2	2.2	2.2	2.7	2.2	2.2	2.7	2.7	2.2	2.2	2.7	2.7	2.8
	0.6	2.3	2.8	3.3	2.8	2.3	2.8	3.3	2.8	2.3	2.9	2.8	2.8	2.8	3.3	3.3	2.8
	0.8	2.4	3.4	3.4	3.4	2.4	3.4	3.4	3.4	2.9	3.4	3.4	3.4	2.9	3.4	3.4	3.4
	1.0	3.0	3.5	4.0	4.0	3.0	3.5	4.0	4.0	3.0	3.5	4.0	3.5	3.0	4.0	4.0	3.5
	1.2	3.6	4.1	4.1	4.1	3.6	4.1	4.1	4.1	3.6	4.1	4.1	4.1	3.6	4.1	4.1	4.1

		$\theta_2 = 0°$			$\theta_2 = 15°$				$\theta_2 = 30°$				$\theta_2 = 45°$			
	D (m)	$\theta_1 = 0°$	15°	30°	45°	$\theta_1 = 0°$	15°	30°	45°	$\theta_1 = 0°$	15°	30°	45°	$\theta_1 = 0°$	15°	30°	45°
H = 10 m	0.2	1.6	2.3	2.8	2.3	1.3	1.8	2.3	2.3	1.8	2.3	2.3	1.8	2.3	2.8	2.3	1.3
	0.4	1.7	2.7	2.7	2.7	2.2	2.3	2.7	2.7	2.2	2.7	2.7	2.2	2.7	3.2	2.7	2.2
	0.6	2.3	2.8	3.3	3.3	2.3	2.8	3.3	3.3	2.8	3.3	2.8	2.8	2.8	3.3	3.3	2.8
	0.8	2.4	3.4	3.9	3.9	2.9	3.4	3.4	3.4	2.9	3.4	3.4	3.4	3.4	3.9	3.9	3.4
	1.0	3.0	3.5	4.0	4.0	3.0	3.5	4.0	4.0	3.0	4.0	4.0	4.0	3.5	4.0	4.0	3.5
	1.2	3.6	4.1	4.6	4.6	3.6	4.1	4.0	4.0	3.6	4.1	4.1	4.1	4.1	4.6	4.6	4.1

		$\theta_2 = 0°$			$\theta_2 = 15°$				$\theta_2 = 30°$				$\theta_2 = 45°$			
	D (m)	$\theta_1 = 0°$	15°	30°	45°	$\theta_1 = 0°$	15°	30°	45°	$\theta_1 = 0°$	15°	30°	45°	$\theta_1 = 0°$	15°	30°	45°
H = 15 m	0.2	1.6	2.3	2.8	2.8	1.8	1.8	2.3	2.3	1.8	2.3	2.3	1.8	2.3	2.8	2.8	1.3
	0.4	1.7	2.7	3.2	3.2	2.2	2.2	2.7	2.7	2.7	2.7	2.7	2.7	3.2	3.2	2.7	2.2
	0.6	2.3	3.3	3.3	3.3	2.3	2.8	3.3	3.3	2.8	3.3	2.8	3.3	3.3	3.3	3.3	2.8
	0.8	2.4	3.4	3.9	3.9	2.9	3.4	3.9	3.9	3.4	3.4	3.4	3.4	3.4	3.9	3.9	3.4
	1.0	3.0	4.0	4.5	4.5	3.0	3.5	4.0	4.5	3.5	3.0	4.0	4.0	4.0	4.5	4.0	3.5
	1.2	3.6	4.1	4.6	5.0	3.6	4.1	4.6	4.6	3.6	4.1	4.1	4.1	4.1	4.6	4.6	4.1

(Continued)

Table 9.5 (Continued) Anchor Blockson Rock Foundation – Base Length (m)

		$\theta_2 = 0°$			$\theta_2 = 15°$				$\theta_2 = 30°$				$\theta_2 = 45°$			
	D (m)	$\theta_1 = 0°$	15°	30°	45°	$\theta_1 = 0°$	15°	30°	45°	$\theta_1 = 0°$	15°	30°	45°	$\theta_1 = 0°$	15°	30°	45°
$H = 20$ m	0.2	1.6	2.3	3.3	3.3	1.8	1.8	2.8	2.8	2.3	2.8	2.3	2.3	2.8	3.3	2.8	1.3
	0.4	1.7	2.7	2.6	3.1	2.2	2.2	3.2	3.2	2.7	2.7	2.7	2.7	2.6	3.1	3.2	2.2
	0.6	2.3	3.3	3.8	3.8	2.8	2.8	3.3	3.3	2.8	3.3	2.8	3.3	3.8	3.8	3.3	2.8
	0.8	2.4	3.4	3.9	4.4	2.9	3.4	3.9	3.9	3.4	3.9	3.4	3.9	3.9	4.4	3.9	3.4
	1.0	3.0	4.0	4.5	5.0	3.0	3.5	4.0	4.5	3.5	4.0	4.0	4.0	4.5	5.0	4.5	3.5
	1.2	3.6	4.1	5.1	5.1	3.6	4.1	4.6	5.1	4.1	4.6	4.1	4.6	4.6	5.1	4.6	4.1

		$\theta_2 = 0°$			$\theta_2 = 15°$				$\theta_2 = 30°$				$\theta_2 = 45°$			
	D (m)	$\theta_1 = 0°$	15°	30°	45°	$\theta_1 = 0°$	15°	30°	45°	$\theta_1 = 0°$	15°	30°	45°	$\theta_1 = 0°$	15°	30°	45°
$H = 25$ m	0.2	1.6	2.8	3.3	3.3	1.8	1.8	2.8	2.8	2.3	2.8	2.3	2.3	3.3	3.8	2.8	1.3
	0.4	1.7	2.7	3.1	3.1	2.2	2.2	3.2	2.6	3.2	3.2	2.7	2.7	3.1	3.1	3.2	2.2
	0.6	2.3	3.3	3.8	4.3	2.8	2.8	3.3	3.8	3.3	3.3	2.8	3.3	3.8	3.8	3.3	2.8
	0.8	2.4	3.4	4.4	4.4	2.9	3.4	3.9	4.4	3.4	3.9	3.4	3.9	4.4	4.4	3.9	3.4
	1.0	3.0	4.0	4.5	5.0	3.0	3.5	4.0	4.5	4.0	4.0	4.0	4.0	4.5	5.0	4.5	3.5
	1.2	3.6	4.1	5.1	5.6	3.6	4.1	4.6	5.1	4.1	4.6	4.1	4.0	5.1	5.1	5.1	4.1

Notes: Block height ≥ 2.0 D; Block base width $B = 3.0$ D. except in the marked region. where it should be 4.0 D.

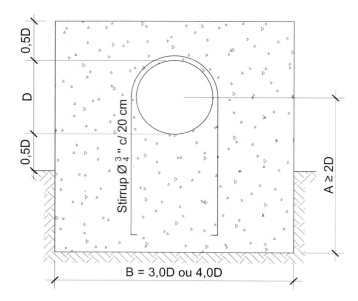

Figure 9.9 Anchoring block surrounded by concrete. Details.

9.4.3 Waterhammer

By definition, waterhammer is the pressure fluctuation in a penstock caused by the change in flow velocity after the opening or closing of the valve or the flow control gate of the turbines.

This operation creates a direct pressure wave near the valve/gate that propagates upstream toward the reservoir at a positive a velocity (waterhammer wave speed). Upon reaching the reservoir, this wave is reflected, but the reservoir pressure height (h_p) is not changed by transient penstock pressures.

The reflected wave has the same magnitude but opposite signal to the direct wave that left the valve seconds before.

In deducing the fundamental equations of waterhammer, it is considered the elasticity of the conduit as well as the compressibility of water under the action of pressure change.

The dynamic equilibrium condition requires that the unbalance force acting on any element of the net mass be equal to the product of the mass by acceleration. Thus, Newton's second law is satisfied. In addition, the condition established by the continuity equation must also be satisfied.

By simultaneously solving the two equations, the waterhammer equations are obtained. The mathematics for solving Saint–Venant's equations is complex and not within the scope of this book.

For details, it is recommended to consult, among others, the publications: Hydraulique Technique (Jaeger, 1954), Chambres d'Equilibre (Stucky, 1958), and Waterhammer Analysis (Parmakian, 1983).

Hydraulic conveyance design

From a design point of view, hydraulic transients are potentially a problem if:

$$\frac{LV}{H} > 3.3 \, \text{m/s} \tag{9.17}$$

where:
L = penstock length (m);
V = penstock flow velocity (m/s);
H = net head (m).

If the condition defined by Equation 9.17 is met, further investigations into the possibilities of waterhammer should be made and the overpressure estimated as shown below.

Calculations of hydraulic transients are complex and are usually done by computer programs that contain details of turbine response time, valve and gate closing, junctions, surge tank, etc.

There are maximum and minimum acceptable pressures in the penstock, and the program must be able to ensure that the limits are not exceeded. The program is used to optimize the design as well as to study the effects of an equilibrium chamber in reducing this pressure wave.

As defined by CBDB/Eletrobras (2003), the penstock shall be designed to resist, in each section, the internal pressure corresponding to the maximum positive waterhammer defined by the piezometric line AB in Figure 9.10. The waterhammer in the valve section must not exceed 50% of the static reservoir load (height D-Turbine).

Negative waterhammer caused by the sudden opening of the valves, or resulting from the fluctuation that occurs after the positive waterhammer, defined by line EG in Figure 9.11, should not cause negative pressure at any point in the penstock (point K).

Waterhammer should be calculated numerically or graphically using Allievi's formula (Equation 9.27) as presented below.

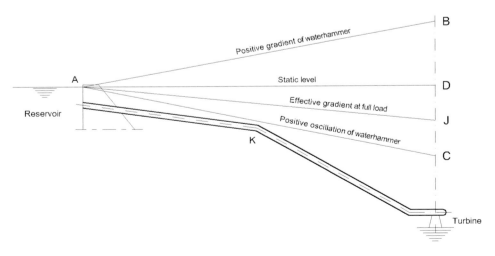

Figure 9.10 Positive waterhammer gradient (CBDB, 2003).

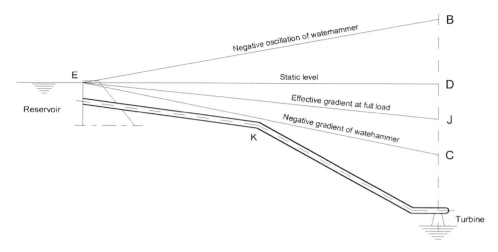

Figure 9.11 Negative waterhammer gradient – sudden open of the valves (CBDB, 2003).

The critical time of the penstock, i.e., the time it takes for the pressure wave to travel through the entire penstock from the closing device to the reservoir or surge tank and back to the starting point is defined by:

$$T_c = \frac{2L}{a} \tag{9.18}$$

where:
  T_c = critical time of the penstock (s);
  L = length of the penstock (m);
  a = waterhammer wave speed (m/s).

As the turbine closing time is generally much longer than the critical time of the penstock, there is always the possibility of closing from a partial opening that completes in a time virtually equal to T_c. Therefore, using the Allievi or Joukowsky formula, the waterhammer for partial closure should be calculated to define the maximum overpressure to which the conduit will be subjected. This maximum overpressure will be equal to:

$$h_{máx} = \frac{a \Delta V}{g} \tag{9.19}$$

where:
  $h_{máx}$ = water column height overpressure (m);
  a = waterhammer wave speed (m/s);
  ΔV = flow velocity for partial flow whose closing time is equal to T_c (m/s);
  g = gravity acceleration (m/s^2).

If the condition defined by Equation 9.18 is met, as stated above, the overpressure shall be estimated, for a closing time $T_f \leq T_c = 2L/a$, as shown below.

9.4.3.1   Overpressure calculation due to instant closing

There is always a variation in pressure, waterhammer, associated with a rapid change in speed. The relationship between these variations, given by the basic physics of linear momentum, is known as the Joukowsky Equation (1898) or Allievi Equation (1902), presented below (Equation 8.20). The deduction, presented by Stucky (1958) in Chambres d'Equilibre, is transcribed below.

Consider a conduit of diameter D, thickness e (Figure 9.12). Let there be a stretch of flow with length x and velocity V, subjected to zero pressure.

Over the upstream face, acts an overpressure γh. The stretch has the length shortened of Δx, and the tube expands from ΔD.

The overpressure h is calculated using the living force theorem, which states:

* Living half-force variation = External forces working

$$\text{Living half force} = \frac{1}{2} \frac{\pi D^2}{4} \frac{\gamma x}{g} V^2 \tag{9.20}$$

$$\text{Stretch crushing work} (\Delta x) = \frac{1}{2} \cdot \left(\gamma h \cdot \frac{\pi D^2}{4} \right) \cdot \gamma h \frac{x}{K} \tag{9.21}$$

i.e., stretch crushing work $(\Delta x) = \frac{1}{2} \times \underline{\text{force}} \times \Delta x$.

$$\text{Pipe expansion work} = \frac{1}{2} \cdot \left(\frac{\gamma h Dx}{2} \right) \left(\frac{\gamma h Dx}{2exE} \right) \cdot \pi D \tag{9.22}$$

(which is measured by pipe wall expansion work = $1/2 \, x$ traction within the tube x specific elongation x pipe perimeter), where:

γ = specific water weight = 1,000 kg/m^3;
K = volumetric modulus of elasticity of water = 21.4×10^8 N/m^2;
E = Young's modulus of elasticity of the penstock material; for steel 2.10×10^{11} N/m^2.

Substituting these values into the living force equation and simplifying by $\dfrac{\gamma \pi D^2 x}{8}$, we find

$$\frac{V^2}{g} = h^2 \gamma \left(\frac{1}{K} + \frac{D}{e\,E} \right) \tag{9.23}$$

or

$$h = \frac{V}{\sqrt{\gamma g \left(\dfrac{1}{K} + \dfrac{D}{e\,E} \right)}} \tag{9.24}$$

For a given conduit, overpressure h is proportional to velocity V in the case of an instantaneous closing.

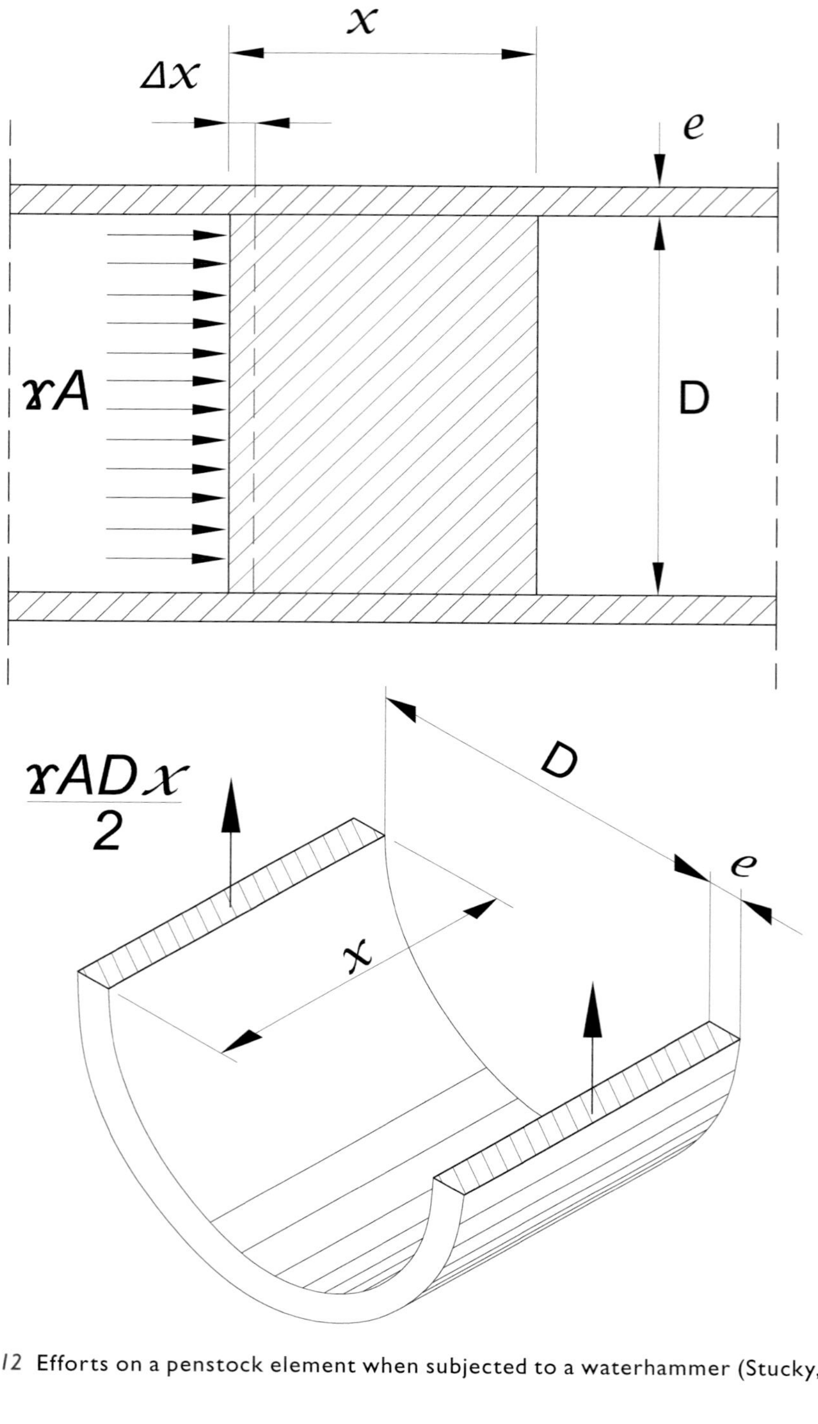

262 Design of Hydroelectric Power Plants – Step by Step

Figure 9.12 Efforts on a penstock element when subjected to a waterhammer (Stucky, 1958).

The speed a is calculated with the support of the amount of motion theorem:

- Variation of the amount of movement over a period of time $t =$ Sum of the external forces.

$$V\left(\frac{\gamma}{g}\frac{\pi D^2}{4}x\right)\frac{1}{t} = \gamma\,h\,\frac{\pi D^2}{4} \tag{9.25}$$

or

$$a = \frac{x}{t} = \frac{h\,g}{V} \tag{9.26}$$

or

$$h = \frac{a\,V}{g}, \text{(Joukowsky or Allievi)} \tag{9.27}$$

Substituting h in Equation 9.24, we find:

$$a = \sqrt{\frac{g}{\gamma\left(\frac{1}{K} + \frac{D}{e\,E}\right)}} \tag{9.28}$$

where:

$a =$ waterhammer wave speed (m/s);
$\rho =$ density of water $= 1{,}000$ kg/m^3;
$K =$ volumetric modulus of elasticity of water $= 21.4 \times 10^8$ N/m^2;
$D =$ penstock diameter (m);
$C_e =$ coefficient for reducing expansion in pipe joints;
$E =$ Young's modulus of elasticity for the penstock material, for steel $= 2.10 \times 10^{11}$ N/m^2;
$e =$ penstock thickness (m).

The speed depends only on the characteristics D, e, and E of the pipe and the volumetric modulus of elasticity K of the water.

Equation 9.20 shows that overpressure h is independent of conduit length L. This surprising finding results from the hypothesis of instantaneous closure. In fact, overpressure h depends on the conduct length.

Observation I

Stucky (1958) pointed out that if the conduit is infinitely rigid, $E = \infty$, Equation 9.28 becomes:

$$a = \sqrt{\frac{K}{\rho}} = \sqrt{\frac{g\,K}{\gamma}} \tag{9.29}$$

264 Design of Hydroelectric Power Plants – Step by Step

Table 9.6 Values of "a" for a Steel Penstock

H (m)	a (m/s)
1,500	1,250
1,000	1,180
500	1,080
250	950
100	830
50	750

which is the known value of the velocity of propagation of an elastic wave in a fluid, which here is the velocity of sound in water (approximately 1,400 m/s). This is the upper limit of a (waterhammer wave speed).

Observation 2

If the conduit is steel, the thickness will depend on the pressure. Speed can be considered as a function of the head only (H). According to Stucky (1958), Allievi provides the following values for a steel penstock (Table 9.6).

These numbers show that the wave has velocities between the limits of 1,250 and 750 m/s. The average speed is in the order of 1,000 m/s.

9.4.3.2 Calculation of overpressure (h) due to gradual closure without surge tank

The speeds mentioned above are much higher than those allowed in the design criteria for concrete and steel penstocks. Considering, for example, $a = 1,000$ m/s and $V = 7.0$ m/s and using Equation 8.20, the overpressure h is 700 m water column (or 700 t/m^2).

Penstocks to resist such overpressures would have prohibitive dimensions. A first means of reducing these pressures is to proscribe instantaneous (or quick) closures and allow only progressive and linear shutter maneuvers, i.e., flow proportional to time, and disregarding the effect of varying pressure on flow.

- Gradual closing – fast (Allievi): $Tf \leq 2L/a$

    Figure 9.10 shows, for gradual closure, the different stages of progression of the overpressure and depression waves. In the case of instantaneous close, the maximum overpressure is given by the following value:

$$h = \frac{\Delta p}{\rho g} = \frac{\rho a \Delta V}{\rho g} = \frac{aV}{g} \qquad (9.30)$$

h water column height overpressure (m);
ΔV change in velocity caused by the opening or closing of the turbine, or a rejection;
ρ density of water (kg/m^3);

a waterhammer wave speed (m/s);
g gravity acceleration (m/s^2).

It is found that the overpressure in the obturator does not decrease and that part of the conduit is relieved (Figure 9.13). This relief (within the calculation assumptions) extends $a \cdot T_f/2$ in length and the overpressure ranges from zero to h the $a \cdot T_f/2$ distance from the inlet.

In this case, the medicine is insufficient. If the gallery is 5 km long, a closing time of at least 10 seconds would be required ($Tf = 10,000/a = 10,000/1,000 = 10$ s).

- Gradual Close – slow: $Tf > 2\ L/a$

For slow closing, the maximum overpressure is given by the following value.

$$\Delta H = \frac{2LV}{gT_f}\,(\text{Michaud Formula}) \tag{9.31}$$

Slow gradual closing allows for an appreciable reduction in overpressure. However, very long closing times must be considered. It has been seen earlier that the overpressure is in the hundreds of meters for instantaneous closure.

For gradual closure to be acceptable without special reinforcement of the casing, the overpressure must not exceed the order of 10–30 m. This will require the closing time (T_f) to be 10–30 times longer than $T_c = 2L/a$. If the gallery is 5 km, this will lead to closing times of 100–300 s, which are too slow and not acceptable for the turbine.

The introduction of a surge tank in the adduction hydraulic circuit is a radical, safe remedy and, therefore, is the solution adopted to reduce sub-pressures by closing as slowly as possible. It is recommended to consult, among other references, Stucky (1958).

Most of the penstocks are classified as thin walled, i.e., $D/e > 25$. The coefficient of expansion Ce for these penstocks according to Chaudhry (1979, apud Gulliver and Arndt, 1991) is given in Frame 9.1.

For thick-walled and underground penstocks, according to CBDB/Eletrobras (2003), the expressions proposed by Jaeger (1954), presented below, may be used.

Thick underground penstock

Frame 9.1 C_e Values (Chaudhry, 1979)

Conduit Type	C_e
With frequent joints	$C_e = 1.0$
Anchored against longitudinal movements	$C_e = 1 - v^2 (*)$
Downstream anchored against longitudinal movements	$C_e = 1.25 - v$
Unlined tunnel	$C_e = e/D$, $E = G$ rock stiffness module
Steel lined tunnel	$C_e = E/E + G(D/e)$

(*) Poisson's ratio and modulus of elasticity are given in Table 9.4.

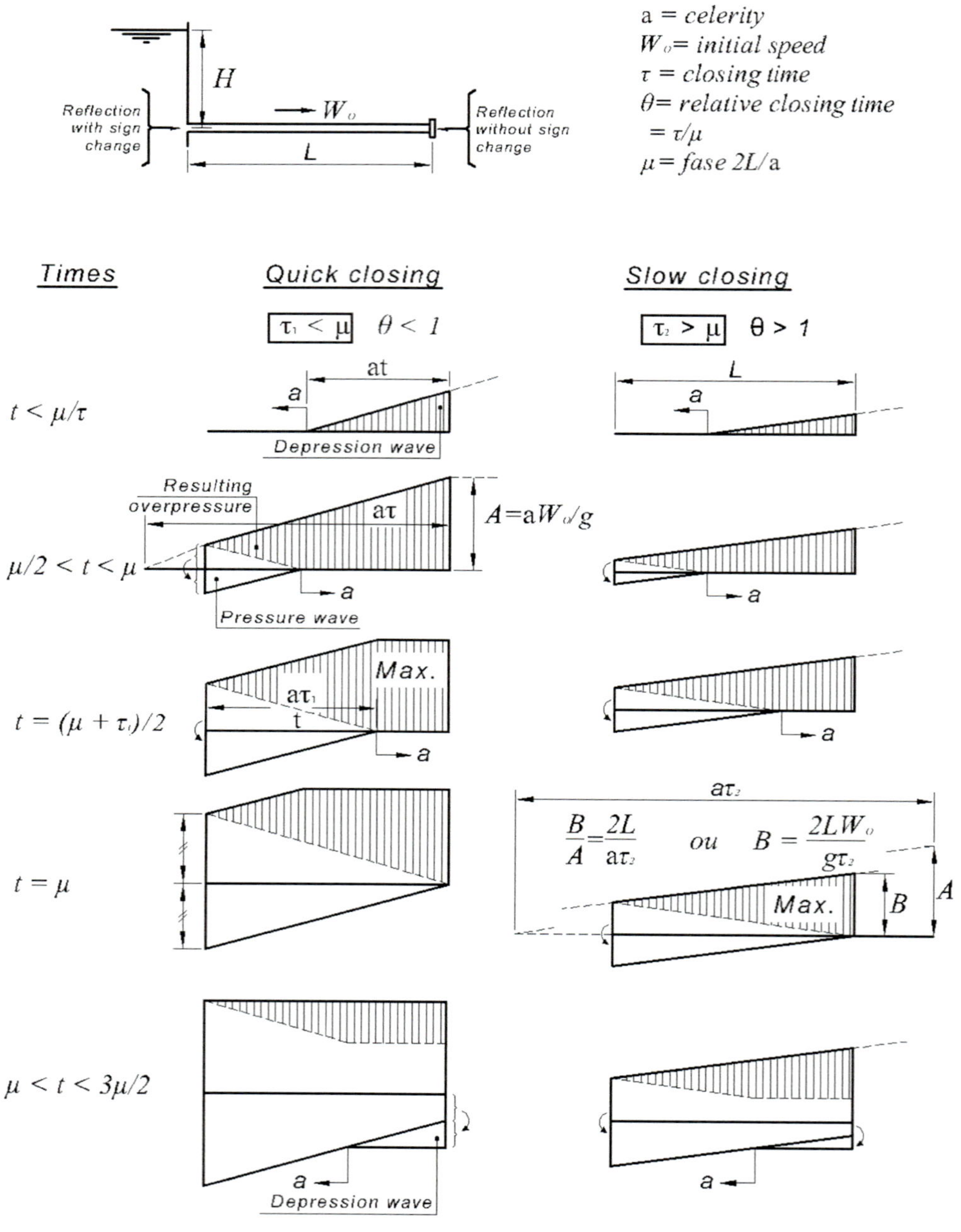

266 Design of Hydroelectric Power Plants – Step by Step

$a = celerity$
$W_o = initial\ speed$
$\tau = closing\ time$
$\theta = relative\ closing\ time$
$= \tau/\mu$
$\mu = fase\ 2L/a$

Reflection with sign change

Reflection without sign change

Times | **Quick closing** | **Slow closing**

$\boxed{\tau_1 < \mu}\ \ \theta < 1$ | $\boxed{\tau_2 > \mu}\ \ \theta > 1$

$t < \mu/\tau$

at — Depression wave

$\mu/2 < t < \mu$

Resulting overpressure — $a\tau$ — $A = aW_o/g$ — Pressure wave

$t = (\mu + \tau_1)/2$

Max. — $a\tau_1$ — t

$t = \mu$

$\dfrac{B}{A} = \dfrac{2L}{a\tau_2}$ ou $B = \dfrac{2LW_o}{g\tau_2}$ — Max. — $a\tau_2$

$\mu < t < 3\mu/2$

a — Depression wave

Figure 9.13 Progressive closing waterhammer (Stucky, 1958). Each diagram is a velocity profile at a given time t. The closure is assumed to be linear. The effect of pressure variation on flow is neglected.

$$a = \sqrt{\dfrac{1}{\rho\left[\dfrac{1}{K}+\left(\dfrac{2}{E}\right)\left(\dfrac{b^2+c^2}{c^2-b^2}\right)\right]}} \qquad (9.32)$$

where:

a = waterhammer wave speed (m/s);
ρ = density of water = 1,000 kg/m³;
b = inner diameter of the penstock (m);
c = outside diameter of the penstock (m);
K = volumetric modulus of elasticity of water = 21.4×10^8 N/m²;
E = Young's modulus of elasticity for the penstock material (N/m²) (Table 9.7);

– Underground penstock in sound rock

$$a = \sqrt{\dfrac{1}{\rho\left[\dfrac{1}{K}+\dfrac{2}{E_r}\right]}} \qquad (9.33)$$

where:

ρ = density of water = 1,000 kg/m³;
K = volumetric modulus of elasticity of water = 21.4×10^8 N/m²;
E_r = rock elastic modulus (N/m²) (Table 9.4) (Figure 9.14).

Table 9.7 Elasticity Modules and Poisson's Coefficients: Tubes of Various Materials

Material	Elasticity Modules E (GN/m²)	Poisson's Coefficients (v)
Aluminum alloys	68–73	0.33
Fiber cement	24	-
Bronze	78–110	0.36
Cast iron	80–170	0.25
Concrete	14–30	0.10–0.15
Copper	107–131	0.34
Glass	46–73	0.24
Lead	4.8–17	0.44
Light steel	200–212	0.27
Plastics		
ABS	1.7	0.33
Nylon	1.4–2.75	
Perspex	6.0	0.33
Polyethylene	0.8	0.46
Polystyrene	5.0	0.4
Rigid PVC	2.4–2.75	-
Rocks		
Granite	50	0.28
Limestone	55	0.21
Quartzite	24.0–44.8	-
Sandstone	2.75–4.8	0.28
Schist	6.5–18.6	-

References: Chaudhry (1979, apud Gulliver and Arndt, 1991).

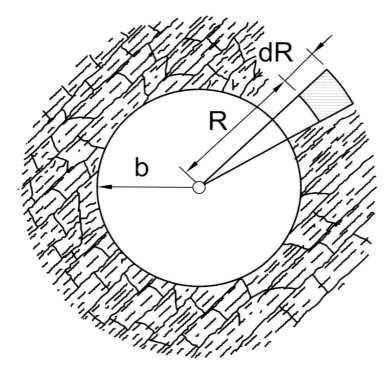

Figure 9.14 Conduit or penstock in sound rock – without steel lining (CBDB/Eletrobras, 2003).

Steel lining conduit

$$a = \sqrt{\frac{1}{\rho\left[\dfrac{1}{K} + \dfrac{2b}{E \cdot e}(1-\lambda_3)\right]}} \qquad (9.34)$$

$$\lambda_3 = \frac{\dfrac{b^2}{E \cdot e}}{\dfrac{b^2}{E \cdot e} + \dfrac{c^2 - b^2}{cE_c} + \dfrac{(v_r+1)b}{v_r E_r}} \qquad (9.35)$$

where:
 ρ = density of water = 1,000 kg/m^3;
 b = inner radius of concrete lining (m),
 c = external radius of concrete lining (m);
 e = plate thickness (m);
 v_r = rock Poisson's coefficient;
 K = volumetric modulus of elasticity of water = 21.4 × 10^8 N/m^2;

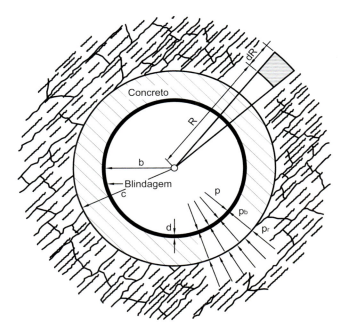

Figure 9.15 Steel lining conduit (CBDB/Eletrobras, 2003).

E = Young's modulus of elasticity = 2.10×10^{11} N/m^2;
E_c = modulus of elasticity of concrete (N/m^2) (Table 8.4);
E_r = modulus of elasticity of rock (N/m^2) (Table 9.7) (Figure 9.15).

9.5 TUNNEL

9.5.1 General design criteria

Tunnels have been used in HPPs under increasing loads, approaching 1,000 m. In Brazil the use has intensified in the last 25 years. They are designed and built to work normally without significant pressurization.

The following is a summary of the design and executive methods of these devices, based primarily on references: CBDB/Eletrobras (2003), Brekke and Ripley (1987) and Edvardsson and Broch (2002).

9.5.1.1 Tunnel alignment

The alignment of the power tunnel should represent the shortest connection between the water intake and the powerhouse, wherever possible. Figure 9.16 illustrates the evolution of this alignment.

In profile the tunnel must be plotted, so that the highest point is always safely below the piezometric line in the most unfavorable case, i.e., when the water level

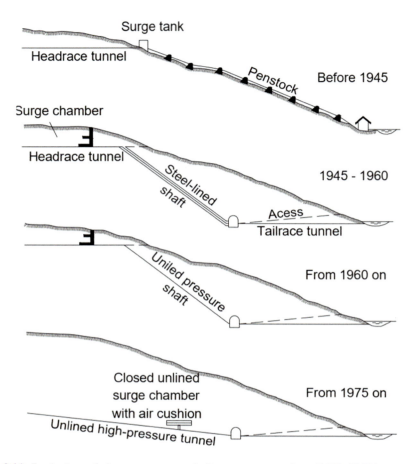

Figure 9.16 Evolution of the power tunnel alignment over time (NIT, 1993).

of the reservoir and the surge tank (if any) reaches the minimum level. In the low-pressure stretch (headrace tunnel), between the water intake and the surge tank, the slope should be low, 1.0%–2.0%, so that there is a laminar flow. This slope will take into account the water drainage aspects. In the high-pressure stretch (shaft), between the surge tank and the valve upstream of the turbine, the dive angle should be appropriate to the need for rock cover. Inclined shafts have advantages over vertical shafts. In general, the maximum slope should be limited to 10%–12%. When the geometry of the layout requires it, the steep sections should be concentrated to small extensions, as they will require differentiated construction methods.

Considering the quality of the rock, in the stretch where the minimum rock cover criterion is met, in principle, no tunnel lining should be provided. Lining will only be required in the stretch that is located where rock cover is insufficient due to geological/constructive requirements. The length of the armored section upon arrival at the powerhouse and, eventually, in other localized sections, should meet the Norwegian minimum rock cover criterion, which will be presented later in this chapter.

9.5.1.2 Covering criteria

In order to reduce the geological risk, the project should be designed based on detailed knowledge of the geology, topography, geo-mechanical characteristics of the rock and its stress state (estimated). Design should be done by a team with proven experience, including hydraulic engineer, engineering geologist, and rock mechanics expert.

Determining the rock top and rock density is very important. It is based on a thorough program of geological-geotechnical investigations and laboratory testing, as described in Chapter 6 of this book. This program should include: direct drilling along the tunnel; walk on guideline and surface geological mapping; and geophysical surveys along the axis.

In addition to geological investigations, it is important to develop a geo-mechanical classification of the rock mass. For example, the criteria of Barton (1974) or Bieniawski (1984) presented in Hoek (2007) can be used. Palmstrom and Broch (2006) presented an update of this classification (Figure 9.17).

The Excavation Support Ratio (ESR) depends on the purpose for which the excavation was designed, as described in Frame 9.2. For details, see Hoek (2007).

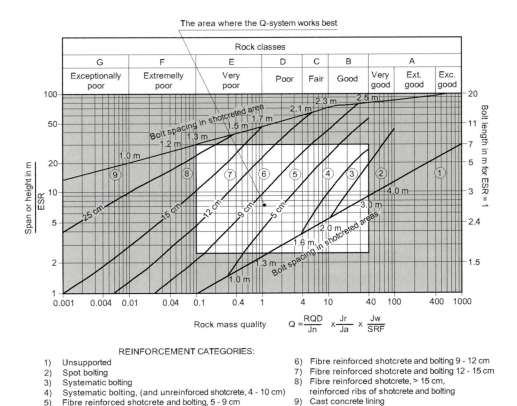

Figure 9.17 Limitations in the Q rock support diagram. Outside the unshaded area supplementary methods evaluations/calculations should be applied – Update: Palmstrom and Broch (2006).

272 Design of Hydroelectric Power Plants – Step by Step

Frame 9.2 Excavation Support Ratio (by Excavation Category)

Excavation Category		ESR
A	Temporary mine openings	3–5
B	Permanent mine openings, water tunnels for hydro power (excluding high-pressure penstocks), pilot tunnels, drift and headings for large excavations	1.6
C	Storage rooms, water treatment plants, minor road and railway tunnels, surge chambers, access tunnels	1.3
D	Power stations, major road and railway tunnels, civil defense chambers, portal intersections	1.0
E	Underground nuclear power stations, railway stations, sport and public facilities, factories.	0.8

Hoek (2007).

After defining the profile and characteristics of the rock, where there is sufficient coverage to ensure containment and sealing conditions, the tunnels are designed unlined over most of their length. In the final stretch, where the cover (rock weight) is insufficient to withstand the internal hydrodynamic pressure, they are shielded which substantially enhances the design.

These conditions make it possible to avoid hydraulic jacking of the rock mass. In design, this is the most important consideration to be made from a tunnel safety perspective. Hydraulic jacking could have disastrous consequences – see cases reported by Broch (1984), Sharma (1991) and Brekke and Rippley (1987).

In addition, it is emphasized that it is very important to provide for the operational phase, as a criterion, the filling and controlled emptying of the tunnel.

The tunnel shall be designed to withstand the maximum internal pressure due to extreme operating conditions of the plant for maximum turbine flow. To do so, one must first meet the Norwegian criterion of minimum rock cover (Figure 9.18), advocated by Bergh-Christensen and Dannevig (1971, apud Broch 1984).

This criterion, for pre-feasibility studies, is translated by two pocket rules:

$$\gamma_r \, h \, \cos\alpha > F \, H \, \gamma_a \tag{9.36a}$$

$$\gamma_r L \, \cos\beta > F \, H \, \gamma_a \tag{9.36b}$$

where:

h = studied point depth (m);
L = shortest distance from the surface to the studied point (m);
α = tunnel inclination;
β = average slope of the valley slope;
H = maximum static water pressure head in the section under study (m);
γ_r = density of rock (t/m^3).
γ_a = density of water = 1.0 t/m^3;
F = safety coefficient adopted for the pressure considered = 1.3.

For nearly horizontal terrain, adopting cos $\beta \approx 1.0$, expression 9.29b is reduced to the traditional criterion, $\gamma_r L > F H \gamma_a$, where the vertical stress associated with the rock cover should always exceed the water pressure at any point in the tunnel.

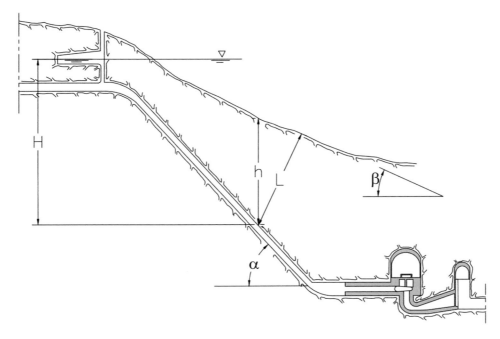

Figure 9.18 Definition of Minimum Coverage Criteria. NIT (Nielsen and Thidermann, 1993).

The pocket rules used in pre-feasibility studies represent a simplification of boundary equilibrium methods that take into account only the forces of gravity. To take into account the stress state in the tunnel valley, especially the topographic stresses, the Norwegian Institute of Technology – NIT (1992) has developed a model to simulate the most real tunnel design situation using the Finite Element Method. The basic principle of the method for this purpose is to find a location where all parts of the tunnel satisfy the following equation, where σ_3 is the minor main stress.

$$\sigma_3 > H \, \gamma_a \tag{9.37}$$

The NIT, Nielsen and Thidemann (1993), has developed a chart (Figure 9.19) for standard design of unlined pressure tunnels, covering valleys with slopes between 14° and 75° and a wide range of stress configurations and rock mass properties. To make the dimensionless model, the static water pressure was expressed as a ratio H/d, where d is the depth of the valley. Solid curves pass points where the internal water pressure in the tunnel is less than the lowest main tension in the surrounding rock mass. The dashed lines represent the curves of the same pressure σ_3. The use of this chart is illustrated by the example:

A. Valley: bottom at elevation 100 m and the top at elevation 600 m; therefore, $d = 500$ m;
B. reservoir WLmax: elevation = 390 m; therefore $H = 290$ m;
C. $H/d = 0.58$.

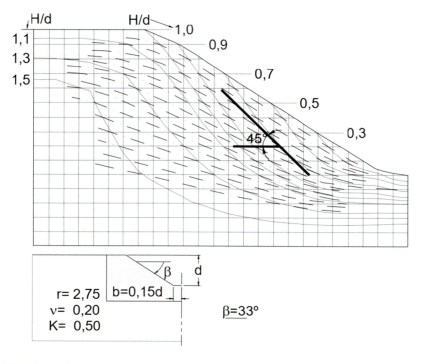

Figure 9.19 Unlined tunnel design chart (Nielsen and Thidemann-NIT, 1993, apud Bjorlykke/Semer-Olsen).

Therefore, at all points below the 0.58 line, the tension σ_3 exceeds the tunnel water pressure and there will be no hydraulic jacking problem. For a safety factor $F = 1.3$, the critical line will be $1.3 \times 0.58 = 0.75$. In Figure 9.19, the probable final location of a tunnel is placed, with $\alpha = 45°$. In cases with very high head of water and complex geology, the model should be adjusted to the topographic and geological conditions as close to reality as possible (see NIT, 1993, for 900 m water columns – for example).

It is noteworthy that the static water load is used in unlined tunnel designs. The dynamic load, usually 5%–10% higher, is used to size the lining where needed. This is because the dynamic load only occurs for a few minutes and it is assumed that the influence of the hydraulic jacking is negligible.

With respect to water leaks, tunnels with sectioned concrete and concrete lining are treated as unlined because this type of lining always has cracks.

9.5.2 Criteria for hydraulic tunnel dimensioning

- Design flow, maximum speed and section
  In principle, the tunnel should be sized for maximum turbine flow. The maximum flow velocity shall be around 2.0 m/s for uncoated tunnels, 3.0 m/s for projected concrete lined tunnels and 4.5 m/s for structural concrete lined tunnels.

The tunnel section has been adopted in arc-rectangle for constructive reasons. Where required, the shield shall be circular. The estimation of the economic diameter of the conduit must be made using the formula of Sarkaria (1979) presented in Section 9.3.2.

In Design Guidelines for Pressure Tunnels and Shafts, Brekke and Ripley (1987) specify a minimum diameter of 3.0 m that can be excavated economically. They also specify a maximum diameter of 10 m, provided that the conditions of the rock massif allow it. According to them, for larger tunnels, the civil works contractor resort to the top heading and bench excavation process, which increases the cost and time of construction. For the stretch under pressure (shafts), they specify a $D = 2.50$ m (minimum).

- Head loss

According to CBDB/Eletrobras (2003), localized head losses resulting from contractions, expansions and curvatures should be estimated according to Sections 3.7.3 and 3.7.4 as already presented in Chapter 8.

Continuous head losses are estimated using the Darcy-Weisbach formula, Equation 9.38, which is applicable for the entire roughness range typically found in excavation methods (drilling/blasting and TBM – Tunneling boring machine).

$$h_f = f \frac{L}{D} \frac{V^2}{2g} \tag{9.38}$$

where:
   h_f tunnel head loss (m);
   f Darcy-Weisbach head loss coefficient;
   L tunnel length (m);
   D tunnel diameter; base or height of arc-rectangle section (m);
   V average tunnel flow velocity (m/s), $V = Q / (\pi D^2/4)$;
   Q flow (m³/s);
   g gravity acceleration (m/s²).

The coefficient of pressure drop f is a function of wall roughness, of tunnel diameter and of flow velocity. The coefficients of strength for unlined tunnels should be obtained from Graphs 224-1/5 to 224-1/6 (HDC, 1973).

The coefficient f can simply be estimated by the expression:

$$f = 124.58 \, n^2 / D^{0.333} \tag{9.39}$$

where:
   n Manning coefficient, which varies as a function of tunnel roughness.

Table 9.8 presents suggested values for the Manning coefficient.
- Wall roughness

As stated earlier, unlined tunnels have been widely used in HPPs as river diversion tunnels and as power tunnels where the rock is good.

Determining the roughness of tunnel walls requires special consideration. The technique used by USACE is presented in HDC Chart 224-1/6 (Figure 9.17).

276 Design of Hydroelectric Power Plants – Step by Step

Table 9.8 Manning Coefficient (CBDB/Eletrobras, 2003)

Lining	n
Unlined	0.015–0.041 (excavation method function)
Concrete	0.013
Steel	0.010

This technique assumes that, in unlined tunnels, the resistance coefficient is independent of the Reynolds number because of the large value usually obtained for relative roughness.

Thus, the Von Kármán-Prandtl equation for completely rough flow, based on the sand grain Nikuradse data should apply. In terms of the Darcy coefficient, the tunnel diameter Dm and the equivalent diameter of the sand grain Ks, the following equation can be used:

$$\frac{1}{\sqrt{f}} = 2\log\left(\frac{D_m}{K_s}\right) + 1.14 \tag{9.40}$$

The value of the overbreak K in unlined tunnels is given by:

$$K = D_m - D_n = \sqrt{\frac{4}{\pi}}\left(\sqrt{A_m} - \sqrt{A_n}\right) \tag{9.41}$$

where:

K = overbreak (mm);

D_m and D_n are equivalent diameters based on the areas A_m and A_n shown in Figure 9.17.

The relative roughness of the tunnel can be expressed as follows:

$$\frac{D_m}{K} = \frac{1}{1 - \sqrt{\dfrac{A_n}{A_m}}} \tag{9.42}$$

The dimension of K is approximately twice the average thickness of the overbreak and therefore is a parameter similar to the diameter of the Nikuradse K_s sand grain.

- Resistance coefficient

  In the graph of Figure 9.20 [HDC (1973), Figure 9.17], resistance coefficient data are plotted from 42 projects in Sweden, Norway, Australia, Malaysia, and the USA, published since 1953.

This data was analyzed in accordance with Equation 9.41 and converted as required for tabulation and plotting using Equation 9.43, which expresses the relationship between the Darcy coefficient f, the Manning roughness coefficient and the tunnel diameter D_m.

Hydraulic conveyance design 277

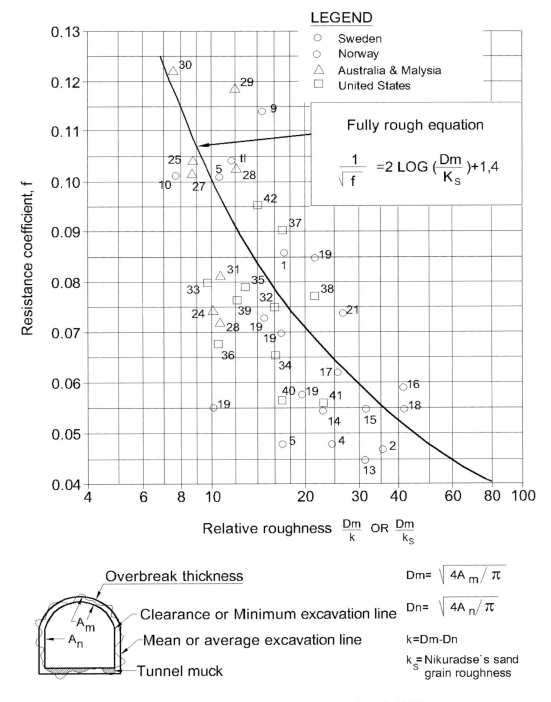

Figure 9.20 Unlined rock tunnels – Resistance coefficient (HDC, 1973).

278   Design of Hydroelectric Power Plants – Step by Step

$$f_m = \frac{185n^2}{\sqrt[3]{D_m}}$$
(9.43)

The relationship between f, D_m and the diameter of the sand grain K_s of Nikuradse, expressed by Von Kármán-Prandtl Equation 9.39, is also shown in Figure 9.20.

The data correlate well with the theoretical curve and indicate that K (Equation 9.40) is a reasonable measure of tunnel roughness. The use is conditioned by the fact that the data presented are in terms of average tunnel excavation areas – average "as built" (A_m). This aspect deserves attention because in projects, before construction, what is known is An (nominal area) as a function of design dimensions D_m, without the effects of overbreak.

9.5.3   Design application

- Preliminary design
    The mean value of the Manning coefficient in the tabulated values is 0.033 (HDC, 1973) and is based on the average area (A_m). This value can be used in preliminary design and economic analysis.
- Final design
    Once the mean area (Am) is established, the curve of the graph in Figure 9.20 can be used in the final design. An estimate of the overbreak K, or relative roughness (D_m/K), is required to assess head losses. K should be assessed on the basis of experience gained from tunnel works in geologically similar areas or on the basis of data presented in the HDC (1973).

Detailed studies of local geology, as well as contractors' experience, are useful in estimating the expected total roughness. After commencement of the work, overbreak should be measured to calculate the value of (D_m/K) to check the design assumptions.

Low values of f should be used in calculations to determine the stability of the tunnel and the pressure wave levels in the surge tank during a load rejection. High values of f should be used in calculations to determine pressure wave levels (surge) during a load take-up (as defined in the Section 9.5).

The loss of charge to be assumed in the tunnel design is an economic issue and should be understood as a portion of waived energy. It is estimated, as presented above, as a function of tunnel diameter and wall roughness. Optimal solution involves analyzing various hypotheses of alternative alignments, various tunnel diameters, with and without coating – full or partial (sections). Two aspects should be highlighted:

- the most economical section will not be dictated by the maximum permissible flow velocity, as the optimal sizing will be dictated by proper analysis of the head loss (renounced power generation), which should be shown to be at low percentage;
- the minimum tunnel section will be dictated by the most economical underground excavation method.

Hydraulic conveyance design 279

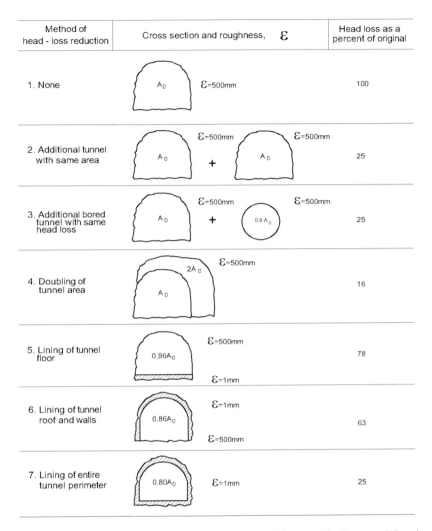

Figure 9.21 Unlined tunnels – Methods for reducing head losses (Gulliver and Arndt, 1991).

If the pressure losses are high, the following solutions are used to reduce them: tunnel lining; increased section of the tunnel; excavation of another tunnel – see Figure 9.21 (Gulliver and Arndt, 1991). For each of these alternatives, an estimate of the head loss reduction is also presented.

Lower roughness values are obtained by excavating the tunnel using the TBM. According to Gulliver and Arndt (1991), the roughness ranges from 60 to 600 mm for the conventional excavation method (drilling and blasting) and from 10 to 15 mm for tunnels excavated using TBM. Table 8.7 presents values suggested by the University of Trondheim (Norway) for the Manning coefficient. The suggested roughness Ks is observed to be between 3.0 and 4.5 mm (Table 9.9).

Table 9.9 Manning Coefficients (Stutsman, 1987; University of Trondheim, 1984)

Tunnel – Excavation Method	$A = 8\ m^2$	$A = 20\ m^2$	$A = 50\ m^2$
1-Drill and blast			
Manning coefficient (n)	0.0305	0.0295	0.0289
Manning number (M=1/n)	32.8	33.9	34.6
2-TBM (schistose rock)			
Roughness K_s (mm)	4.5	4.5	4.5
Manning coefficient (n)	0.0159	0.0162	0.0165
Manning number (M = 1/n)	62.7	61.8	60.5
3- TBM (homogenous rock)			
Roughness K_s (mm)	3.0	3.0	3.0
Manning coefficient (n)	0.0152	0.0157	0.0160
Manning number (M=1/n)	65.9	63.5	62.4

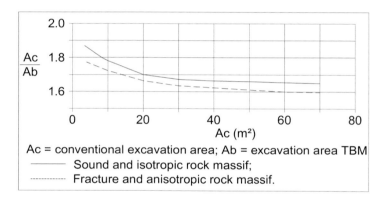

Figure 9.22 Hydraulic equivalence between unlined tunnels excavated by conventional and TBM methods (Skjeddedal, 1984).

The Darcy-Weisbach equation shows that losses increase with increasing roughness and decrease with increasing diameter.

Then there may be a hydraulic equivalence with respect to the pressure losses:

i. for a smaller diameter, tunnel with a softer surface should be obtained using TBM in the excavation; and,
ii. for a larger diameter, tunnel with the roughest surface should be obtained by conventional excavation drilling/blasting.

This equivalence occurs when the ratio between the excavated areas by the conventional process (Ac) and the process using the TBM (Ab) is in the range $1.6 < Ac/Ab < 1.8$ (Figure 9.22).

Figure 9.23 shows examples of the quality of tunneling.

Hydraulic conveyance design 281

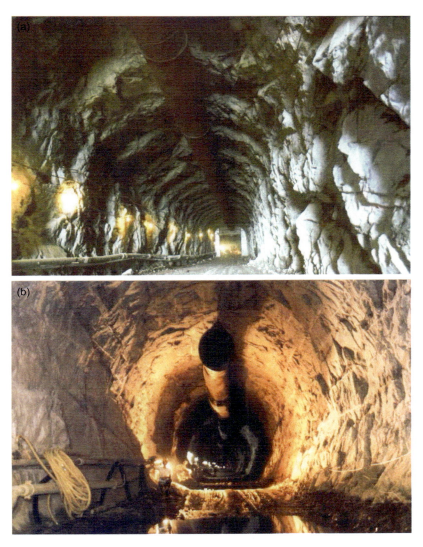

Figure 9.23 (a–b) Examples of excavation. (a) bad detonation; (b) good detonation (Hoek, 2007).

9.5.4 Assumptions for tunnel lining dimensioning

When confinement is inadequate or where the rock mass is very permeable, it will result in excessive infiltration, and it will be necessary to case the tunnel (steel or concrete lining). This measure makes the project very expensive.

The excavation progress should be accompanied by an experienced field geologist who will classify the rock mass using Barton's (1974) criterion, considering: rock type;

282 Design of Hydroelectric Power Plants – Step by Step

an evaluation of its geological-geotechnical parameters (fracture degrees, alteration, coherence and hydraulic conductivity); and analyzing the structural aspects (faults, relief joints, fracture conditions and intrusions). The types of shoring, treatment and containment in each stretch will be a function of this classification.

The dimensioning of the case thickness should consider two situations:

- the case shall fully comply with the maximum internal head load with the tunnel in operation at each point calculated by the difference between the upstream water level and the tunnel floor height, plus the overpressure coefficient = 1.3;
- the case shall additionally meet the condition prevailing in the tunnel emptying operation when the external pressures of the natural water table or the artificial sheet created by the tunnel operation act in the opposite direction, i.e., the liner crushing.

To relieve the external crushing pressures in the lined stretch, the following aspects must be considered: drainage system of the water around the section; prediction of radial injections between unlined and lined stretch.

The case shall be able to withstand the minimum crushing pressure corresponding to the injection pressure of the rock mass to shell shielding process, typically 3.0 kg/cm^2, corresponding to a 20 m water column plus a safety factor of 1.5 (20 m = 2.0 kg/cm^2). In the tunnel outlet section, the shield must be checked for maximum pressure.

9.6 SURGE TANKS

Waterhammer are transient pressure waves that occur in the penstock (or tunnel) when the valve is closed quickly, or the flow control distributor or when there is a load rejection in the system. Pressure waves are reflected back in the surge tank. For this to occur, the portion of the duct to be used in the analysis of the hydraulic transients must be between the turbine and the surge tank.

Surge tanks are used to dampen waterhammer and also to meet the frequency regulation requirements of turbine-generator sets. These two problems are all the more acute as the lengths of the ducts (L) increase, in relation to the gross head of the plant (H).

The surge tank also acts as a power reserve during a turbine load shedding and provides the water for starting the machines. The water mass is accelerated or decelerated in the duct and the pressure wave amplitudes of the circuit are then reduced by the actuation of the surge tank.

The surge tank directs the regulating characteristics of a turbine, and hence, the length of the penstock will be of prime importance in defining the starting hydraulic inertia time.

Hydraulic transients can have a devastating effect on control structures. They are estimated using computer programs that contain details of turbine response, valve or gate shutoff, junctions and so on.

9.6.1 Types of surge tanks

The surge tanks types are shown in Figure 9.24.

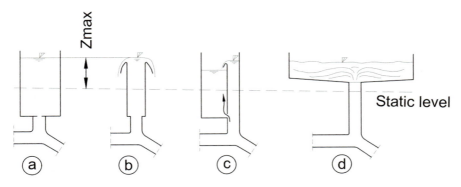

Figure 9.24 Surge tanks types (Stucky, 1958).

a. Orifice surge tank
It has a strangulation in its cylindrical base, where the hole section is smaller than the tank section and creates a significant head loss. It is cylindrical if there is no hole in the base.
b. Spillway surge tank
It is the surge tank that, by evacuating a certain volume of water, eliminates its position energy.
c. Differential surge tank
Combination of the orifice surge tank with the spillway surge tank. It is the surge tank composed of a simple surge tank on the outside and an internal surge tank whose base section is the strangulation hole area. The heights of the vertical walls will be equal, and there is still communication between the parties.
d. Expansion tank
Simple surge tank that allows water to expand in a horizontal gallery (tank).

9.6.2 Criteria used in inventory studies (Canambra)

In Inventory Studies conducted by the Canambra Engineering Company in the 1960s, a simplified criterion was used to predict the need for a surge tank. It was considered that in isolated systems requiring good frequency regulation, chimneys would be necessary if $L/H > 4$.

This criterion remained for large interconnected systems when the plant was to contribute to regulation. If the planned plant was small in relation to the system and was designed to provide additional base power, it could operate at $L/H > 10$ without a surge tank.

In Brazil, Inventory Studies CBDB/Eletrobras (2003) recommends to foresee surge tanks in the following cases: for small hydropower plants with reservoirs for WLmax when $L/H > 10:1$; for small hydroelectric plants when $L/H > 6:1$; for large hydroelectric plants when $L/H > 4:1$.

284 Design of Hydroelectric Power Plants – Step by Step

9.6.3 Canambra criteria

Micheaud's classic formula for estimating overpressure due to waterhammer suggests the inclusion of flow velocity in the duct to extend the previous criterion:

$$\Delta h = \frac{2VL}{gT_e}$$

(9.44)

Δh = net overpressure in meters of water column (m);
V = average conduit velocity for maximum flow (m/s);
L = conduit length (m);
T_e = effective closing time (s) and g = gravity acceleration (m/s^2).

Adopting T_e = 5s and g = 9.81 m/s^2, the above equation turns into $LV=25H$. Surge tank will be required if $LV>25H$. This criterion corresponds to that used by Canambra in Inventory Studies for V = 6.25 m/s, demonstrating the possibility of slowing down to lower speeds.

Waterhammer can be further reduced by increasing the machine's closing time (T_e), as specified in CBDB/Eletrobras (2003). The waterhammer can also be reduced by increasing the closing time (T_e) of the machine. However, a longer shutdown time corresponds to a higher turbine overspeed (speed increase over the nominal speed that occurs when the load is suddenly removed and the turbine continues to operate under regulator control). For satisfactory conditions of regulation of an isolated unit, it is desirable that the overspeed be maintained around 45%.

9.6.4 Rotating masses inertia

Turbine regulation conditions are closely linked to the inertia of water in the hydraulic circuit and its relationship to the inertia of the rotating masses. Due to its inertia, the turbine generator set has a flywheel effect that can be expressed by the mechanical starting time (T_s), defined as the time in seconds to accelerate the rotating mass from zero to the nominal rotation n.

$$T_s = \frac{WR^2 n^2}{67,000 P}$$

(9.45)

where:
T_s = unit acceleration time (s);
WR^2 = unit inertia effect reported by suppliers (kgf.m^2);
n = synchronous rotation (rpm);
P = full power of the unit (hp).

The higher the value of Ts, the greater the natural stability of the group. Generator and turbine WR^2 values should be obtained from equipment suppliers. As an initial guideline, the following expressions proposed by USBR (1976) may be used:

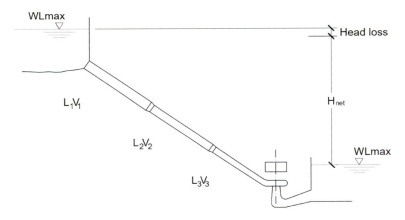

Figure 9.25 Stability control for a simple system (NIT, 2003).

turbine: $WR^2 = 1{,}000 \, (P/n^{1.5})^{1..5}$ (kgf.m^2);
generator: $WR^2 = 15{,}000 \, (KVA/n^{1.5})^{1.25}$ (kgf.m^2).

With the opposite effect, the water inertia acts in the "penstock-spiral case- draft tube" system (Figure 9.25). This inertia may be expressed by the transient hydraulic time, T_p, corresponding to the time required to accelerate the water mass from zero to the maximum velocity, V, under the action of the net head H_{net}.

This parameter, T_p, in many publications is called TW (or Tw).

$$T_p = Tw = \frac{\sum LV}{g \cdot H_{net}} \qquad (9.46)$$

or

$$T_p = Tw = \frac{(Lr + Ls + Lp) \cdot V}{g \cdot H_{net}} \qquad (9.47)$$

where:
  Lr = surge tank length (m);
  Ls = half-length of spiral case (m);
  Lp = penstock length (m);
  V = average speeds in the penstock, spiral case and draft tube at full head (m/s);
  H_{net} = net head (m).

According to the "Selecting Hydraulic Reaction Turbines", EM nº 20 USBR (1976), the units where:

$$T_s \geq 2(T_p)^2 \qquad (9.48)$$

for all plant head conditions will have good regulating conditions.

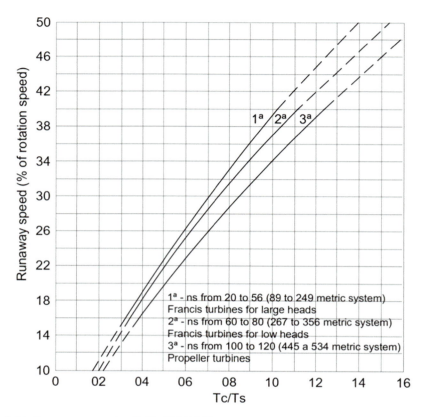

Figure 9.26 Runaway speed × Tc/Ts (CBDB/Eletrobras, 2003).

This criterion is not dimensionally homogeneous and does not refer to the servomotor operating time, on which the waterhammer and the runaway speed of the machine depend. Overspeed may be obtained from the graph in Figure 9.26 as a function of the relationship between (T_c), total servomotor action time, and (T_s), mechanical transient time, and turbine specific rotation. The total servomotor action time will be equal to the effective closing time plus 0.25–1.5 seconds:

$$T_c = T_e + (0.25 - 1.5) \text{ seg.} \tag{9.49}$$

To account for the effect of overpressure due to waterhammer, the value obtained in the graph in Figure 8.23 should be multiplied by the ratio, $(1 + T_p/T_e)$.

9.6.5 Interconnected system operation

The "waterhammer regulation" involves several complex theoretical and practical aspects, with conflicting economic repercussions that are difficult to assess. On the one hand, the greater the inertia of the turbine generator set, the better the system stability and regulation conditions, as stated in the CBDB/Eletrobras (2003).

On the other hand, an increase of inertia with respect to the minimum required by the machine characteristics will represent an increase in the cost of the generator design, cranes and powerhouse civil structures.

Increasing machine speed does not result in considerable inertia gain and is generally limited by turbine conditions normally selected to operate at high speeds near the upper limit. The larger the machine size in relation to the system, the more important the inertia requirements will be. In general, a unit that provides about 40% or more of the load should be treated as an isolated unit. Also, if the plant can be temporarily isolated from the system, for example by a transmission line accident, its operating conditions become more critical and should be considered.

Allowable frequency variations strictly depend on the type of load. The stricter the specifications, the greater the inertia required for the hydraulic circuit and the higher the costs of utilization.

In large low-head turbines with short penstocks, the distributor closing time may be limited by the non-rupture condition of the water column in the draft tube. In this case, very fast closures can rupture the water column which is followed by a positive waterhammer on the turbine's guiding blades.

Maximum depression can be assessed by Micheaud Equation 9.44, where ε ranges from 1.0 to 1.7, according to draft tube length: 1.0 for short draft tubes and 1.7 for long draft tubes. Reducing the flow velocity in the penstock reduces the inertia needs of the group, but it will increase the cost of the powerplant.

The graph in Figure 9.27 makes it possible to evaluate the frequency regulation conditions of the group, taking into account most of these aspects (Gordon, 1961): the

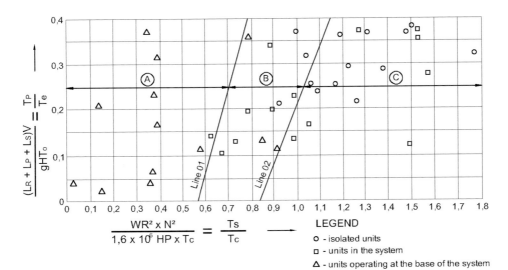

Figure 9.27 Criteria for adopting surge tanks (Gordon, 1961; CBDB/Eletrobras, 2003).

T_c/T_p ratio was placed on the horizontal axis – T_c is the total servomotor action time and T_s is the mechanical transient time; the T_p/T_e ratio was placed on the vertical axis – T_p is the hydraulic transient time and T_e the effective regulator time.

9.6.6 Surge tank need – summary

The following is a summary of a surge-tank need:

- Surge tanks should be adopted if the corresponding waterhammer reduction effect results in a more economical "adduction hydraulic circuit – turbine" alternative, or to meet the overspeed limitations required by turbine-generator set regulation;
- No surge tank will be required at the plant to the right of line 1 of Figure 9.27, considering the natural inertia of the turbine-generator set;
- If the point is between lines 1 and 2 for the group's natural inertia, no surge tank is required at the plant if its stake in the interconnected system is less than 40%, taking into account the smallest system to be met by power plant;
- If the point is to the left of line 2, a surge tank in the plant will be required, or the turbine-generator set WR^2 will be increased to allow the plant to participate in system frequency control.

The alternative without surge tank and without participation of the plant in the frequency control, if any, will be economically confronted for the selection of the final utilization scheme (increase of the tunnel section area; decrease of the closing time of the turbine distributor – T_f).

9.6.7 Minimum dimensions of the surge tank

The surge tank must satisfy the following conditions (CBDB/Eletrobras, 2003):

- Sparre condition, to ensure reflection and to provide satisfactory protection to the adduction conduct upstream of the surge tank: $A \geq a$, where A = surge tank cross-sectional area (m^2) and a = bifurcation adduction cross-sectional area with the surge tank (m^2), as defined in Figure 9.28.

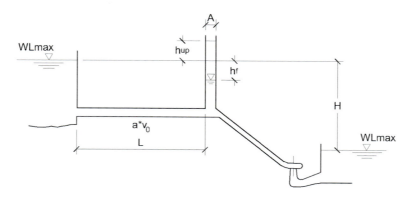

Figure 9.28 Surge tank. Thoma criteria (NIT, 2003).

- Thoma condition to ensure surge tank stability: $A >$ Amin of the surge tank being:

$$A_{MIN} = \frac{V_0^2}{2g} \frac{a \cdot L}{hf \cdot (H - hf)} \tag{9.50}$$

where:
$A =$ surge tank area (m^2); $A_{MIN} =$ surge tank minimum area – Thoma section (m^2);
$V_0 =$ flow velocity for normal flow (m/s);
$a =$ adduction duct cross-sectional area (m^2);
$L =$ adduction duct length (m);
$H =$ gross head (m);
$g =$ gravity acceleration (m/s^2);
$hf =$ head loss in the adduct duct to the surge tank for nominal flow (m).
$hup =$ surge tank level rise (m).

As a criteria condition A>Amin is adopted. If the Thoma condition is met with a small margin of safety, the issue of surge tank stability should be examined in greater detail, taking into account the head losses at the surge tank inlet and the penstock, as well as the curves of turbine efficiency. The maximum height of the surge tank level rise (hup) for a full load rejection, assuming no friction, can be estimated by the expression:

$$hup = \sqrt{\frac{aL}{gA}} \cdot Vo. \tag{9.51}$$

Stability shall also be verified by simulating finite load variations, equal to 10% and 20% of the nominal power of the group, initially operating at 80% of its nominal power. Still according to these criteria, the surge tank should always be sized for the hypothesis of rapid rejection of the maximum flow, corresponding to the total shutdown of the plant when under full load, to the maximum normal level (positive oscillation) and the minimum normal level (negative oscillation) of the reservoir.

The dimensioning for intake maneuvers will be made according to the situation of the plant in relation to the system:

- for the plant isolated or serving more than 40% of the system, the dimensioning should be to predict the rapid opening of zero to full flow, considering the minimum reservoir level;
- for power plant in interconnected system, responsible for less than 40% of the load, the design shall provide for the rapid opening of 50%–100% of the maximum flow, considering the WLmin.

9.7 POWERHOUSE

In HPPs, basically two types of structure have been used: external powerhouse, which may be indoor or outdoor, and underground powerhouse. The standard design types are shown in Figure 9.29. In the powerhouse, areas should be provided for: auxiliary electrical and mechanical equipment (electrical and mechanical galleries); erection

290 Design of Hydroelectric Power Plants – Step by Step

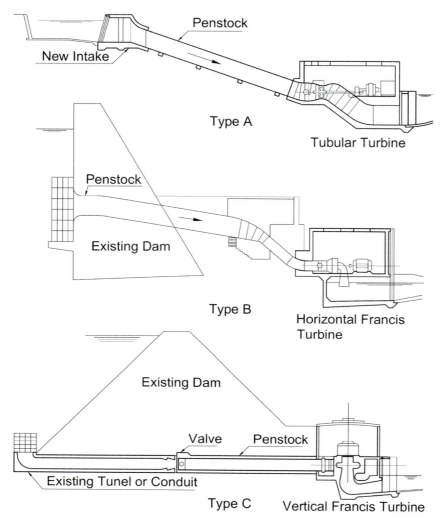

Figure 9.29 (a)- Standard design types (Gulliver and Arndt, 1991). Type A-Canal drop layout: will occur where an existing site contains a drop localized. Type B and C–Concrete or embankment dam layout: may be constructed downstream of the dam. (b) Standard design types (Gulliver and Arndt, 1991). Type D – Dam/powerhouse type layout: construction of the dam and powerhouse simultaneously.

(Continued)

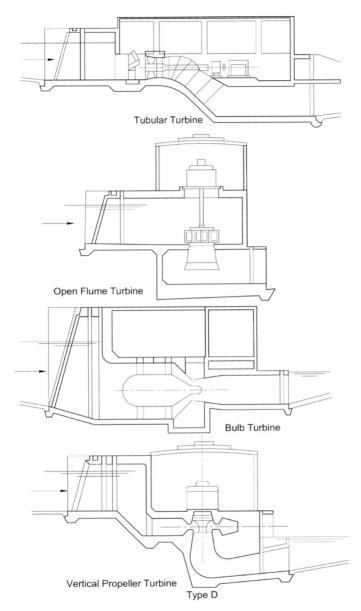

Figure 9.29 (Continued) (a)- Standard design types (Gulliver and Arndt, 1991). Type A-Canal drop layout: will occur where an existing site contains a drop localized. Type B and C—Concrete or embankment dam layout: may be constructed downstream of the dam. (b) Standard design types (Gulliver and Arndt, 1991). Type D – Dam/powerhouse type layout: construction of the dam and powerhouse simultaneously.

292 Design of Hydroelectric Power Plants – Step by Step

bay, whose dimensions must be defined by the equipment supplier; and COU-plant operation center. The transformers may be installed inside or outside the powerhouse, depending on the particularities of each case.

The main elevations of the powerhouse are defined taking into account the remarkable downstream and submergence water levels of the turbine and, consequently, the draft tube, which determines the definition of the powerhouse foundation level and the drainage galleries. The quality of the flow rating curve is of utmost importance in fixing these elevations, such as the transformer floor level.

The clearance of the cross-section should be as small as possible and therefore the span of the overhead cranes or gantries. However, the location of the downstream wall is fixed by the size of the spiral case. It is convenient to leave a passage between the generator shaft and the superstructure pillars.

The height of the superstructure or gantry crane depends on the maximum height of the above-ground winch hook, required so that the largest part (turbine rotor with shaft or generator rotor) can climb over already assembled machines or in operation.

For the assembly/disassembly the designs must consider the beams for the definitive crane(s). It is noted that some equipment and parts may be supplied pre-assembled.

Stairs should be designed in comfortable proportions and shapes. Workers often have to rush through them, carrying heavy and cumbersome parts in emergencies. Spiral and sailor ladders should only be designed for places where there is no proven option. External access should be defined considering the main floor level of the assembly area, taking topographic aspects of the site into account along with admissible ramps for the transport equipment and the layout of the downstream works (e.g., substation).

The stability of the structure should be checked for current loading cases. The downstream wall sizing should, in some cases, consider the support of the transmission line output structure.

The following figures show some examples of powerhouses, standard designs used routinely around the world. It's worth repeating that the author's main objective is to illustrate the theme for young engineers in the beginning of their profession (Figure 9.30).

Balbina HPP, 250 MW. Uatumã river, Amazon Region (Figure 9.31).

Itumbiara HPP, 2,082 MW. Paranaíba river. Minas Gerais, Southeast region (Figure 9.32 and 9.33).

Emborcação HPP, 1,192 MW. Paranaíba river, Minas Gerais, Southeast region (Figures 9.34–9.36).

9.7.1 Outdoor powerhouses

9.7.1.1 Powerhouse at the foot of the dam

Tucuruí HPP (Tocantins river, Amazon region)

The first stage powerhouse of Tucuruí HPP, Figure 9.37a at the bottom, with 3,960 MW, houses 12 Francis turbines of 330 MW each. The cross section is presented in Figure 9.2.

The second stage, Figure 9.37a in front, 4,125 MW, houses 11,375 MW machines. The length of the two powerhouses is approximately 1,000 m.

Hydraulic conveyance design 293

Figure 9.30 Balbina HPP. Indoor powerhouse.

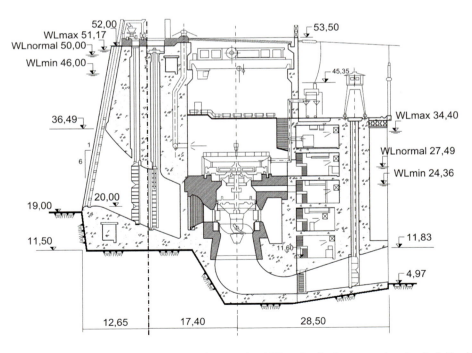

Figure 9.31 Balbina HPP. Indoor powerhouse. $L = 160$ m (including erection bay). 5 Kaplan turbines 50 MW each (CBDB, 2000).

Figure 9.32 Itumbiara HPP (2082 MW), Paranaíba river. Owner – Furnas. Operarating since 1980 (unit 01).

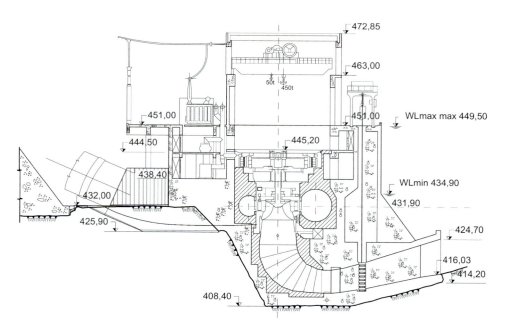

Figure 9.33 Itumbiara HPP. Indoor powerhouse in the foot of the dam. $L = 189$ m (including erection bay). 6 Francis turbines 380 MW each (CBDB, 2000).

Hydraulic conveyance design 295

Figure 9.34 Emborcação HPP. 1,192 MW. Paranaíba river. CEMIG (CBDB, 2000). Operarating since 1982.

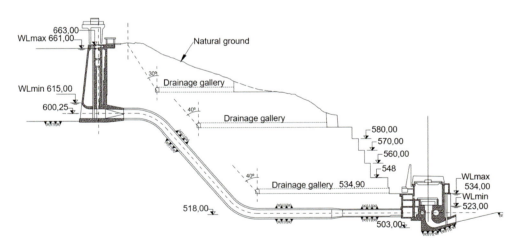

Figure 9.35 Emborcação HPP – Paranaíba river. Powerhouse downstream from the dam. Powerhouse length = 96 m, excluding the erection bay (CBDB, 2000).

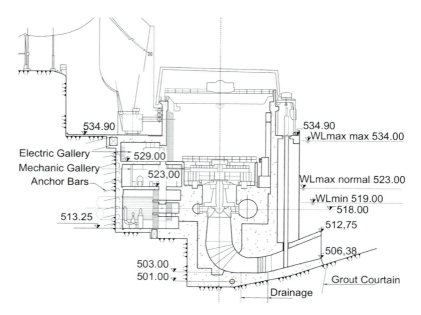

Figure 9.36 Emborcação HPP. Powerhouse detail (CBDB, 2000). $L = 96$ m (excluding erection bay).

Figure 9.37 (a–d) Tucuruí HPP powerhouse.

9.7.1.2 Powerhouse as part of the dam

Karakaya HPP, Euphrates river (Turkey)

Powerhouse, incorporated to the dam, with an installed capacity of 1,800 MW, housing six Francis 300 MW machines. The powerhouse is under the spillway (ski jump to $Q = 18,000$ m^3/s). The dam is a gravity concrete with a maximum height of 180 m. and volume $\sim 2 \times 10^6$ m^3 (Figures 9.38–9.40).

9.7.1.3 Powerhouse downstream of the dam

Barra HPP. Brazilian Aluminum Company (CBA)

The Barra HPP, in the following figures, is located on the Juquiá-Guaçu river, in Tapiraí, Ribeira do Iguape river basin, approximately 120 km from the city of São Paulo.

The construction of the plant took place from 1982 to 1986. The reservoir, with the WLmax at elevation 402 m, floods an area of 2.0 km^2. The live storage is around 25.5×10^6 m^3 for a depletion of 17 m.

The Barra HPP has only one 40 MW Francis machine designed for a gross head of 100 m and turbine flow of 33.10 m^3/s 85% of the time. The minimum gross head is 79 m.

The useful capacity for flow regularization is 25.5 10^6 m^3. Flow is returned to the river by a 410 m long tailrace tunnel of 4.80 m diameter (Figure 9.41).

In the longitudinal profile of the circuit, after the powerhouse, there is a surge tank (25 m long and 4.8 m diameter).

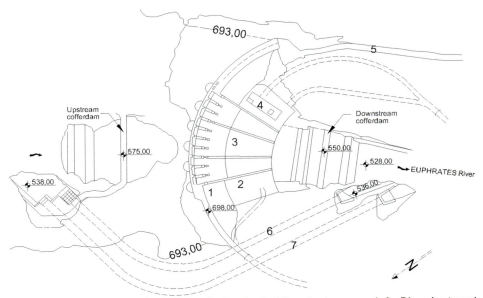

1 - Dam 2 - Power house 3 - Spillway 4 - Service building 5 - Acess road 6 - Diversion tunnel, bottom outlet after construction 7 - Diversion tunnel, closed after construction

Figure 9.38 Karakaya HPP. Plant (Stutz, 1979).

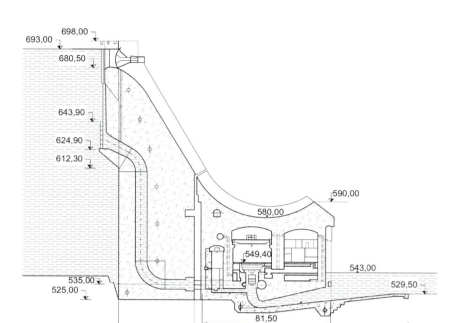

Figure 9.39 Karakaya HPP. Typical section.

Figure 9.40 Karakaya HPP.

Hydraulic conveyance design

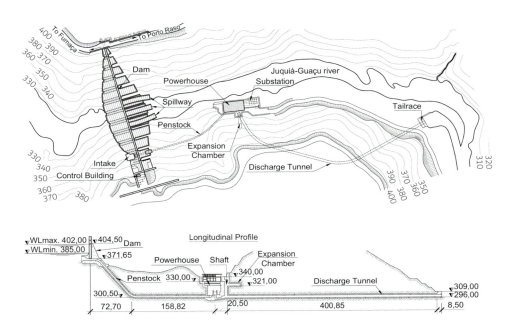

Figure 9.41 Barra HPP (CBA, 1993).

The dam is a gravity-relieved dam with elevation crest 404.50 m, 10 elements, 258 m long and 94.85 m high. The abutment blocks are of massive gravity (Figure 9.42).

The spillway, with capacity of 800 m³/s, has three 13.50 m × 3.50 m tilt gates. The sill is at elevation 398.50 m. The bottom outlet, located in the fifth element of the dam, has a capacity of 250 m³/s. The sill is at an elevation of 322.30 m and has a floodgate of 2.40 × 3.00 m (Figure 9.43).

9.7.2 Underground powerhouses – examples

Serra da Mesa HPP, Tocantins river (Goiás State).

The layout was presented in Chapter 3. The underground powerhouse has three 400 MW Francis machines each. The WLmax is at El. 460.00 m; and the tailrace WL is at El. 334,50 m. The gross head is 125.5 m. The structure is 149 m long, 29 m wide and 70 m high. The surge tank is 69 m long and 77 m high. The pressure tunnels have a diameter of 9 m and a length of 126 m. The tailrace tunnel is 500 m long (Figures 9.44 and 9.45).

CERAN Complex, Antas river (Rio Grande do Sul State, South of Brazil)

The CERAN Antas river Energy Complex is composed by Castro Alves HPP, Monte Claro HPP and 14 de July 14 HPP. Total installed capacity is of 360 MW. The proprietary consortium is formed by CPFL, CEEE and DESENVIX.

While the Monte Claro HPP began operations in December 2004, the other two plants began operating in December 2008. The main characteristics of the powerhouses are shown in Table 9.10 (Figure 9.46).

Figure 9.42 Barra HPP (40 MW) Downstream view. Note spillway

Figure 9.43 Barra HPP Powerhouse.

Hydraulic conveyance design 301

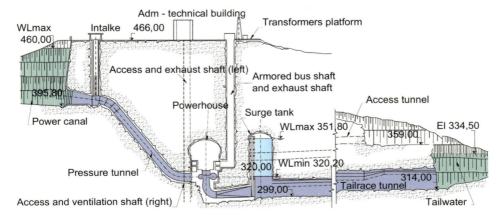

Figure 9.44 Serra da Mesa HPP. Underground powerhouse.

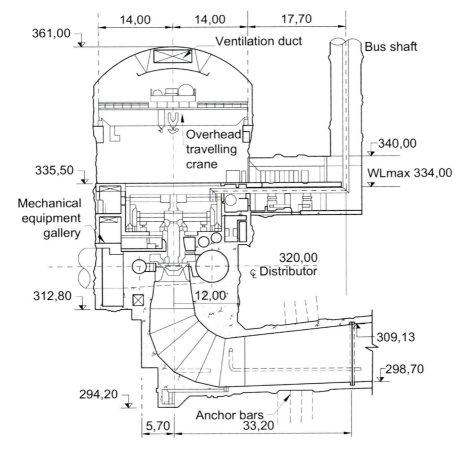

Figure 9.45 Serra da Mesa HPP. Underground powerhouse (CBDB, 2000).

302 Design of Hydroelectric Power Plants – Step by Step

Table 9.10 CERAN Complex (CBDB, 2009)

	Castro Alves	Monte Claro	July 14
WLmax reservoir (m)	246.36	156.50	110.15
WLnormal reservoir (m)	240.00	148.00	104.00
WLmax. downstream (m)	148.00	106.60	70.60
NAmin. downstream (m)	-	104.68	68.40
WLmax. max. downstream (m)	161.30	131.00	90.82
Powerhouse – type	underground	shaft	underground
Block width (m)	12.10	22.00	20.00
Chamber length (m)	14.70	20.10	22.40
Penstocks – quantity	3	1	2
Penstock diameter	4.00	12.50×12.50	8.80×9.75
Average length (m)	180	1,140	220
Turbines – type	3 Francis	2 Kaplan	2 Kaplan
Turbines – power (MW)	44.6	67.1	51.8
Rotation (rpm)	300	150	171.40
Q (m³/s)	58.5	191.4	180.3
Reference head (m)	83.6	37.7	31.7

9.8 TAILRACE

At the turbine outlet there is the draft tube and downstream there is the tailrace channel (or tunnel) through which the effluent flow is returned to the river (see Figures 9.1, 9.44, 9.47–9.49, 9.52 and 9.53). The tailrace channel has free surface. In some cases of tunnel, this may be pressurized if the turbine requires back pressure to prevent cavitation, as in Paulo Afonso IV, São Francisco river (Figure 9.49) (Figure 9.50 and 9.51).

Depending on the length of the draft tube, it is sometimes necessary to introduce a surge tank downstream to ensure that transient flows at the distributor opening or closing do not cause problems in the generator circuit, as in the case of Serra da Mesa HPP, Tocantins river – Figures 9.44 and 9.45.

The length of the channel is highly variable and depends on the type of layout. Proper hydraulic design of the transition from draft tube to tailrace channel is important to minimize pressure losses in mechanical turbine design.

The geometry sizing of this channel should be performed using the Manning formula (Section 8.1). The design will be conditioned by the type and dimensions of the powerhouse as well as the distance to the river. The flow in the channel for the maximum turbine flow should be fluvial, with low speed.

For free surface channels, the width is commonly variable along their length. At first the width will be equal to that of the powerhouse. At the confluence with the river, the width at the end of the canal should be large enough not to introduce any control over the flow.

The slope of the channel will also vary depending on the difference in elevation between the bottom of the draft tube and the bottom of the river. Initially, gentle ascending ramps such as 1 (V): 6 (H) or 1 (V): 10 (H) should be adopted according to the draft tube geometry.

Hydraulic conveyance design 303

Figure 9.46 (a–b) July 14 HPP. (CPFL, 2007). Powerhouse excavation. Note arrival of the tunnels.

304 Design of Hydroelectric Power Plants – Step by Step

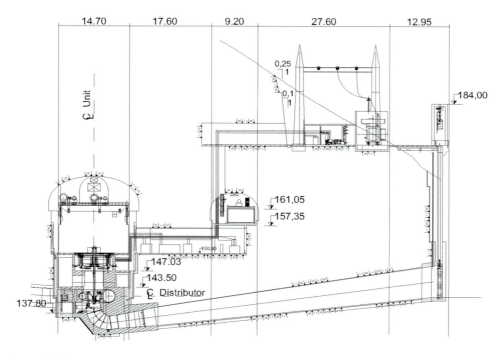

Figure 9.47 Castro Alves HPP – Underground powerhouse (CBDB, 2009).

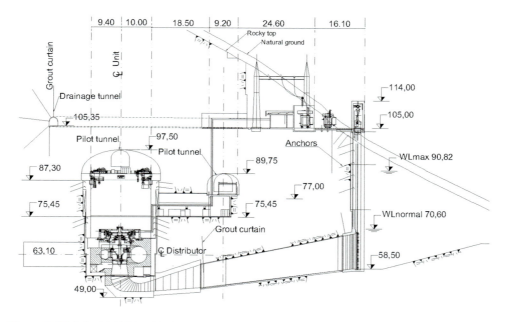

Figure 9.48 July 14 HPP – Underground powerhouse (CBDB, 2009).

Hydraulic conveyance design 305

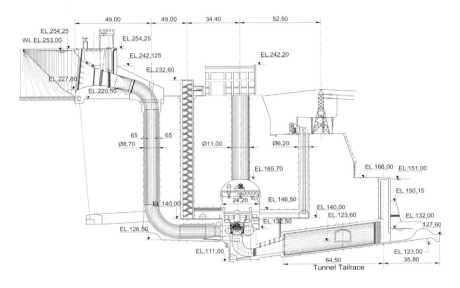

Figure 9.49 Paulo Afonso IV HPP. Underground powerhouse. (MBD, 1982).

Figure 9.50 Paulo Afonso IV HPP. Underground powerhouse (FGV).

Figure 9.51 Paulo Afonso IV HPP. Underground powerhouse (Cetenco Eng. S. A.).

Figure 9.52 Tucuruí HPP. Powerhouse 1 and 2 and tailrace channel. Power output: 8,125 MW: 12 units of 330 MW, 2 units of 20 MW and 11 units of 375 MW. Ski jump spillway – 110,000 m³/s: 23 radial gates 20 × 21 m. $L = 560$ m. H dam = 80 m. L dam structures = 8,100 m.

Figure 9.53 Flaming Gorge HPP – 3 units; Power output 150 MW. Green river. Utah. Tailrace channel. Construction: USBR 1958–1962. $H = 153$ m. $L = 392$ m. Tunnel spillway – 820 m^3/s.

Chapter 10

Mechanical equipment

10.1 GATES AND VALVES

The construction of gates originated from irrigation, water supply and river navigation work. In ancient times, water was dammed by small dams and diverted to irrigation canals. Excess water was discharged over the dam.

Movable dams were built with movable gates which could be opened to give way to excess water, allowing greater flexibility of operation to hydraulic works.

This book addresses the gates used in the headrace canal (or lower pressure tunnel) and spillways of hydroelectric plants. Only the basic characteristics important for selecting these devices will be presented. For details, it is recommended to consult the manufacturers and suppliers.

10.1.1 Preliminary considerations

Headrace canals, low-pressure tunnels, and high-pressure tunnels typically use gates and bottom valves to protect the penstocks and turbines. Spillways are usually provided with surface gates, but there are bottom outlets. Prior to device selection, basic hydraulic requirements such as maximum load and discharge must be defined. These requirements provide the basis for determining the dimensions and number of conduits required.

The dimensions of the penstock, or penstocks, are the starting point for the selection of gates and valves, as the various types have maximum load limitation (maximum head). The load limitation may be due to seals or possible hydraulic problems (cavitation trends and damage, head losses, required maintenance and cost). Size is usually limited by manufacturing and transportation problems.

Even if the turbines have their own closing devices (distributor), it is prudent to provide for another circuit closing mechanism. The USBR (1976, 1974) has as a rule to provide at least two separate conduits, each with a gate or gate valve. This arrangement, though is more expensive, ensures a high degree of freedom to control discharges. In addition, it enables maintenance and repair work without stopping operation when the gate becomes inoperative (which is what happens most often).

The conditions under which the conduits will discharge must also be established in the gate or valve selection process. These conditions include determining whether the discharge will be thrown into the air (ski jump) or submerged, when a dissipation basin may be required.

DOI: 10.1201/9781003161325-10

Besides that, the degree of spray that adjacent electrical installations can tolerate must be determined. For conduits with air discharge gate and valve designs, we need to be sure that they will never be submerged when operating the plant (turbined flows increase the downstream level and, if there is an error in the tailwater curve, can drown the outlet of the conduit).

In many cases, the discharge conduit of a dam is a key aspect of safe structure operation. For this reason, gates and valves for these conduits must be selected and arranged to operate safely hydraulically. Security is paramount and cannot be compromised by the economy.

After these considerations, the following is a summary of the important design points for the gates and valves. The terminology used in describing the various types of closure is presented below.

- Gate: it is a device composed of a board, with stop and wake, that moves to control the flow. Contains fixed parts and maneuvering mechanism. The seals fixed to the wall are usually made of rubber profiles.
- Valve: is a closing device in which the element of this action remains axially fixed with respect to the flow path and is rotated or moved longitudinally in order to control the flow.
- Gates or Guard Valves: these devices operate fully open or closed and act as a secondary safety device to cut the flow in case the primary device becomes inoperable. Guard gates are usually operated under pressure-balanced, non-flow conditions except for emergency closures. Gates and regulating valves can be used to control reservoir outlet flow or to control transient pressures that occur during turbine operation.
- Throttling Gates or Valves: these devices operate under high pressure to throttle the flow and control discharge rates, which are freely discharged, or at relatively low backpressures. This does not include high throttle back pressure. They can also control transient pressures during turbine operation.
- Bulkhead Gates or Stoplogs: these are installed at the entrance of the hydraulic circuit to isolate it for inspection or maintenance, and are always operated in pressure balance. They cannot be considered as guard gates.

Many gates and valves designs have been prepared over the years. The strongest (rugged) and simple are usually the best. Some commonly used gates and valves with proven safety are shown in the figures tables that follow. In these tables, a brief summary of the characteristics of each one of them is provided, by the type of function to be performed in the outlet: flow regulation (throttling gates or valves) or guard (guard gates or valves). Gates and valves with high flow coefficients and hydraulic efficiencies deserve the first consideration, but selection should be made considering all other factors listed.

10.1.2 Gates

Initially it should be noted that for detail on this matter it is recommended consulting Davis (1952) and Erbisti (2004). It should be noted also that each country has its norms and standards to be obeyed.

10.1.2.1 Types of gates

There are several types of gates: flap, cylinder, stoplog, slide, caterpillar, miter, roller, segment/radial, sector, stoney, drum, bear-trap, fixed-wheel and visor gates, as illustrated in Figures 10.1–10.4 (Davis, 1952). There are also fuse gates, a simple, robust and safe system to increase storage and daily generating capacity as well as spillway capacity (Figures 10.5 and 10.6).

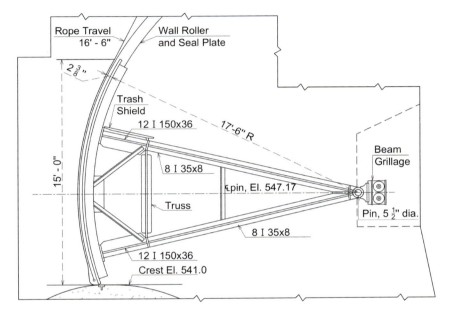

Figure 10.1 Tainter or radial gate (Davis, 1952).

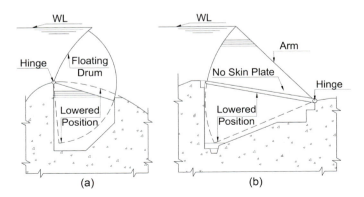

Figure 10.2 Drum gate (Davis, 1952).

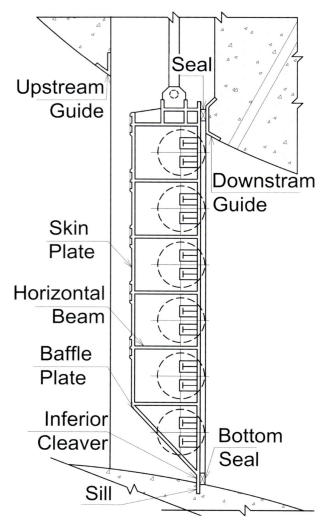

Figure 10.3 Sliding gate Fixed-wheel gate/bulkhead gate (Davis, 1952).

10.1.2.2 Gate classification

According to their features, the gates may be grouped based on various manners. Among others, the following classification criteria may be listed: purpose, movement, water passage, leaf composition, location and skin plate (Table 10.1).

10.1.2.3 Selection of the type of gates

The choice should be based on an analysis of all factors capable of influencing equipment performance, cost, quality and reliability, such as: operational safety; lower

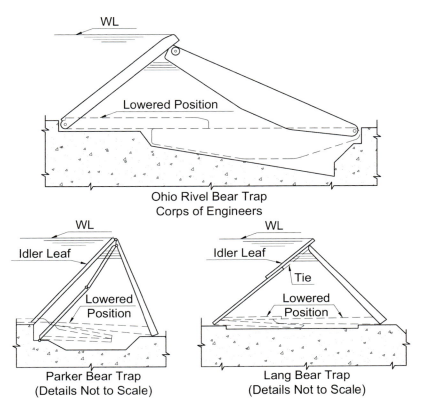

Figure 10.4 Bear-trap gate (Davis, 1952).

supply weight; simplicity of operation; ease of maintenance; structural requirements (slots, chambers, guides, etc.); magnitude and direction of the forces transmitted to the concrete; capacity of the maneuvering mechanism; and ease of transport and assembly.

In the selection of the gate, we also consider the experience acquired in projects carried out, as well as the experience of the manufacturers. Table 10.2 lists the applicable types of gates by project structure.

10.1.2.4 Usage limits

The continuous development of hydropower plants in the last 50 years of the last century has caused the application limits of each type of gate to increase considerably (span, height and hydrostatic load).

For example, at Itaipu HPP, 14 segment gates with 20 m span and 21.34 m height were installed, designed for a water height of 20.84 m in relation to the sill (see Figure 10.7).

At Tucuruí HPP, 23 gates of 20 m span and 21 m height were installed (see Figure 10.8).

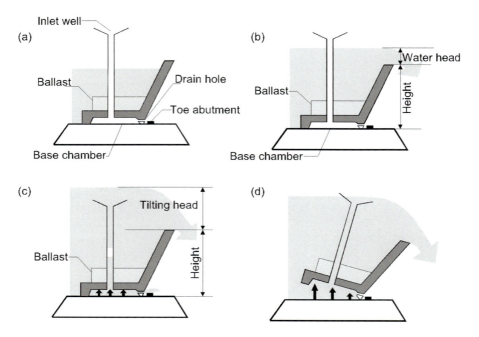

a) The water level is increasing;
b) Fusegate acts as an ungated spillway;
c) The water head reachs to the predetermined elevation (design tipping head), and
d) The fusegate tips.

Figure 10.5 Fuse gate (Alla, 1996).

The current limits of gate size and heads are presented by Erbisti (2004) through a series of tables and graphs, head on sill×gate area and gate height×gate span, as follows:

- high-pressure gates: segment, fixed-wheel, caterpillar and slide
    head on sill up to 250 m×gate area up to 250 m^2;
- underflow spillway gates: slide, roller, fixed-wheel, Stoney and segment (or radial)
    gate height up to 25 m×gate span up to 60 m;
- overflow spillway gates: bear-trap, flap, sector and drum
    gate height up to 15 m×gate span up to 60 m;
- underflow and overflow spillway gates: fixed-wheel with flap, segment with flap and double-leaf fixed-wheel
    gate height up to 20 m x gate span up to 60 m.

10.1.2.5 Outlet discharge coefficients

- Penstock throttling gates
    The discharge coefficients of the throttling gates vary between 0.6 and 0.8; for guard gates, they range from 0.9 to 1.0 (Kohler, 1969). These coefficients vary

Mechanical equipment 315

Figure 10.6 Fuse gate. Shongweni dam. South Africa (Hydroplus). Spillway width = 126.2 m; new capacity = 5,000 m³/s; quantity of fuse gates = 10; gate height = 6.5 m; width = 9.73 m; increased discharge capacity = 235%.

Table 10.1 Gate Classification

Classification	Type	Utilization
According to the function	Service	Service gates are used to permanently regulate flow or water levels. Examples: spillway gates; bottom discharge gates; lock gates; and automatic flood control gates.
	Emergency	Emergency gates are used sporadically to interrupt water flow in conduits and channels; they are generally designed for normal operation in an open or closed position. Examples: intake gates; gates installed upstream of penstocks service valves; Kaplan turbine draft tube gates; and outlet service gates.
	Maintenance	Maintenance gates are operated with standing water only, and their main function is to allow the penstock or channel to be emptied for proper access and maintenance of the main equipment (turbines, pumps or even other gates). Example: stoplog.

(Continued)

316 Design of Hydroelectric Power Plants – Step by Step

Table 10.1 (Continued) Gate Classification

Classification	Type	Utilization
According to the movement of the gate	Translation	Sliding: slide, stoplog and cylinder. Rolling: fixed-wheel, caterpillar and Stoney.
	Rotation	Rotation gates: flap, miter, segment, sector, drum, bear-trap and visor.
	Translation-rotation	Roller is the only gate that performs a combined motion of rotation and translation.
According to the water passage in relation to the leaf position	Over the leaf–flap	Flap, sector, bear-trap and drum gates; in the opening operation, they move down around the articulation axis located on the sill, permitting the water passage over the gate.
	Under the leaf–slide	Slide, caterpillar, roller, segment, fixed-wheel, visor and Stoney gates move upwards, making possible the flow of water under the gate.
	Over and under the leaf	Mixed and double gates permit discharge alternately over and under the leaf, according the operational requirements.
According to the gate leaf composition	Plain	The leaf has only one element.
	Mixed	The main leaf has, at the top, a flap gate. Many applications are known of segment, fixed-wheel, roller and Stoney combined with flap gates, mainly in Europe.
	Double	The leaf comprises two movable elements. The lowering of the upper element permits discharge over the plate, while the lower element can be lifted to discharge as an orifice. Both elements are raised for passage of the maximum flow. Fixed-wheel and segment gates are the only known types of double-leaf gates.
According to the location	Surface	All types of gates can be used on crest works
	Bottom	Only a few gates are applied on submerged installations, namely fixed-wheel, segment, caterpillar, slide, stoplog, cylinder and Stoney. According to the water over the sill, gates are usually classified as: • low-head gates: up to 15 m; • medium-head gates: from 15 to 30 m; • high-head gates: over 30 m. However, the criteria based on the water head are subjective and change according to the technological evolution.
According to the skin plate shape	Plain	Slide, caterpillar, fixed-wheel, Stoney, stoplog and bear-trap gates.
	Radial	Segment, sector, drum, visor, cylinder and roller gates. Flap and miter gates may have a flat or curved skin plate. Reverse segment gates, very common in Germany, most times have a flat skin plate.

Table 10.2 Applicable Types by Structure/Objective

Structure	Gate Type
Intake	Fixed-wheel, slide, caterpillar, segment and cylinder.
Spillway	Segment (compression or traction), flap, fixed-wheel, sector, drum, segment with flap, fixed-wheel gate with flap and double-leaf fixed wheel hook gate.
Outlet	Fixed-wheel, slide, caterpillar and segment.
Lock gates	Miter, slide and segment (with trunnions mounted on horizontal and vertical axes)
Lock aqueducts	Fixed-wheel, slide and segment traction.

Figure 10.7 Itaipu spillway: 14 segment gates, 20 m span × 21.34 m height.

according to the specifics of each project as they depend on the characteristics of the approaching and leaving conduit flow lines, which depend on the shape of the crest and the position of the trunnion (Figure 10.9A).

- Conduit Tainter gates

For the discharge coefficients of bottom conduits Tainter gates (or radial, or sector), see Figure 10.9B (HDC 320-1). The curves shown are based on the equation given by R. von Mises, obtained from the Garrison tunnel model. For details, see HDC (1973) and USBR – DSD (1974).

318 Design of Hydroelectric Power Plants – Step by Step

Figure 10.8 Tucuruí spillway: 23 segment gates, 20 m span × 21 m height.

In Figure 10.10 and in the Table 10.3, the terms "bonneted and unbonneted" gates appear. Bonneted slide gates are used to regulate runoff on outlet works. Encapsulated, they are designed and manufactured to be embedded in concrete (except switching equipment). They are often used in conjunction with a downstream gate and an upstream emergency shutter. They have been manufactured with maximum dimensions of 3 m × 3 m and are used for high heads – over 150 m (Rodney Hunt) (Figures 10.11 and 10.12).

Figure 10.13 shows the head loss coefficients for partially open circular and rectangular gates (Miller, 1978). For other gates it is recommended to consult.

10.1.2.6 Discharge coefficients – spillways segment gates

The flow coefficient depends on the characteristics of the approaching and leaving orifice flow lines, which depend on the shape of the crest, the gate radius and the position of the trunnion.

Figure 10.14 (HDC chart 311-1) presents a series of model and prototype data for various spillways crest shapes and radial gate designs for non-submerged flows.

The data presented are based mainly on tests with more than three spans in operation. Discharge coefficients for a single span should be lower because of lateral contractions. At the time, they had no data available to present. For details, see HDC (1973).

It is recommended to consult Erbisti (2004) for more information on the other gate design items: stress and strain, maneuvering, drive systems, building materials, fences, fabrication, transportation and assembly.

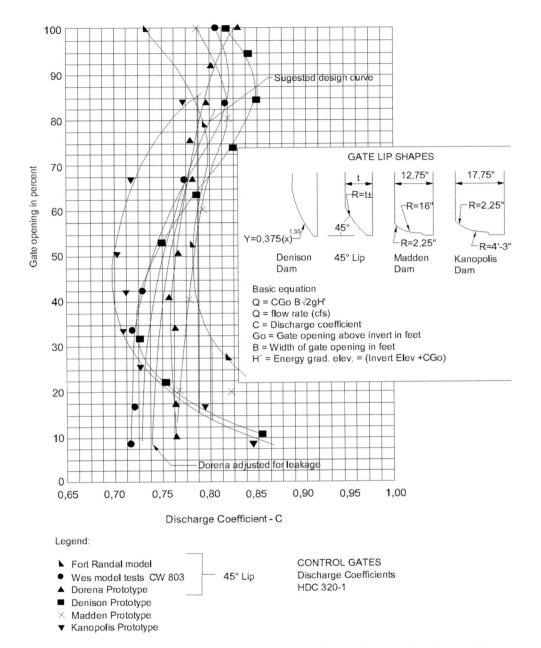

Figure 10.9A Control gates. Discharge coefficients. Slide and caterpillar (tractor) gates (HDC, 1973).

320 Design of Hydroelectric Power Plants – Step by Step

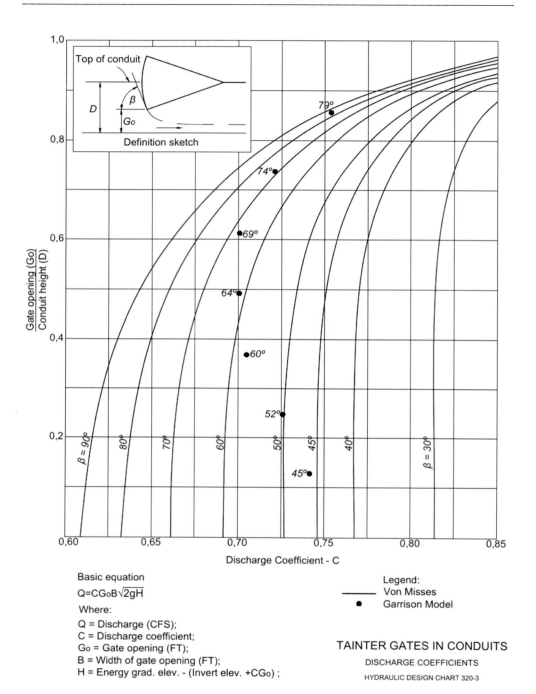

Figure 10.9B Tainter gates in conduit. Discharge coefficients (HDC, 1973).

Mechanical equipment 321

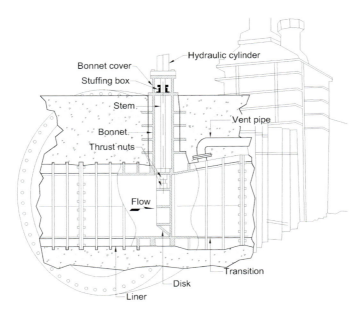

Figure 10.10 Bonneted slide gate (Rodney Hunt).

Figure 10.11 Bonneted gate (Rodney Hunt).

Figure 10.12 Bonneted gate detail.

10.1.3 Valves

In this item, the valves are described briefly, without any detail. Needle valves are used at the outlets exits to control discharges normally under extremely high loads. They are designed to discharge into the air, thereby eliminating the opportunity for cavitation within the penstock. Although they have been used in a number of dams, such as the Hoover dam in the Colorado river (Nevada/Arizona). They have been superseded by more economical and efficient valves, such as cone-fixed and hollow jet valves.

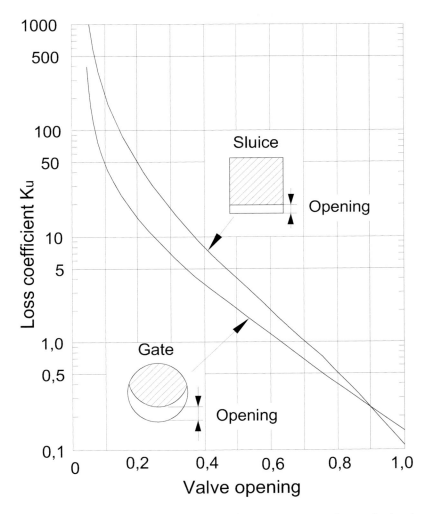

Figure 10.13 Head loss coefficients for partially open rectangular and circular gates (Miller, 1978).

Fixed-cone valves, or Howell-Bunger, are widely used to flow regulation. In addition, they are a very good energy dissipator and very good aerator. The hollow-jet valve is essentially half a needle valve, with the needle turned in such a way that it moves upstream on closing. The ball or spherical valve (Figure 10.15) consists of a large sphere in a housing with a cylindrical bore of the same size as the duct. When the valve is open, the cylindrical bore is aligned with the duct. Turning the ball 90° closes the valve. The head losses on a ball valve are negligible, while the costs are generally higher than those of a butterfly valve. The butterfly valves (Figures 10.16 and 10.17) are the most used as emergency valves upstream of turbines, where penstocks are long, or where emergency shutoff valves are used in outlet works. The two dominant valves can be compared as shown in Table 10.3. For details, see Tables 10.4–102.6 (Kohler, 1969).

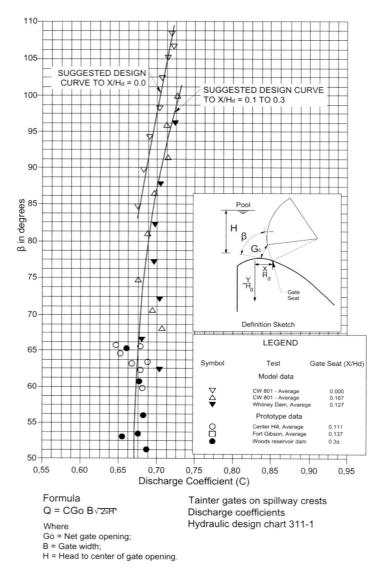

Figure 10.14 Tainter gates on spillway crests. Discharge coefficients (HDC, 1973).

The example adopted at the Funil HPP, 216 MW, Grande river, in the municipality of Resende, Rio de Janeiro, is shown below. The Funil dam was built between 1961 and 1969 and it is 85 m high and 385 m long at the crest. The gross head is 77.83 m on the intake sill. The intake has three caterpillar gates, 4.50 m wide and 6.20 m high. The plant has two tunnel spillways, one on each bank, with 4,400 m^3/s of total discharge capacity. The plant has also a 3.50 m diameter Howell Bunger valve (Figure 10.18), with a flow capacity that varies from 190 m^3/s to 285 m^3/s.

Mechanical equipment 325

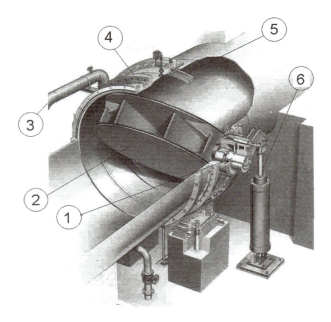

Figure 10.15 Ball or spherical valve with rotor (Vinogg, 2003): (1) valve housing, (2) valve rotor, (3) maintenance seal, (4) main seal and (5) ring piston servomotor.

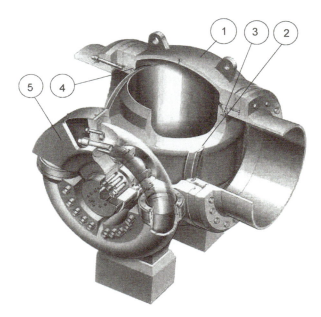

Figure 10.16 Large biplane butterfly valve (Vinogg, 2003): (1) trunnion assembly, (2) biplane disc (lattice), (3) pipe for by-pass, (4) valve housing with main seal, (5) movable seal assembly and (6) servomotor.

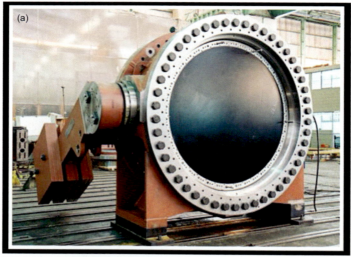

Figure 10.17 Butterfly valves: (a) closed valve and (b) open valve in assembly.

Table 10.3 Comparison of Dominant Guard Valves (Vinogg and Elstad, 2003)

	Butterfly Valve	Spherical Valve
Usual head (H) range (m)	Up to 200–300 m	Up to the highest heads
Usual diameter (D) range (m)	Up to 5.0–6.0 m	Up to 3.0–4.0 m
Max. relative size H (m) and D (m)	Up to 1,200 (1,500)	Up to 2,000 (2,500)
Head loss coefficient	0.2, 0.5	0.02, 0.05
Discharge coefficient[a]	0.7, 0.8	1.0

[a] The coefficients are approximate and may vary depending on the characteristics of the projects. It is recommended to consult additionally the HDC, Charts 330 (1973).

Table 10.4 Throttling Gates (Kohler, 1969)

Service Classification	Throttling Gates				
Schematic diagram					
Name	Unbonneted slide gate	Bonneted slide gates		Jet-flow gate	Top seal radial gate
		High pressure type	Streamlined type		
Max head (approx.)	75'	200'	500'	500'+	200'–500'
Discharge coefficient (a)	0.6–0.8	0.95	0.97	0.80–0.54	0.95
Submerged operation	No	No	Yes (1)	Yes (1)	No
Throttling limitations	Avoid very small discharge	Avoid very small discharge	Avoid very small discharge	None	None
Spray	Minimum	Minimum	Minimum	Small	Minimum
Leakage	Small	Small	Small	None	Small to moderate
Nom. size range (b)	To 8' wide & 12' high	To 8' wide & 9' high	To 10' wide & 20' high	10'–20' dia.	To 15' wide & 30' high
Availability	Commercial std. (1)	Special design	Special design	Special design	Special design
Maintenance required	Paint	Paint	Paint (1)	Paint	Paint and seals (1)
Notes	1. Gates are readily available from several commercial sources. They are not on off-the-shelf item, however.		1. Air vents required 2. Use of stainless steel surfaced fluid ways will reduce pointing requirements and cavitation damage hazard	1. Air vents required	1. Seal replacement in 5–15 years is probable depending on design and use
a. Coefficients are approximate and may vary somewhat with specific designs					
b. Size ranges shown are representative and are no limiting					

(*)Original table, no update.

Table 10.5 Throttling Valves (*Kohler, 1969)

Service Classification	Throttling Gates				
Schematic diagram					
Name	Fixed-cone valve	Hollow-jet valve	Needle valve	Tube valve	Sleeve valve
Max head (approx.)	1,000′	1,000′	1,000′	300′	
Discharge coefficient (a)	0.85	0.70	0.45–0.60	0.05–0.55	0.80
Submerged operation	Yes (1)	No (1)	No	Yes	Yes (1)
Throttling limitations	None	Avoid very small discharge	None	None	None
Spray	Very heavy	Moderate	Small	Moderate (1)	None
Leakage	None	None	None	None	None
Nom. size range (b)	6″–108′ dia.	30″–108′ dia.	10″–96″ dia.	36″–96″ dia.	12″–24″+dia. (2)
Availability	Commercial std. (3)	Special design	Special design	Special design	Special design
Maintenance required	Paint	Paint	Paint (1)	Paint	Paint
Notes	1. Air vent required 2. Spray rating will change to moderate if a downstream hood is added 3. Valves are not stock items but std. commercial units are available	1. Submergence to axis of valve is permissible	1. If water operation is used disassembly at 3–5 years interval for removing scale deposits is usually necessary	1. Spray is heaviest at openings of less than 35%. At larger openings the rating would be better than moderate	1. Valve is designed for use only in fully submerged conditions 2. Large sizes seem feasible and will probably be developed

a. Coefficients are approximate and may vary somewhat with specific designs

b. Size ranges shown are representative and are no limiting

(*)Original table, no update.

Table 10.6 Guard Gates and Guard Valves (*Kohler, 1969)

Service Classification	Guard Gates						Guard Valves		
Schematic diagram	See Figure 10.3 for diagrams								
Name	Slide gates			Ring-follower gate		Wheel-mounted or roller-mounted gates		Butterfly valve	Spherical and plug valves
	Unbon-neted	High pressure	Stream						
Max. head (approx.)	100′	250′	500′	500′		500′		750′	1,500′+
Head loss (a)	(1)			Negligible				None to small (1)	Negligible
Leakage	*	*	*	None		Small to moderate			None
Nom. size range (b)	*	*	*	36′–120″ dia.		To 10′ wide & 30′ high		12″ to over 12″ dia.	12″ to over 10″ dia.
Used as guard unit for	(2)	(2)	(2)	All types of circular conduit throttling gates and valves		Top seal radial gates and other square or rectangular units		All types of circular conduit throttling gates and valves	
Availability	*	*	*	Special design		Special design		Special design (2)	Std. and special (1_
Maintenance required	*	*	*	Paint		Paint and rubber seals		Paint and seals (3)	Paint
Notes	(*)Data in Figure 10.3					1. Normally wheel-mounted gates are used except for high heads		1. Rubber-seated valves have no leakage, while new metal seats will have some leakage.	1. Sizes to about 24″ are fairly standard. Large sizes and high pressures are special
c. Head losses are approximate and may vary with specific designs	1. Head loss coefficients will vary from about 0.2–0.4 depending on entrance							2. Sizes to 36″ or 18″ are fairly standard. Larger sizes and high pressures are usually special.	
d. Size ranges shown are representative and are not limiting	2. Usually used with a similar type throttling gate. Sometimes used for other types							3. Metal seals may require periodic adjustment	
	3. Used closely coupled with a similar throttling gate. See Figure 10.3								

(*)Original table, no update.

Figure 10.18 (a–b) Downstream view of Funil HPP. Working Howell Bunger valve detail.

10.2 TURBINES

This chapter provides a summary of the project as well as criteria for choosing a turbine for a hydroelectric plant for a given waterfall and flow, based on Vinogg-Elstad (2003) and Engevix (1982). Details on the subject can be found in Gulliver and Arndt (1991), as well as in other references cited at the end of this chapter.

10.2.1 Generalities

The turbine converts the hydraulic energy {Q.H – flow (kinetic) and head (potential)} to mechanical energy (T.n, motor torque and rotation) on the turbine shaft.

The flow is controlled by opening of the distributor. The turbine shaft is directly connected to a generator that converts mechanical energy into electrical energy (E.I, electrical voltage and electrical current). The rotational speed n (rpm) of the turbine-generator set must be the synchronous speed corresponding to the system frequency (f) of alternating current (in Brazil, 60 Hz):

$$n = 60 \cdot f/p \tag{10.1}$$

where p is the number of pole pairs of the generator.

The highest possible speed is generally desired because the turbine-generator set will be smaller, the unit will require less space, the dimensions of the powerhouse will be reduced and the plant will be cheaper.

Mechanical equipment 331

A hydraulic turbine is usually tailored to fit a liquid head and design flow. There are two types that are based on two different principles of energy conversion, action turbines and reaction turbines, as described below.

10.2.1.1 Action turbines

They are those in which all hydraulic energy entering them is converted to kinetic energy in the stationary parts in front of the rotor. The rotor of this turbine works out of the water (operating near atmospheric pressure) which does not allow it to fully utilize the energy of the installation as the free height above the rotor cannot be used for power generation. The Pelton turbine is the most common type in this category and is used for very high heads.

10.2.1.2 Reaction turbines

They are those whose rotors work submerged in water, being subjected to a downstream back pressure, which allows full use of the energy of the installation.

In these turbines, only part of the hydraulic energy entering them is converted to kinetic energy in the stationary parts after the rotor. The conversion of hydraulic energy to mechanical energy in the rotor can be divided into two: the impulse action caused by the change in direction of input speed to the rotor output and the reaction contribution caused by pressure drop across the rotor. In the draft tube some of the kinetic energy at the rotor outlet is converted to potential energy.

The Francis, Kaplan and bulb turbines belong to this group.

The field of application for each type of turbine is shown in the following graphs as a function of flow (m^3/s) and net head (m) (Figure 10.19).

10.2.2 Design conditions and data

The nominal data of a turbine is usually: H_n = net head (m); n = rotation (rpm); and P = turbine output power (kW) for net head H_n. The turbine design is based on the best head-flow efficiency point to optimize its hydraulic performance for a wide range of flow and head. This point, depending on the type of turbine, may be in the range of 60%–95% of total flow. The structural design of the circuit surrounding the turbine is made for the maximum hydrostatic pressure plus the maximum pressure transient due to water hammer in the penstock aiming at the safety of the plant.

In choosing the turbine, in addition to the parameters mentioned, we also use the K Factor – a practical comparative dimensionless factor that defines the technical evolution in turbine manufacturing:

$$K = n_s (H)^{0.5} \tag{10.2}$$

$$n_s = n P^{0.5} / H^{1.25} \tag{10.3}$$

where:

n = rotation (rpm); P = turbine output power (kW); H = head (m); n_s = specific speed.

332 Design of Hydroelectric Power Plants – Step by Step

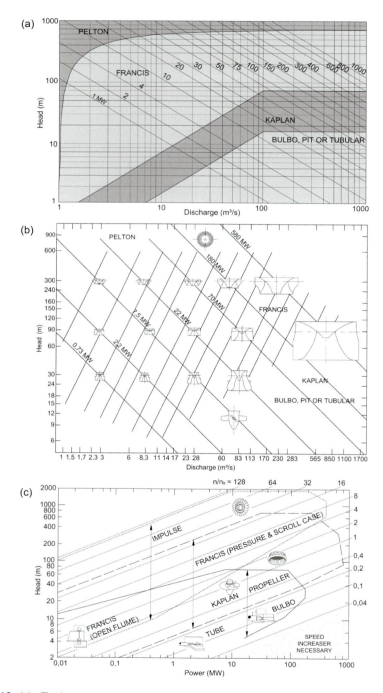

Figure 10.19 (a) Turbine type selection. Graph flow × net head (Vinogg and Elstad. 2003). (b) Turbine type selection. Graph flow × net head (Miranda, 1982). (c) Turbine type selection. Graph flow x net head (Gulliver and Arndt, 1991).

Mechanical equipment   333

Figure 10.20 shows the turbine rotor configuration according to the specific speed. The K factor can be adopted based on manufacturers' worldwide experience:

- K factor, K for Francis turbines: $2.100 < K < 2.300$;
- K factor for Kaplan turbines: $2.400 < K < 2.800$.

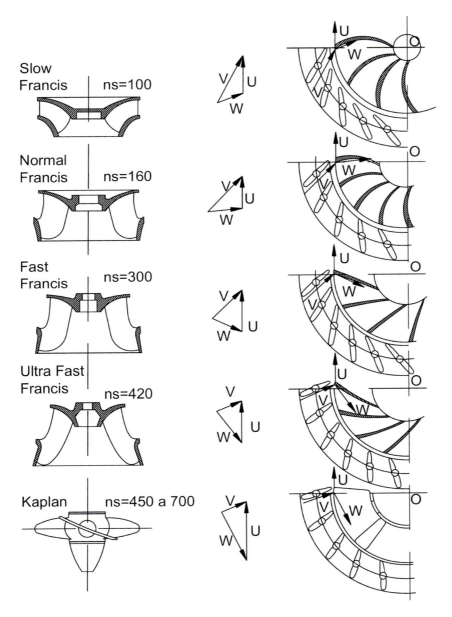

Figure 10.20 Turbine rotor configuration according to the specific speed (Miranda, 1982).

334   Design of Hydroelectric Power Plants – Step by Step

The increase of the specific velocity of the factor K, under the same head (H), implies the appearance of the inconveniences listed below (Miranda, 1982):

- greater possibility of cavitation, which implies the need to install the turbine in a lower position;
- higher speed in the turbine passages, which causes premature wear;
- reduction in turbine performance due to thicker profiles to the distributor and, sometimes, also of turbine rotor;
- the yield curve as a function of flow or power loses its flattened characteristic as the specific speed increases;
- appearance of undesirable pressure and power fluctuations over a wider range of turbine's operating field;
- higher frequencies and amplitudes of pressure fluctuation, as well as vibrations of hydraulic origin.

Preliminary specific speed calculation

$$\text{Francis turbines:} \; n'_s = 2,200/(H)^{0.5} \tag{10.4}$$

$$\text{Kaplan turbines:} \; n'_s = 2,600^{1.25}/(H)^{0.5} \tag{10.4a}$$

Nominal rotation determination

$$n' = n_s \cdot H^{1.25} / P^{0.5} \tag{10.5}$$

$$\text{number of poles} = 7,200 / n' \tag{10.6}$$

$n = 7.200$ / number of poles – (Equation 10.1)

$$\text{Specific speed corrected:} \; n_s = n \cdot P^{0.5} / H^{1.25} \tag{10.7}$$

$$\text{Corrected } K \text{ value:} \; K = ns / H^{0.5} \tag{10.8}$$

Rotor diameter determination

In the Basic Design, a preliminary diameter calculation is used according to the following formula. The final diameter must be calculated by the turbine manufacturer.

$$\text{Francis turbines:} \; D = 84.5 \, Ku \, H^{0.5} / n \, (\textbf{discharge diameter}) \tag{10.9}$$

$$\text{Kaplan turbines:} \; D = 84.5 \, Ku \, H^{0.5} / n \, (\textbf{rotor outer diameter}) \tag{10.9A}$$

where:

Ku = velocity coefficient = peripheral speed (m/s)/speed of water vector (m/s), whose values can be obtained from Figure 10.21 as a function of the specific velocity.

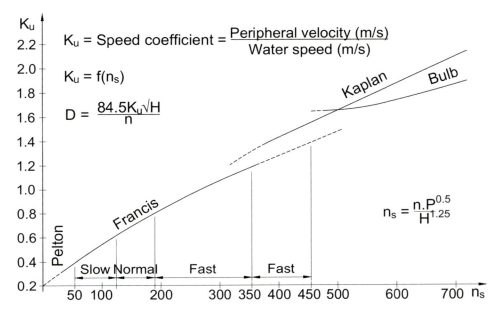

Figure 10.21 Velocity coefficient × specific velocity (Miranda, 1982).

10.2.3 Turbine efficiency and plant efficiency

As shown below, the turbine efficiency is defined by the relationship between the output and input power of this equipment. Figure 10.22 shows typical turbine efficiency curves.

$$\eta_T = \frac{P_T}{gH_n\rho Q} \tag{10.10}$$

where:
P_T = output, is the mechanical power delivered by the turbine on the axis.

The input power is the available hydraulic power for the turbine, which is the net specific hydraulic energy of water gH_n (J/kg) multiplied by the mass of the water flow ρQ (kg/s), where ρ (kg/m³) is the density of water

Relative discharge is the ratio between the discharge at the point under study and the discharge at the maximum yield point of the turbine. If the operating point being studied is the maximum yield point, the relative discharge is equal to 1.

Turbine efficiency considers turbine losses, but there are also other losses in the "water to wire" generation process:

- head losses in the hydraulic circuit before and after the turbine;
- generator losses, which reduce the mechanical power at the turbine output;
- transformer losses, which reduce the output power of the generator compared to what arrives at the transmission line.

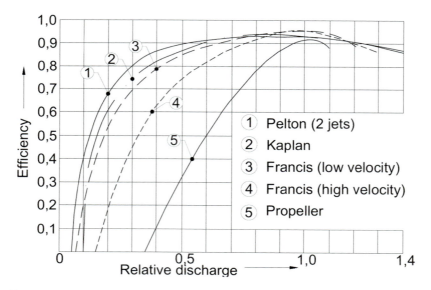

Figure 10.22 Turbine efficiency × relative discharge (Vinogg and Elstad, 2003).

Plant efficiency can be expressed by the following relationship:

$$\eta_{usina} = \frac{P_{Trafo}}{gH_{bruta}\rho Q} \tag{10.11}$$

Extra losses included in plant efficiency are not proportional to flow or output power. Therefore, the shape of the plant's efficiency curve differs considerably from the efficiency of the turbine and this has to be taken into account at the planning stage.

10.2.4 Turbine equation

The turbine converts hydraulic power to mechanical power P. Water flow creates a torque T at the turbine rotor running at an angular velocity ω.

Referring to Figure 10.23, this can be written as follows:

$$P = T\omega \tag{10.12}$$

where:
$\omega = 2\pi n / 60$ is the angular velocity of the turbine (rad/seg);
n is the rotation (rpm).

The torque is

$$T = F r \tag{10.13}$$

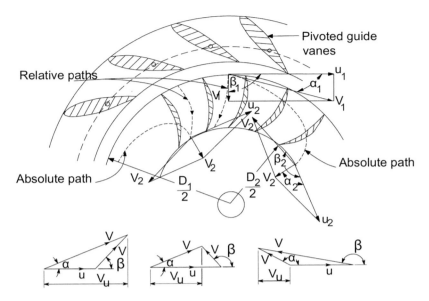

Figure 10.23 Definition sketch for radial flow turbine runner and typical velocity triangles (Gulliver and Arndt, 1991).

where:
  F is the sum of all circumferential forces acting on the rotor blades;
  r is the effective radius.

Thus, mechanical power can be expressed by the equation:

$$P = Fr\omega \tag{10.14}$$

As,

$$U = r\omega \tag{10.15}$$

is the circumferential velocity of the blades, the equation becomes

$$P = FU \tag{10.16}$$

From basic hydraulics, the simple form of momentum principle establishes:

$$F = \rho Q \Delta V \tag{10.17}$$

where ΔV is the change in velocity vector.

Then, it is possible to calculate the force F by acting in the circumferential direction on the rotor blades. This brings out the classic Euler equation:

$$F = \eta \rho Q (U_1 \cdot V_{C1} - U_2 V_{C2}) \tag{10.18}$$

338 Design of Hydroelectric Power Plants – Step by Step

where:

- subscripts 1 and 2 refer to the turbine inlet and outlet speeds, respectively;
- C is the circumferential component of velocity;
- $V_{C1} = V_1 \cos\alpha_1$
- $V_{C2} = V_2 \cos\alpha_2$
- η is the hydraulic efficiency of the turbine.

At the best efficiency point, V_{C2} should be zero, which means that the rotor must have "absorbed" all circumferential flow forces. There is only one component of the velocity, V_2, remaining in the flow when it leaves the turbine rotor.

10.2.5 Hydraulic similarity and speed number

The turbine's hydraulic design is based on rotation n, net drop H_n and flow at the best efficiency point $*Q$ (* means the best efficiency point). A dimensionless number, the speed number $*\Omega$, characterizes the hydraulic design of the turbine. Turbines with the same speed number are geometrically similar if they have a great design.

$$*\Omega = \omega \frac{*Q^{0.5}}{(2gHn)^{0.75}} \tag{10.19}$$

For Pelton, Kaplan and bulb turbines, the speed number is also often applied to the maximum flow point, or nominal flow rate. It is usually written $°\Omega$ (° means full flow).

The speed number is used as a parameter for turbine rating. Each turbine type is found within a range of speed number values, as shown in Table 10.7. These numbers are not absolute, but they give a good indication of a safe design.

Low-speed turbines (Pelton) are hydraulically characterized as "low-speed turbines" because the circumferential rotor speed U (shells) is relatively low compared to the free jet speed $V_0 = (2gH_n)^{0.5}$. U_{PELTON} is 50% of V_0, approximately.

High-speed turbines (Kaplan) are "high-speed turbines" because the circumferential rotor speed is high compared to the free jet speed calculated from the net drop $V_0 (2gH_n)^{0.5}$. U_{Kaplan} is in the range of 120%–130% of V_0. This shows that low-head turbines (and small V_0) are "speeded up" relative to V_0 in order to achieve a reasonable size, which improves machine economics.

Table 10.7 Speed Numbers

Turbine	Minimum Speed Number	Maximum Speed Number	Speed Number Type
Pelton, one jet	0.05	0.08	$°\Omega$
Pelton, four jets	-	0.15	$°\Omega$
Pelton, six jets	-	0.19	$°\Omega$
Francis	0.19	1.50	$*\Omega$
Kaplan	1.20	3.20	$°\Omega$
Bulb	2.10	4.30	$°\Omega$

Mechanical equipment   339

This feature of relative hydraulic speed, "high" and "low" turbine speeds, should not be confused with turbine n rotation, because they rotate slowly, whereas Pelton currently rotates faster than the "faster" Kaplan due to different net heads H_n and, consequently, different free jet speeds V_0.

10.2.6   Specific numbers

In order to compare turbines, it is common practice to convert speed and flow to velocity number n_{11} and flow number Q_{11} of a "unit turbine" with 1 m head and 1 m rotor diameter output:

$$n_{11} = n D / H_n^{0.5} \tag{10.20}$$

and

$$Q_{11} = Q / \left(D^2 H_n^{0.5} \right) \tag{10.21}$$

where:

n, D and H_n are given to the turbine (model or prototype) before conversion to the unit turbine.

This speed and flow specific numbers are also used in the new IEC code for model test acceptance, but there the term head has been replaced by specific hydraulic energy (E). The numbers are called "speed factor" and "flow factor".

Previously, several other numbers were also used:

- specific speed n_s used as a "unit turbine" with 1 hp output and 1 m head;
- specific speed n_q used as a "unit turbine" with 1 m³/s flow and 1 m head.

These specific numbers are usually based on nominal power and nominal flow, i.e., maximum power and maximum flow, causing some confusion when other specific numbers are related to the best efficiency point. This becomes even worse when English UK or US units are used for the "unit turbine". Caution is required when using Manual's data.

10.2.7   Operation out of design head

Power P and flow Q, in a net head H different from design head H_0, can be calculated using the "affinity laws", when power P_0 and flow Q_0 at design drop are known:

$$P = P_0 \left(H / H_0 \right)^{1.5} \tag{10.22}$$

and

$$Q = Q_0 \left(H / H_0 \right)^{0.5} \tag{10.23}$$

These equations are generally valid when the net operating head is relatively close to the design head, within a range of ± 2%. Within this range, the efficiency level

340 Design of Hydroelectric Power Plants – Step by Step

Table 10.8 Best Efficiency Point Regarding Maximum Power Point *Q/°Q

Turbine	Speed Number	*Q/°Q
Pelton	0.08	0.65
Francis, high head	0.20	0.75
Francis, low head	1.50	0.90
Kaplan	1.30	0.65
Bulb	3.10	0.60

remains unchanged. If the head deviation is large, the rotor flux vectors will not adjust with the blade angles and efficiency will be reduced. The turbine performance diagram ("the shell curves") shows how efficiency varies with flow and net head. The top of the hill is the point of highest operating efficiency. Isolines show the decreasing steps from the top. Examples of these curves, which are provided by the manufacturers, can be found in the references cited. The location of the best efficiency point, relative to the maximum output power point at design head, depends on the turbine type and speed number (Table 10.7). The point of maximum output power is usually determined by the cavitation limit described below. Typical values are shown in Table 10.8.

10.2.8 Runaway speed

When the electrical load on the generator is turned off for some extraordinary reason, the turbine speed increases reaching runaway speed, and the turbine regulator or protective device initiates turbine shutdown. If the turbine, due to a control system failure, does not shut down, the distributor will remain fully open and the unit will quickly reach its speed limit, while the hydraulic torque on the rotor will be reduced to almost zero because the flow will not adjust to distributor guide vane angles.

Turbine torque will then balance the friction losses on the machine. This speed limit, the runaway speed depends on the turbine type and design (speed number). Rough estimates for runaway speed are presented in Table 10.9.

Table 10.9 Runaway Speed

Turbine	Runaway Speed (%)
Pelton	175
Francis, high head (low velocity number)	155
Francis, low head (high velocity number)	190
Kaplan (on-cam[a])	200
Kaplan (off-cam[b]*)	300

a **On-cam**: The combination of the distributor guiding vanes and the rotor blades of a Kaplan turbine or bulb is working.
b **Off-cam**: Conjugation not working; in this case, the rotor blades have no brake action causing the runaway speed to increase.

Mechanical equipment 341

For the Kaplan turbine, the highest runaway speed occurs when the automatic coordination of the distributor guiding vane opening×rotor blade angle is off-cam, and the distributor guiding vans remain fully open, while the rotor blade angle is small ("closed" rotor).

The turbine and generator must be designed mechanically at this high speed, even though it should not happen when the control and protection system work. High centrifugal forces are critical for both the turbine wheel and generator rotor, and acts as a well support system.

There are safety barriers that will close the upstream flow of the turbine if the control system fails.

All Pelton turbines and most small Francis turbines with head above 50–100 m have a valve at their inlet which is designed to close at full flow.

Low-drop Francis turbines, Kaplan turbines, and bulb turbines have an emergency gate in the water inlet or draft tube instead of valve. The gate normally closes by gravity with no extra force requirements, but, in some cases, it has to be closed manually. The closing of the gate may take time, but the duration of the uncontrolled speed situation is limited.

10.2.9 Hydraulic thrust

Ideally, hydrostatic forces acting on a turbine rotor should be only circumferential and create torque driving the rotor.

The Pelton turbine has symmetrical shells that split the jet into two halves, and has virtually no axial hydraulic force.

Reaction turbines, however, can have considerable axial force from pressures acting on the rotor blades, crown and rotor belt of the Francis turbine, and on the rotor blades and hub of a Kaplan turbine.

The calculation of these forces is complex but very important for sizing the thrust bearing. Methods are used to reduce hydraulic thrust such as surge tank pipes connected to the suction pipe or holes in the rotor crown of Francis turbines by connecting the high-pressure zone at the rotor inlet to the low-pressure zone.

The thrust force increases with the speed number. A Kaplan turbine has rotor axial flow, and in an uncontrolled runaway speed situation, the full hydraulic head can act axially on the "closed" rotor and cause a higher thrust force. Hydraulic thrust is defined based on the reduced turbine model test.

10.2.10 Suction height and cavitation

The suction height Zs of a reaction turbine is the vertical distance between the turbine center and water level of the tailwater (Figure 10.24A). The "turbine center" is the center of the distributor guide vanes for the Francis turbine and the wheel center for the Kaplan turbine. If the center of the turbine is above the water level of the tailwater channel, the suction height is positive.

It is considered an advantage to have the machine center above water level as the risk of flooding is limited only to high flood situations. Additionally, exhaust pumps are required only to access the draft tube and the need for the gate is reduced. A simple stoplog solution is sufficient, but it is rare to have a positive suction height.

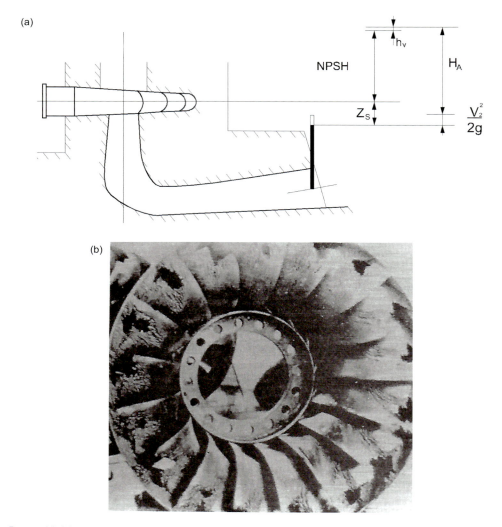

Figure 10.24 (a) Suction height (Z_s) and NPSH for reaction turbines (Vinogg and Elstad, 2003). (b) Cavitation erosion in a turbine runner (Gulliver and Arndt, 1981).

To have a smaller, cheaper turbine-generator unit, the rotational speed must be as high as possible. This requires deepening the turbine setting because of the high flow velocities in it. Cavitation limit dictates turbine speed and setting. Deeper setting enables high speeds without cavitation problems.

Cavitation occurs at the rotor outlet, where speed is high and local pressure may drop to vapor pressure in run-off sections. Vapor bubbles are formed in the low-pressure regions – where the water vaporizes. Vapor bubbles are transported by the flow and collapse when they reach a high-pressure zone immediately downstream giving off a lot of negative energy (suction). When collapse occurs near the steel lining there is damage to the surface, called cavitation erosion (Figure 10.24B).

Mechanical equipment 343

Stainless steel is more resistant to cavitation erosion than ordinary carbon steel. The introduction of stainless-steel rotors was therefore a big step in the field of hydraulic turbine technology. However, stainless steel also has its limit to cavitation erosion. In order to avoid these problems, the turbine rotor is usually seated below the water level of the tailrace where the pressure is high. This is an important procedure for generating unit design optimization. Deeper setting increases the volume and cost of the excavation, but the unit will have a higher speed and that too at lower cost.

Instead of suction height, the term NPSH, Net Positive Suction Head, has become more common. NPSH is the absolute hydraulic load (minus the load due to vapor pressure) at the suction tube outlet (draft tube) in relation to the machine reference level:

$$NPSH = H_A - h_v - Z_s \frac{V_2^2}{2g} \tag{10.24}$$

where:

H_A = atmospheric pressure head ($\sim$10 m);
h_V = vapor pressure (negligible);
Z_s = suction height (positive when the tailwater level is above the turbine center);
V_2 = average speed in the draft tube.

10.2.11 Cavitation limits

The performance of reaction turbines under cavitation operating conditions is studied in the laboratory according to IEC specifications.

The pattern and intensity of cavitation and the resulting reduction in efficiency can be observed and measured in the model and thus critical operating conditions are quantified.

The results of the turbine model cavitation tests can be converted to the prototype by the dimensionless parameter:

$$\sigma = NPSH/H_n \tag{10.25}$$

In order to have a cavitation free flow, the condition $\sigma \geq \sigma_{crit}$ has to be met. A simplified calculation of cavitation limits for Francis and Kaplan turbines states that:

$$\sigma_{crit} = 0.2 * \Omega^{1.7}. \tag{10.26}$$

A bulb turbine requires a deeper setting than the Kaplan turbine with the same speed number for two reasons:

- the straighter suction tube of a bulb turbine has less pressure drop than the Kaplan turbine elbow draft tube. Less pressure head on the low-pressure side means low pressure on the bulb turbine rotor output if the setting is the same;
- the suction height of the horizontal turbine bulb well is referred to the center of the turbine, but the upper half of the wheel operates above the center of the turbine, where the pressure is low.

10.3 PELTON TURBINES

10.3.1 Application range

The Pelton turbine (Figures 10.25 and 10.26) is the dominant type for heads over 300 m and for smaller heads when flows are low, on the order of 2 m³/s or less.

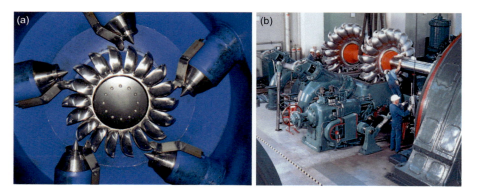

Figure 10.25 (a–b) Pelton turbines. (a) mechanical-eng.comb) en.wikepedia.org.

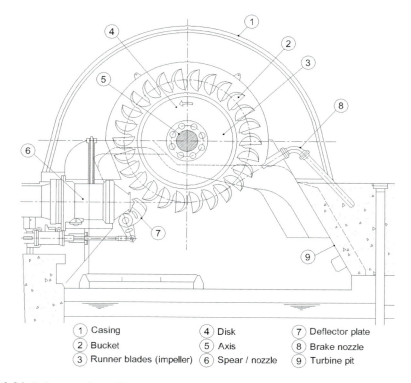

① Casing ④ Disk ⑦ Deflector plate
② Bucket ⑤ Axis ⑧ Brake nozzle
③ Runner blades (impeller) ⑥ Spear / nozzle ⑨ Turbine pit

Figure 10.26 Pelton turbine (Brazil; see NBR 6445/1987).

Mechanical equipment   345

10.3.2   Basic principle

The jet from the injectors hits the rotor tangentially to the shell pitch circle and is divided by the center edge. This edge is usually the first part of the shell that enters the jet and its shape is very important to prevent cavitation. The circumferential velocity of the shells at the diameter step is close to 50% of the jet velocity.

The flow through the nozzle is regulated by means of a conical needle. The needle needs to regulate flow slowly to prevent a high pressure increase as a function of the penstock extension.

To prevent unacceptable speed increase in large load rejections, a jet deflector is installed at the nozzle outlet. In such cases, the deflector moves rapidly toward the jet and deflects it of the rotor. The needle slowly moves to the position where the flow adjusts to the new reduced load while the deflector diverts from the jet.

Multi-jet turbines have an arrangement for automatic selection of the number of jets in operation to achieve the highest possible efficiency at partial loads.

10.3.3   Dimensions

Depending on the flow and head, Pelton turbines are designed with a number of injectors ranging from 1 to 6. The decisive parameter is the speed number. Increasing the number of speeds implies increasing the number of jets.

Single or twin jet machines generally have a horizontal axis. With more than two injectors, it requires a vertical axis that allows it to properly displace the water discharge from the rotor. Rotor diameters are usually between 1.0 and 5.0 m.

The number of shells ranges from 18 to 24 for modern Pelton wheels. The maximum diameter of the D_{JET} jet shall not exceed 10% of the wheel diameter pitch. The width of the shell relative to the D_{JET} shall not exceed $W/D_{JET} = 3.4$. Wider shells reduce efficiency.

10.3.4   Performance data

The Pelton turbine has a good efficiency for a wide flow range. Peak efficiency is between 91% and 92% for a good hydraulic design. The net head of a Pelton turbine is defined as the load above the wheel center elevation for vertical-axis units, or the average jet elevation for horizontal-axis units. This means that a Pelton turbine has an additional loss, which is not considered for turbine efficiency: the loss is in terms of the drop between the center of the turbine and the leakage channel water level.

The wheel's vertical setting must be well above the maximum downstream water level, at least by one diameter, to prevent splashing of water or foam in the well from causing extra friction losses.

In case of large variations of the downstream water level, a very high setting is required but this reduces the energy and annual generation of the unit.

Some Pelton turbines are equipped with an air compressor to pressurize the wheelhouse and thus reduce the water level in the well during periods of high water level.

It is noted that a 1.869 m head and a 423 MW power turbine were installed at the Bieudron Hydroelectric Power Plant of the La Grande Dixence Complex (Switzerland).

346 Design of Hydroelectric Power Plants – Step by Step

See Chapter 12 for a summary of the accident that occurred with the forced conduit of this plant in December 2000.

10.4 FRANCIS TURBINES

10.4.1 Application range

The Francis turbine, Figure 10.27a, is the dominant type for heads between 60 and 300 m, but it can be used for larger heads. For hydraulic, mechanical, practical and economic reasons, 750 m is considered as the upper limit. In the upper range, between 300 and 750 m, there is an overlap between the Pelton and Francis turbines. In the lower range, there is an overlap between the Francis and Kaplan turbines. The choice of turbine will be determined by the initial investment and operating and maintenance costs. Francis turbine performance is best in the range of 50%–100% of flow. For smaller discharges, efficiency drops and performance is not good.

10.4.2 Basic principle

Flow in the scroll case is guided toward the rotor blades by the pre-distributor fixed blades and the distributor directional vanes. Flow is controlled by the distributor directional vanes. A large vortex created by the scroll case and cascades of the fixed blades and guide vanes enters the rotor. Speed and pressure energy in the vortex are "absorbed" by the rotor, creating torque and speed in the turbine chamber.

All change and total load rejections are normally controlled by the opening and closing movements of the distributor guiding vanes. However, in the case of turbines with long penstocks, there is an unacceptable risk of increased pressure at the turbine inlet due to waterhammer caused by the quick closing of the distributor. In this case, the scroll case should be equipped with a bypass valve to reduce pressure. It opens and discharges water directly into the draft tube as the distributor directional vanes close quickly to reduce turbine generator set speed. When the guiding vanes are closed, the bypass valve opens to smoothly slow down the penstock flow – reducing pressure increase due to waterhammer (Figure 10.28).

10.4.3 Dimensions

Small Francis turbines may have the horizontal axis. Medium and large turbines always have the vertical axis. The rotor may have 9–19 blades, depending on the design and speed number. The rotor of a high-head turbine usually has more blades than a low-head turbine. Rotor diameters are usually between 1.0 and 10 m. Figure 10.29 shows sections across two different Francis rotors. A typical rotor for high-head plants (low speed number), another is a typical rotor for low-head plants (high speed number).

There is an established upper limit for the speed of a Francis rotor. In order to avoid serious cavitation, noise and vibration problems, the circumferential velocity U_2 at the rotor D_2 output (Figure 10.27b) is usually limited to 43 m/s.

Mechanical equipment 347

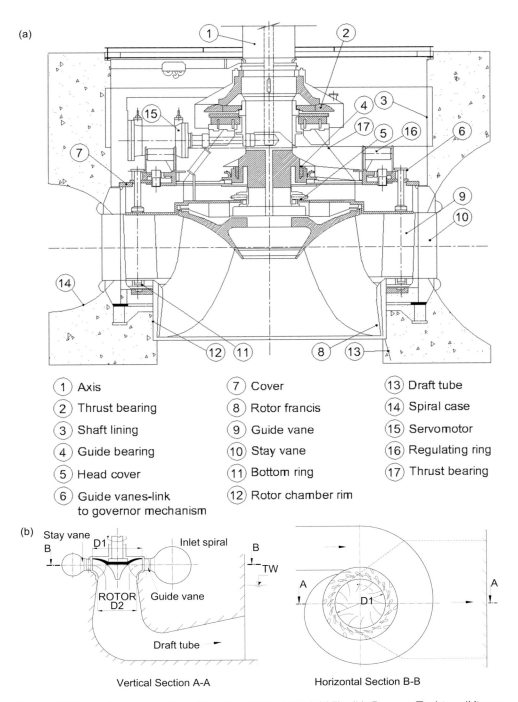

Figure 10.27 (a) Francis turbine (Brazil – NBR 6445/1987). (b) Francis Turbine (Vinogg and Elstad, 2003).

Figure 10.28 Francis turbine view. USBR. (For Francis runner main dimensions, see Figure 10.14.)

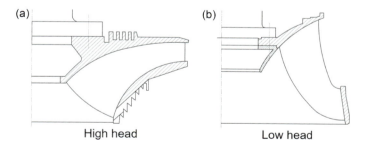

Figure 10.29 Francis rotor sections (Vinogg and Elstad, 2003).

10.4.4 Performance Data

The peak efficiency of a Francis turbine is in the range of 93%–96%, depending on its speed number and size. The best efficiency is obtained for the velocity number of the order 0.6, and the shell curves have different geometry as a function of this number.

10.5 KAPLAN TURBINES

10.5.1 Application range

The Kaplan turbine (Figure 10.30a) is designed to utilize low heads and comparatively high flow rates. For technical reasons, the maximum head considered is around 70 m. There is an overlap between Francis and Kaplan turbines in the 50–70 m range and between Kaplan and bulb turbines in the 15–20 m range.

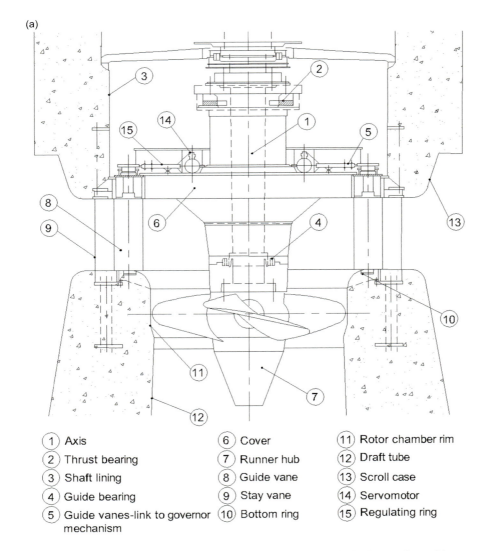

1. Axis
2. Thrust bearing
3. Shaft lining
4. Guide bearing
5. Guide vanes-link to governor mechanism
6. Cover
7. Runner hub
8. Guide vane
9. Stay vane
10. Bottom ring
11. Rotor chamber rim
12. Draft tube
13. Scroll case
14. Servomotor
15. Regulating ring

Figure 10.30 (a) Kaplan turbine (Brazil – NBR 6445/1987). (b) Kaplan turbine (Vinogg and Elstad, 2003). (Figure 10.15 illustrates the main dimensions of the Kaplan turbine distributor.)

(Continued)

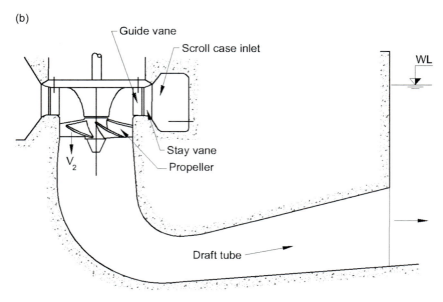

Figure 10.30 (Continued) (a) Kaplan turbine (Brazil – NBR 6445/1987). (b) Kaplan turbine (Vinogg and Elstad, 2003). (Figure 10.15 illustrates the main dimensions of the Kaplan turbine distributor.)

Turbine choice will be strongly influenced by unit size and expected flow variation. The Kaplan turbine performs well in the 30%–100% flow rate range, which is longer than a Francis turbine. For lower flow rates efficiency drops and performance worsens (Figure 10.31).

10.5.2 Basic principle

A Kaplan's scroll case is made of concrete for low heads and steel for high heads. There are pre-distributor fixed blades and distributor guide vanes as for the Francis turbine. The rotor is an axial flow propeller. On Kaplan turbines, the rotor blade is adjustable and it automatically adapts to an optimum angle for any flow under the control of the turbine speed governor.

The optimal combination of distributor vane opening and rotor blade angle is researched in a small model and is improved in field trials. In propeller turbines, the rotor blades are fixed. Because of low heads and large flow rates, the velocity head ($v^2/2g$) at the rotor outlet is a relatively large part of the design head. The draft tube is then carefully designed to recover this problem as much as possible by converting it into pressure on the diffuser.

10.5.3 Dimensions

The main dimensions of a Kaplan turbine are usually smaller than those of a Francis turbine with the same drop and flow. Rotor diameters are usually between 2 and 10 m.

Figure 10.31 Kaplan turbine.

The number of rotor blades is determined from the speed number or gross head: 4 blades are typically used for heads shorter than 30 m; five blades for heads in the range of 35–50 m; and six to seven blades for heads of the order of 70 m.

The clearance between the rotor blade and the rotor chamber wall should be as small as possible (a few millimeters) as water leakage reduces efficiency.

10.5.4 Performance data

The peak efficiency of a Kaplan turbine is in the range of 93%–95%, depending on size and design. Adjustable rotor blade angle provides good efficiency over a larger flow range than a Francis turbine. If the rotor blades are fixed, a peaked efficiency curve is obtained.

10.6 BULB TURBINES

10.6.1 Application range

The bulb turbine (Figure 10.32) uses lower heads, less than 20 m, between 3.5 and 15 m. The power may exceed 40 MW.

The 50 turbines of Jirau HPP and 44 turbines of Santo Antônio HPP, Madeira river (Rondônia), have heads of 15.2 m and unit powers of 75 and 71.6 MW (flows of 549.13 m^3/s and 542.65 m^3/s).

It can be considered a horizontal-axis Kaplan turbine. The main difference is that the generator is encapsulated. However, the bulb turbine will reach the design head limit due to transfer of the concentrated hydraulic load on the bulb that supports the

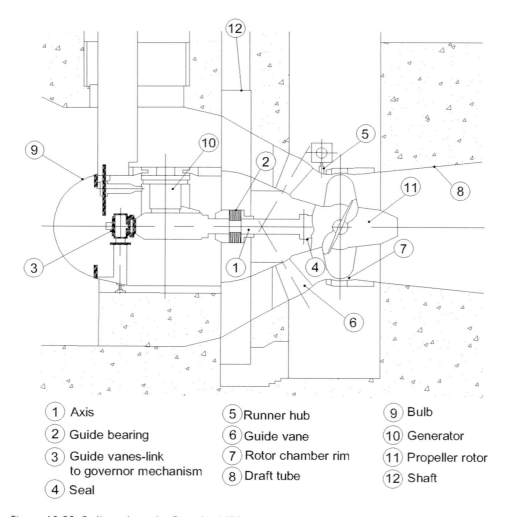

1) Axis
2) Guide bearing
3) Guide vanes-link to governor mechanism
4) Seal
5) Runner hub
6) Guide vane
7) Rotor chamber rim
8) Draft tube
9) Bulb
10) Generator
11) Propeller rotor
12) Shaft

Figure 10.32 Bulb turbine (in Brazil – NBR 6445/1987).

foundation concrete. Therefore, the head is limited to 15–20 m. The regulation range for acceptable performance is 25%–100% of the flow rate.

10.6.2 Basic principle

Water flows axially with minimal changes in flow direction along the bulb. Adjustment of the load is done by the guide vanes of the distributor, as in the Kaplan turbine. Depending on the characteristics of the utilization, bulb turbines with fixed guide vanes or fixed rotor blades can also be used (Figure 10.33).

10.6.3 Dimensions

The bulb turbine has 4 or 5 blades on the rotor, but for very small heads, 3 blades can be used. Rotor diameters are usually between 3.5 and 7.0 m. Tadami HPP (Japan) has machines with 6.70 m in diameter. The Jirau HPP Brazilian machines have diameters of 7.5 m, while Chinese machines are with the diameter of 7.9 m. Those of Santo Antônio HPP have diameters of 7.5 m. The Canoas I HPP machines have a diameter of 5.0 m (181 m^3/s; reference net head of 16.3 m, speed of 138.5 rpm) (Figure 10.34).

10.6.4 Performance data

The efficiency of the bulb turbine is slightly better than that of the Kaplan turbine, due to better axial flow conditions and elbow without draft tube. Shell curves look the same as Kaplan turbine curves.

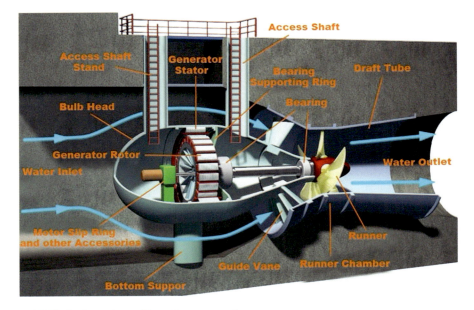

Figure 10.33 Bulb turbine (Dongfang power).

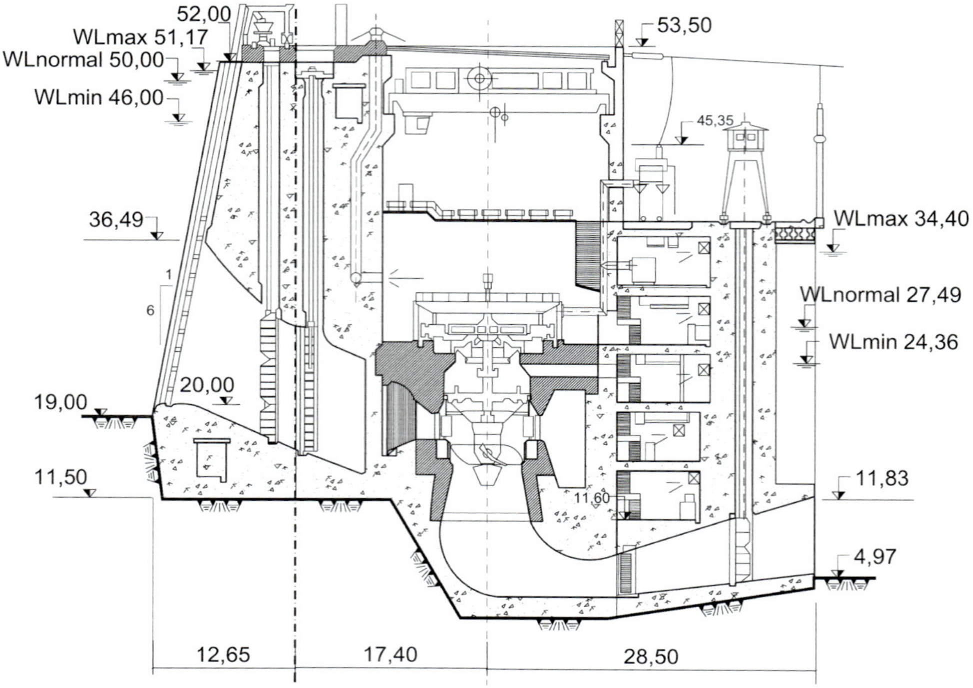

354 Design of Hydroelectric Power Plants – Step by Step

Figure 10.34 Canoas I HPP, Paranapanema river, São Paulo (1989).

10.7 TUBULAR TURBINES

The tubular turbine (Figure 10.35) is used for heads between 5 and 15 m, lower than those of bulb turbines. Figure 10.36 is a typical picture of your installation. For details, see the referenced bibliography, for example Warnick (1984) and Mayo (1979).

10.8 STRAFLO TURBINES

The Straflo turbine (Figures 10.37 and 10.38) is also used for lower heads than bulb turbines. It was developed by L. F. Harza. Units were built and promoted by EscherWyss during World War II under the name Straflo (derived from the words straight + flow), with rim generators.

The generator rotor is installed on the periphery of the propeller turbine rotor and the stator is mounted inside the civil works around the flow passage. The layout shortens the powerhouse, reducing space due to simplified civil works. Only one crane is required for maintenance. The relatively large rotor diameter provides the inherent inertia of the bulb generators, an advantage in operational stability.

Mechanical equipment 355

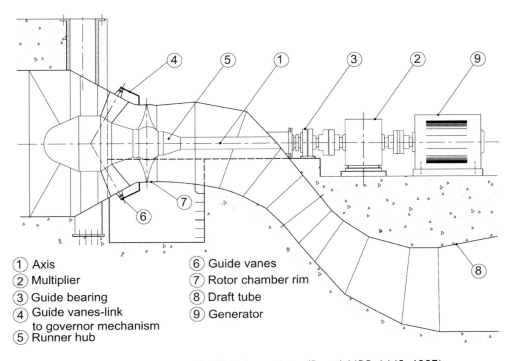

① Axis
② Multiplier
③ Guide bearing
④ Guide vanes-link
   to governor mechanism
⑤ Runner hub
⑥ Guide vanes
⑦ Rotor chamber rim
⑧ Draft tube
⑨ Generator

Figure 10.35 Typical Installation of a Tubular turbine (Brazil-NBR 6445, 1987).

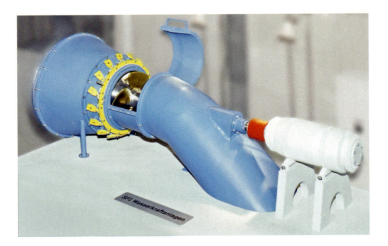

Figure 10.36 Tubular turbine (Voith).

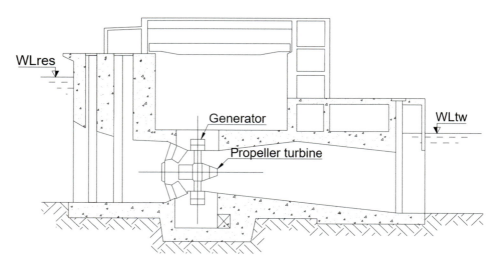

Figure 10.37 Typical installation of a Straflo turbine (rim-generator) (Warnick, 1984).

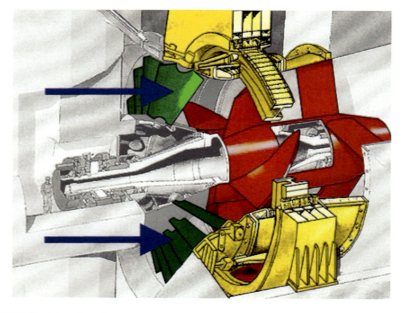

Figure 10.38 Turbina Straflo.

10.9 OPEN FLUME TURBINE

Mention should also be made of low-head facilities (mini-hydro) in which water can be driven through a channel directly into the turbine rotor, as shown in Figures 10.39 and 10.40 (channel or open box). These installations require a vortex device at the channel

Mechanical equipment 357

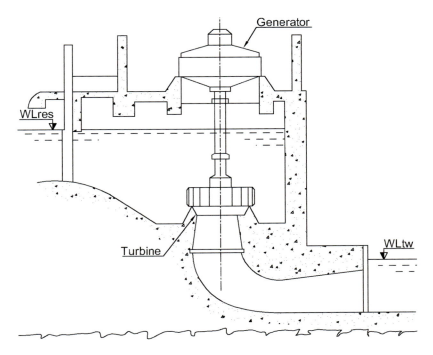

Figure 10.39 Typical installation of an open flume low head plant (Warnick, 1984).

entrance as well as a protective trash rack to contain debris. The upper limit for using this type of installation is that the head should not exceed 6 m.

10.10 TURBINE PERFORMANCE TESTS

10.10.1 Performance guarantees

When calling for a bid, a turbine buyer will usually inform suppliers of plant data: reservoir and downstream levels, hydraulic circuit dimensions, pressure drop, water quality, etc. The contract between the chosen supplier and the buyer will detail the main turbine performance, providing guarantees for the various operating regimes. To determine if warranties are met, acceptance tests are performed. The specifications usually determine the methods of these tests.

The guarantees must cover: the power; the flow rate; the efficiency×flow (or power); suction height limitations; maximum instantaneous runaway speed; maximum/minimum instantaneous pressure; maximum runaway speed in the permanent regime; regulating stability in net head and specified speed; etc. In addition, the cavitation erosion rate and turbine rotor fatigue life must be guaranteed.

There are two ways to determine turbine hydraulic performance: field tests and model tests. Field testing is preferred because it means testing a ready-made delivered

Figure 10.40 Typical installation of an open flume.

turbine. However, field tests for low-head turbines with large sections of water passages are very expensive and take too long. In such cases, laboratory testing is more cost effective.

10.10.2 Field test

Preparations for field testing, as well as efficiency measurement methods, turbine power measurements, pressure drop measurements, measurement equipment, etc., are not within the scope of this book. For this, it is recommended to consult IEC specialized publications.

10.10.3 Model tests

They must be performed in the manufacturer's laboratory according to the traditional standards for this type of testing (see specific references).

10.11 TURBINE CONTROL

The main basic functions of turbine control are as follows:

- maintain the speed and thus the stable and constant frequency grid (approx.), regardless of small load changes when operating on a large interconnected electrical system; for isolated systems adjustments are required.
- maintain generator output power or stable and approximately constant water level, regardless of small load changes when operating on a large interconnected electrical system;
- minimizing speed variations at large load variations, keeping pressure variations in the hydraulic circuit within acceptable limits and ensuring rapid return to permanent operating state.

The basic input data for the control is the speed or frequency grid. The output power of the generator is the command signal to increase or decrease the turbine flow. Figure 10.41 shows a sketch of the speed control principle.

Turbine control can have additional functions such as:

- load limiter for maximum output power and sometimes for minimum output;
- double regulation, such as optimum run-in of blade angle as a function of guide-line vane opening and head for Kaplan and bulb turbines. In addition, there is also an optimal injector combination for Pelton multi-jet turbine; supervision of normal start and stop procedures of the machine, supervision of emergency procedures, etc.

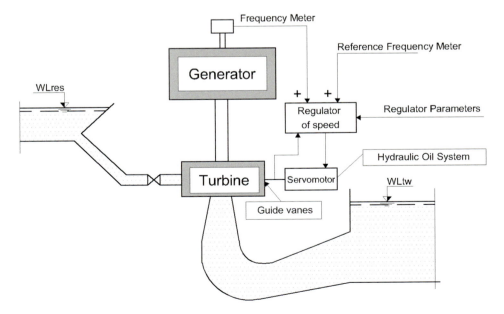

Figure 10.41 Speed controller (Vinogg and Elstad, 2003).

- ensure optimal load distribution across the various plant units;
- complete record of the operation.

10.12 MECHANICAL AUXILIARY EQUIPMENT

The plant's lifting equipment, trash rack cleaners, gantries and overhead cranes are not part of the scope of this book. For this matter, it is recommended to those interested to consult the bibliography and specialists in the subject. Mechanical auxiliary systems are also not within the scope of this book. It should be noted that, normally, the following systems are provided.

- Emptying's system of the units;
- Drainage system;
- Compressed air system;
- Ventilation system;
- Control room air conditioning system;
- Sanitary sewage system and treated water system;
- Units cooling system;
- Service water system;
- Transformer water/oil separator system;
- Fire protection system.

Chapter 11

Electrical equipment

Operation and maintenance

11.1 SYNCHRONOUS GENERATOR

The main reference that was used to support the elaboration of this text was Westgaard et al. (1982) given the clarity and simplicity of this work in exposing the subject. The stability of the electrical power system is outside the scope of this book. However, here is just a reference to IEEE Task Force on Terms & Definitions, "Proposed Terms & Definitions for Power System Stability", *IEEE Transactions on Power Apparatus and Systems*, Vol. PAS-101, No. 7, July 1982, pp. 1894–1898.

11.1.1 Synchronous machines

The following is a brief introduction of the machine as part of the electrical system. Details of synchronous machines as well as asynchronous machines, concepts of frequency and wavelength, three-phase system, voltages and currents, electrical and magnetic circuits, magnetic fields, capacitor and capacitive reactance can be found in the references. In Brazil, the energy in public transmission and distribution networks is three-phase alternating current (AC), with a frequency (f) of 60 Hz (cycles per second). In an interconnected electrical system, such as the Brazilian system, the electrical frequency must be the same for all installations and machines.

Turbine designers always calculate the rotational speed of the machines as previously defined in Section 10.2.1, which must match the synchronous speed that corresponds to the condition:

$$n = 120.f/p \tag{11.1}$$

where:
f = alternating current frequency (Hz); in Brazil, f = 60;
p = number of poles (always in even numbers);
n = turbine generator set rotational speed (rpm).

Thus, $n = 7,200/p$. By varying the values of "p" we find the values of "n".

p	2	4	6	-	-	-	-	-	20
n	3,600	1,800	1,200						360

DOI: 10.1201/9781003161325-11

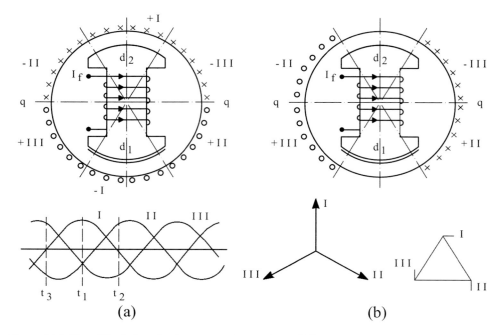

Figure 11.1 Simplified sketch of a two-pole machine (Westgaard et al., 1994).

For low-head power plants, the rotations are lower, which implies generators with a larger number of poles, and vice versa. For high-head power plants, the rotations are higher and, consequently, correspond generators with a smaller number of poles.

In most hydro plants, six poles are more used. These machines have salient poles. Figure 11.1 shows a sketch of the design principle. The two-pole example is chosen because it is easier to follow the space and time angles used. The rotor or pole wheel is supposed to turn counter clockwise. The rotor – exciter – field winding conducts a DC current "I_f". The airgap between the stator and rotor is graded giving a sinusoidal distributed magnetic induction along the stator that bore through its evenly distributed slots.

The stator winding is placed in the slots and divided into three groups or phases, each occupying two times 60° of the circumference. The winding consists of coils with the sides placed in diametrical or near diametrical slots. The direction of the field is from d_1 to d_2. In the position shown, the voltages in the conductors in the upper part of the stator are directed down in the paper plane. This is marked with crosses ("x"). The conductors in the lower half have a voltage in the opposite direction and are marked with dots ("o"). On the terminals of the phases, the geometric sum of the voltages in the coils is measured. The voltage in phase I has its amplitude value, and with this as a reference, phase II is 120, and phase III is 240° delayed as can be seen from Figure 11.1a. Figure 11.1b shows the relation at time t_2 in Figure 11.1a. The voltage in phase I is zero.

The windings of generators in power plants are always Y connected and the neutral is in most cases connected to earth through an impedance – often a pure resistance. For details, see the references cited.

11.1.2 The energy conversion

The energy conversion that takes place in a hydropower plant is sketched in Figure 11.2. The energy in the falling water is converted to mechanical energy of the rotating masses of turbine wheel (1), with the shaft and the generator rotor (3). With a constant head a change in energy output can only be made by changing the water flow. The controlling mechanism is the guide vane and the turbine governor. The valve (2) positions are either open or closed.

In the generator (3), the mechanical energy is converted through the magnetic field to three-phase electrical energy. The exciter circuit consists of the transformer (7), the rectifier (6), a demagnetizing resistance (5) and the slip rings (4). Though not shown here, there is a control circuit for the rectifier and an exciter starting circuit that supplies pole winding from the service battery for some seconds.

Power consumed at the plant is supplied by an auxiliary source (8). To prevent damage by short circuit, a current limiting coil is used (9). Important plants have emergency generators run by diesel motors.

The main generator is electrically connected to the transformer (11), which raises the generator voltage to that of the transmission grid. This connection can be made by insulated cables only for powers smaller than approximately 50 MVA. For higher powers, high conductor sections would be required. In such cases, the use of enclosed

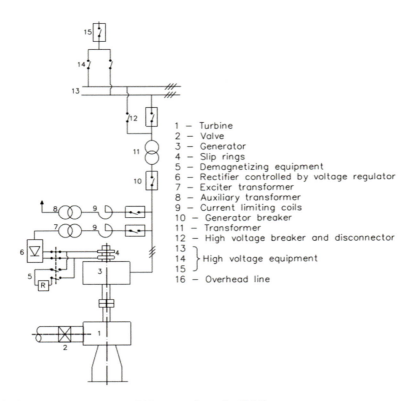

1 — Turbine
2 — Valve
3 — Generator
4 — Slip rings
5 — Demagnetizing equipment
6 — Rectifier controlled by voltage regulator
7 — Exciter transformer
8 — Auxiliary transformer
9 — Current limiting coils
10 — Generator breaker
11 — Transformer
12 — High voltage breaker and disconnector
13
14 } High voltage equipment
15
16 — Overhead line

Figure 11.2 Energy conversion (Westgaard et al., 1994).

364 Design of Hydroelectric Power Plants – Step by Step

power bus is more appropriate. In order to increase the level of reliability against short-circuit defects between phase conductors and phase-to-ground conductors, the three-phase bars are conducted separately within aluminum enclosures.

In large power plants, there is a transformer for every generator. A breaker between the generator and the transformer is not used in every plant. They are then said to be "block connected". In large generators the stored energy is large, and the breaker is needed for safety reasons. It has become more common also in medium size plants to invest in a breaker (10) so as to be able to separate the main units during faults.

When submitted to loading, the generator set and the waterways contain stored energy in the moving water, in the rotating components and in the magnetic field in the generator poles. Then the load should be changes so that such energy gets released. The worst case is a full emergency stop at full load.

Sudden stopping of the valve (2), and consequently of the moving water, creates an increase in the pressure and oscillations in the water column. The way of dealing with this problem is described in Chapter 10, "Mechanical Equipment". The generator inertia is important in limiting the increase in speed. Usually, the turbine consultant or manufacturer decides the necessary size of the rotating mass of the generating set (GD^2).

The energy stored in the magnetic field is supplied by the exciter current. The magnetic flux cannot decrease too quickly because that means high voltages being induced in the pole windings. The only way to reduce the energy is to convert it into heat as $R.I^2$ losses. This is done by the demagnetizing circuit shown in Figure 11.2. Simultaneously the wires from the slip rings are connected to a resistance, and the connection to the rectifier is cut.

The current breakers reach a full opening in about 0.08 s while each of the three-phase currents in the stator winding passes through zero a hundred times per second, i.e., each 0.01 s. With the field present, the opening of an inductive circuit by a breaker is usually followed by an arch between the contacts. The arch may extinguish and re-ignite several times until the field has decreased sufficiently.

There is also energy in the rotating parts. In Francis and Kaplan turbines where the wheel is always submerged, braking starts when the valve is closed. This reduces speed quickly in the beginning of the braking period. To protect the axial bearing all generators should also have a friction brake used from say 70 rpm and below. In Pelton turbines, the water jet is forced away from the wheel which then rotates in the air. For a quicker stop, a special water jet is used to counteract the rotation.

The full emergency stop is brought about only in cases of severe faults in the bearings of the aggregate. All means of braking are engaged when the valve is closed. The friction brakes are also engaged. Brake linings must usually be renewed after a stop like that.

The complete energy system is protected and supervised by several measuring and signaling elements. To protect the generator against over voltages, a surge arrester system consisting of surge arresters and capacitors is installed next to it, in a cubicle called a surge cubicle. Plant operators use computer programs to evaluate signals and command precautionary measures in accordance with the work plan. This plan is prepared based on the experience gained in the operation.

In order to limit the fault current to earth to a value that gets withstand by the generator and at the same time is sensitive to the protection system, the generator is grounded through a grounding system, where the most usual for unitary system (generator connected directly to the step-up transformer) is accomplished through a grounding transformer that is installed in a cubicle called the grounding cubicle.

Electrical equipment: operation and maintenance 365

11.1.3 Generator main elements

The electromagnetic characteristics of synchronous machines can be found in various electrical engineering references. Design principles are fair and do not permit many variations. With the same rated power and speed, the machines have many similarities, independent of the manufacturer. The synchronous generator is composed of the fixed part, the stator, and the rotating part, the rotor (Figure 11.3).

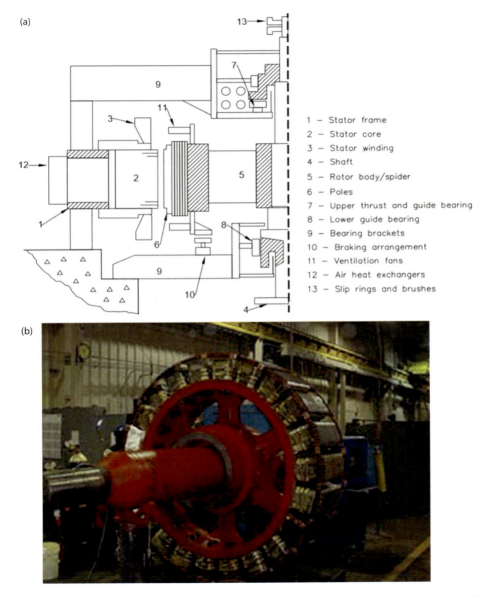

Figure 11.3 (a–b) Salient pole generator – cross section and picture (Westgaard et al., 1994).

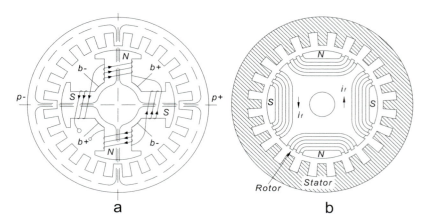

Figure 11.4 Rotor type: (a) salient pole and (b) smooth pole or coiled pole.

The stator frame (1), in welded fabrication, supports inside the stator core, composed of packages of blades that carry the slots to the coils. This housing shall be strongly anchored to the powerhouse structure in order to be able to withstand the twisting moment that assumes its maximum value in the event of a short circuit.

The rotor (5) may be of salient pole (for medium- to low-head, low-speed and large-diameter plants), or smooth or coiled pole (for high-head, high-speed and small-diameter plants), as shown in Figure 11.4.

The salient pole rotor consists of a hub with the axis coupled directly to the turbine axis. It is supported vertically by the thrust bearing (8). To the rotor hub are attached the arms/radius, at which end the magnetic rim/ring is welded.

The rim/magnetic ring is also made up of pressed blades which are provided with slots where the poles with their windings are inserted. This solution is normally used in low speed, large diameter, and with a high number of poles.

In the smooth pole machine, the conductors are mounted in slots and distributed along the periphery. This solution is normally used in high speed, small diameter and with a small number of poles.

The rotor poles, magnetized by direct current, pass through the stator coils, inducing alternating current in them. The number of poles must always be in even numbers as explained in Section 11.1.1 (Equation 11.1). The direct current for magnetization of the poles is generated by the excitation system, which establishes internal voltage of synchronous generator.

11.1.4 Generator rated capacity

Once the output of the plant has been defined (Chapter5), it is possible to give dimensions of:

- the active power of the generator (MVA), dividing the installed power by the generator power factor (fp)

$$MVA = PI/fp \tag{11.2}$$

where fp should be confirmed by generator manufacturer, which defines the specification of this equipment;
- the reactive power that can be delivered by the synchronous generator through the ratio:

$$MVAr = MVA \times \sqrt{\left(1 - fp^2\right)} \tag{11.3}$$

It is the contribution of the synchronous generator to maintain the rated operating voltage when the load is connected;
- generator power is equal to turbine power (PT) multiplied by generator efficiency (η_G):

$$PG = PT\,\eta_G \tag{11.4}$$

The results are then compared to the turbine specifications to analyze the compatibility between them, as the generator shaft power must be supplied by the turbine when it is in its rated operation.

For those interested in developing an HPP in Brazil, please refer to ANEEL Resolution 420, of November 30/2010, which establishes the system for determining the "Power output" and "Net Power" of the generation project, as well as for purposes for the granting, regularization and supervision of electricity generation services. Article 3, paragraph 2, item I states that "Installed Power" shall be defined on the basis of the lowest value between the rated power of the driving equipment (kW) and that of the electric generator (kW). It is defined by the product of the electric power. Apparent (kVA) by the fact of power (fp), both are taken directly from the board approved by the manufacturer for continuous operation.

11.1.5 Dimensioning factors

When bidding for a generator, the buyer provides the following set of information and instructions:

- Rated power in (MVA) and the power factor $-\cos\phi$;
- Rated frequency/speed and speed in runaway conditions (rpm);
- The rotor moment-of-inertia (GD^2);
- The hydraulic load from the turbine (m);
- Temperature of the cooling water;
- Transport factors, i.e., bridges, tunnels and lifting capabilities;
- Per-unit synchronous and transient reactance;
- Testing by IEC and specification of deviations if wanted.

The first rough estimate of the main dimensions is normally based on Poisson's formula:

$$P = C\,D^2 L\,n_N \tag{11.5}$$

where:
- P – rated power (MVA);
- C – Poisson constant;
- D – inner diameter of stator (m);
- L – length of stator core (m);
- n_N – nominal speed (rpm).

Poisson's constant (C) can be estimated using the graph shown in Figure 11.5. Known C has consequently $D^2 L$.

The formula for the natural fly-wheel effect is given as follows:

$$GD^2 \sim k D^4 L \tag{11.6}$$

where:
- k – dependent on rating and speed and the design principles assumed (for a solid cylinder $k = 3.1$);
- D – inner diameter of stator (m);
- L – stator core length (m).

The value of k is dependent on the rotor design:

- Full rotor: $k = 2.6$–2.8 (Pelton and Francis turbines);
- Rotor with rim and Spyder: $k = 0.45 - 0.50 \, n_M^{0.25}$ (Francis and Kaplan turbines)

where n_M = max. overspeed (rpm)
with the highest values for large numbers of poles.

With these relationships, a first estimate of D and L can be made, considering the designer's experience. Some criteria, however, must be observed:

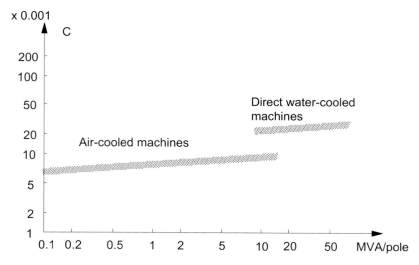

Figure 11.5 Poisson constant C as a function of rating and speed (Westgaard et al., 1994).

- in large machines ($P > 70$ MVA) it is common for the polar pitch to vary from 0.5 to 0.9 (m) and the L/ζ_p ratio to vary from 2 to 3;
- the polar pitch is given by the expression: $\zeta_p = \pi D/p$.

Arbitrating the value for the polar pitch, according to the above recommended range, determine the values of D and L.

Saturation in the present available material limits the maximum flux density in the airgap to about 1.1 Tesla at rated voltage and no load. With the estimates of the diameter, length and number of poles, the magnetic flux per pole is also determined and the next step is to find a reasonable number of slots in the stator. The number of slots must necessarily be a multiple of three in a three-phase machine. The number of slots greatly influences the size and design, and the design prefers having a free choice of voltage in order to optimize the machine.

The optimal voltage depends on the MVA rating of the machine. This is usually 5–6 kV for small ratings and up to 20 kV at 400 MVA. Detailing involves synchronous reactance required by users, but its details are beyond the scope of this book. For such, it is suggested to the interested one to consult the references cited.

Finally, it is noted that the turbine spiral in the longitudinal direction of the powerhouse is larger than the generator, so that the length of the powerhouse depends on the size of the turbine, but the generator determines its width.

Figure 11.6 shows a diagram from which another estimate of the stator outside the diameter of a vertical-axis generator can also be obtained. The diameter should

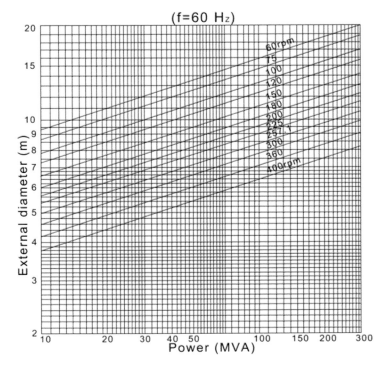

Figure 11.6 Stator diameter of a vertical shaft generator (Schreiber, 1977).

370   Design of Hydroelectric Power Plants – Step by Step

be added to stator with added thickness of the radiators at about 60 cm. Also, a passage should be there to reach the internal diameter of the generator shaft, applied to the pre-dimensioning of the powerhouse. It is also noted that the approximate rotor weight of a generator can be evaluated by the expression:

$$P_R = K \left[\frac{MVA}{n^{0.5}} \right]^{0.74} \tag{11.7}$$

where:

$K = 40$ for horizontal-axis generators and 50 for vertical-axis generators;
P_R = Rotor weight (t):
MVA = rated power (MVA);
n = generator rotation (rpm).

For further details, it is recommended that users consult the references cited as well as other specific electrical engineering or equipment supplier references.

11.1.6   Design principles

The design concepts used for hydropower generators depend on the rated power, speed range and type of turbine. They are shown in Figure 11.7.

The Pelton turbine is used for high-head plants and is characterized by high rated speed, medium overspeed ratios and no axial thrust apart from the weight of the turbine. The turbine wheel runs in the air. The natural braking torque is small and the generator braking equipment must be dimensioned accordingly.

The Francis turbine is used for the intermediate speed range, with relatively low overspeed ratios, but a substantial axial water thrust. The axial thrust requires a sturdy axial bearing with strong brackets and stator frame (Figure 11.3). The overspeed requirements affect the design of the rotor since the shaft must have a high critical speed. The turbine wheel is always submerged and the natural braking torque is considerable.

The Kaplan type gives very high axial thrust which varies considerably with the load. The overspeed ratio may also be high, but since the rated speed is low this will normally not influence the design in any significant way. The turbine wheel is always submerged.

Figure 11.7a shows the main concepts for high-speed machines. They have a combined guide and support bearing on top of the rotor and a guide bearing on the bottom.

Figure 11.7b shows a design with a combined guide and thrust bearing below the rotor and a guide bearing on a separate bridge above. This design is used for lower speeds when the diameter of the stator bore is large and the axial thrust is very high. In this way, these forces are more directly transferred to the foundation and the arms of the bearing bracket are shorter. One disadvantage is the greater difficulty in disassembling bearing parts if that is necessary on site.

Figure 11.7c shows the umbrella type used by low-speed machines where it is possible to use a combined bearing.

The most common rotor designs are shown in Figure 11.8. The stacked plate design is used for large diameters and is often combined with the under rotor thrust bearing in an umbrella design. At the plant, sufficient space is required for mounting the rings.

Electrical equipment: operation and maintenance 371

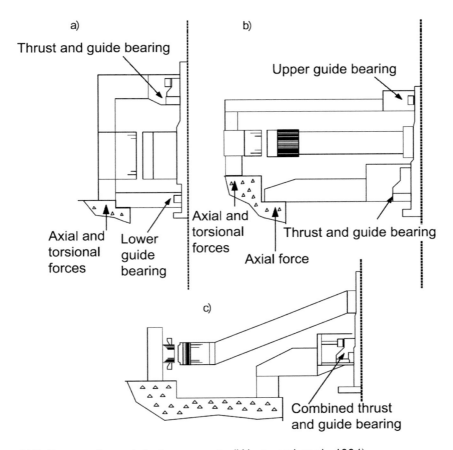

Figure 11.7 Commonly used design concepts (Westgaard et al., 1994).

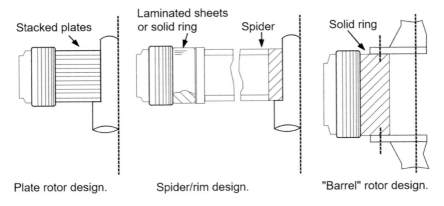

Figure 11.8 Rotor design concepts (Westgaard et al. 1994).

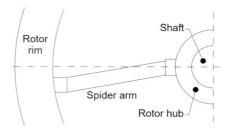

Figure 11.9 Schematic drawing of an oblique element (Westgaard et al., 1994).

The linear thermal expansion of steel is 1.15 mm per meter and 100°C. For large diameters of the stator bore, for instance 10 m, a temperature difference of 50° gives a diameter change of nearly 6 mm.

The ABB company has introduced so-called oblique elements as shown in Figure 11.9 to transform the radial stresses to tangential stresses with a torsional movement of the stressed parts. The same principle is used for the stator.

11.1.6.1 The stator core

The pressure in stator core pressure is a critical factor affecting the life of this equipment as a whole. All commonly used design concepts are based on an elastic compression system, either through the elasticity of the stator frame itself or by using through-bolts in the core yoke.

The natural quantity of "bulges" on the new stator core sheets will gradually be compressed and eliminated through magnetic vibration and thermal cycling, and after a number of years the pressure in the core will decrease. If the pressure disappears, the long-time effects will lead to increased vibration in the core, iron short-circuits with local hot spots and eventually damage of the stator winding insulation. Sufficient pre-stressing during assembly of the core and an initial elastic deformation exceeding the "bulge" height are necessary to avoid these aging effects.

Low pressures in the core may also cause problems at both ends in the outermost part of the core teeth. The variable magnetic field in this region can induce vibrations in the end sheets with material fatigue and breakage as the final result. The method used to avoid this is to have greater thickness in the outermost sheets, a gradual shortening of the teeth length towards the end of the core, as well as adequate pressure in the teeth. Another common problem associated with the stator core is the buckling, especially in low-speed generators with large diameters. The main cause is the difference in thermal expansions of stator core and stator frame. This results in long-term changes in pressure with possible vibrations and deterioration of core sheet insulation. The design elements used to prevent these problems include flexible frame elements that absorb thermal extension of core without lasting deformation. A relatively new way of achieving this is by using oblique elements between the core and the foundation.

With a fractional number of stator slots per pole and phase, a 100 Hz vibration, not present at no load or at rated current in short circuit, may occur in some machines.

Electrical equipment: operation and maintenance 373

The vibration level may range from unpleasant to intolerable. The cause is that evenly distributed magnetic attraction forces between the poles and the stator at no load is modulated by the stator ampere turns. These vibrations only occur where the stator winding has a fractional number of slots per pole and phase. In most cases they can be avoided by correct connections of the winding groups.

11.1.6.2 The stator winding

Stator winding is the most critical part of the generator. It also represents one of the most advanced technologies used in this area. The insulation of high-voltage windings with line-to-line voltages up to 25 kV are exposed to severe dielectric stresses voltages. In addition to these stresses, the unavoidable chemical aging of insulation materials reduces the quality of the insulation over the years of operation. Up to 1960, the carrier tape and resin in the insulation were based on organic materials while present insulation systems are based on synthetic materials which considerably lengthen the life span of winding.

In order to keep the eddy current losses down, the conductors are divided in relatively thin strands. Due to the uneven distribution of the magnetic field across the slot, the strands are transposed in the winding overhang or in the slot itself (the Roebel bar principle).

The testing of insulation is defined according to international standards such as IEC and includes 1 minute exposure to two times the rated line to line voltage plus 3 kV.

When evaluating a stator winding design, it is normal to emphasize the corona protection devices, the slot wedge system and the sidewise retainment of the bar or coil side in the slot.

The main cause of insulation problems is chemical deterioration due to high temperatures and dielectric discharges together with movement/vibration in the slot (Figure 11.10).

11.1.6.3 The poles and pole windings

The most common design for the poles is the one with laminated plate core and stapled end plates. At the feet of the pole, there are round holes to receive the copper rods of the damper winding, which have their ends welded to the end plates. The cores of the poles and windings are very robust elements and rarely cause operational problems. Regular maintenance and cleaning of windings is recommended to prevent short circuits (Figure 11.11).

11.1.6.4 The bearings

Anchor bearing lubrication for vertical shaft units only works at a given machine speed. After completely shutting down the turbine, the machine loses its speed at first quickly, then ever more slowly because of the large inertia of the generator rotor. To avoid damage to the thrust bearing, the generator is braked when the speed has reached half of normal. Modern bearings are very reliable and can run for many years without any problem (Figure 11.12).

374 Design of Hydroelectric Power Plants – Step by Step

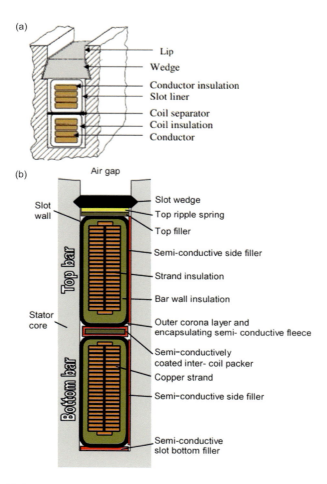

Figure 11.10 (a–b) Stator slot with winding.

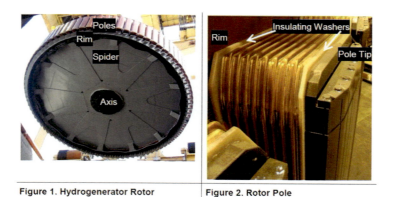

Figure 1. Hydrogenerator Rotor   Figure 2. Rotor Pole

Figure 11.11 Rotor pole.

Electrical equipment: operation and maintenance 375

Figure 11.12 Thrust bearing segments.

11.1.6.5 The cooling system

The losses of a generator, from 1% to 2.5% of its rated power, are transformed into heat that can cause the machine to heat up. To avoid this, a cooling system, water or air is provided. Figure 11.13 shows a sketch of the air-cooling system.

Experience shows that the method of cooling water using hollow conductors is more efficient – Figure 11.13a. For details see, among others, Westgaard (1994). Figure 11.13b shows the main elements of the turbine generator set.

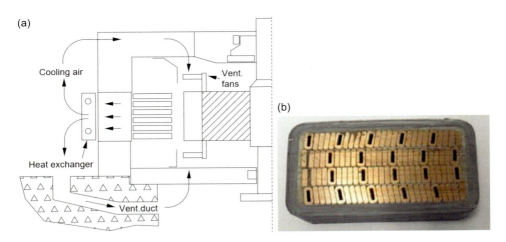

Figure 11.13 (a) Schematic drawing of the air-cooling circuit (Westgaard et al., 1994). (b) Direct water-cooled stator winding bar. Hollow conductor. (c) Main elements of a turbine-generator set. Luka Selak, University of Ljubljana, Slovenia (2014).

(Continued)

376 Design of Hydroelectric Power Plants – Step by Step

(c)

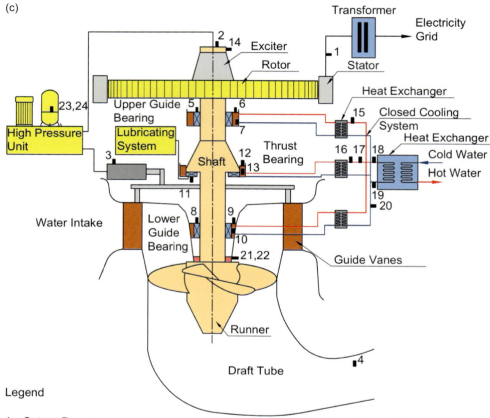

Legend

1. Output Power
2. Displacement of Runner Regulating Oil Distributor
3. Displacement of Guide Vanes Servomotor
4. Net Head, Pressure Sensors
5. Temperature of Upper Guide Bearing Babbit
6. Upper Guide Bearing Oil Level
7. Upper Guide Bearing Oil Temperature
8. Temperature of Lower Guide Bearing Babbit
9. Lower Guide Bearing Oil Level
10. Lower Guide Bearing Oil Temperature
11. Temperature of Thrust Bearing Babbit
12. Thrust Bearing Oil Level
13. Thrust Bearing Oil Level Temperature
14. Rotation Speed
15. Cooling Water Temperature From Upper Guide Bearing
16. Cooling Water Temperature From Thrust Bearing
17. Cooling Water Flow From Thrust Bearing
18. Temperature of Input Water in the Heat Exchanger
19. Temperature of Output Water From the Cooling Loop
20. Pressure of the Cooling Water
21. Water Flow Through the Seal Ring
22. Water Pressure of the Seal Ring Water
23. Oil Pressure in the High Pressure Unit
24. Oil Level in the High Pressure Unit

Figure 11.13 (Continued) (a) Schematic drawing of the air-cooling circuit (Westgaard et al., 1994). (b) Direct water-cooled stator winding bar. Hollow conductor. (c) Main elements of a turbine-generator set. Luka Selak, University of Ljubljana, Slovenia (2014).

11.1.7 Monitoring and instrumentation

The main parameters to be monitored are as follows:

- voltage and current of generator and transformers;
- stator winding temperatures;
- bearing temperature;
- temperature and flow measurements in cold air and water circuits;
- insulator measurements;
- various measurements of oil sump levels, excitation system values, vibrations, etc.

11.1.8 Transport of turbine-generator and assembly

The design of large generators should include the planning of the transportation of their parts to the site, as well as the planning of their assembly. The heaviest parts, the stator and rotor condition the dimensions and capacity of the overhead cranes, with the generator rotor being the heaviest part to move. Likewise, the dimensions of the mounting area are conditioned by the dimensions of the stator and rotor, the rotor being normally all mounted therein and transported entirely to the installation site, placed inside the stator and connected to the turbine shaft. This piece can reach a few hundred tons. The use of two bridges can accelerate assembly, meeting distinct activities of transporting parts in different locations and when coupled together, meet the movement of the heaviest part that is the generator rotor (Figures 11.14–11.16).

Figure 11.14 Tucuruí HPP. Turbine rotor of unit 13, second stage on the road transporter disembarking at the port of Belém – June 13, 2002 (Eletronorte).

Figure 11.15 Tucuruí HPP. Turbine rotor of unit 13, second stage on the Belém-Tucuruí barge-loaded road transporter. June 13, 2002 (Eletronorte).

Figure 11.16 Tucuruí HPP. Overhead crane in service. Unit 13 stator descent, step 2. May 2002 (Eletronorte).

11.1.9 Tests

The number and type of field tests are usually specified by the owner and are stated in the contract documents. The procedures and technical requirements are stipulated in international standards such as IEC and IEEE.

Tests include control of all guaranteed values, such as temperatures, reactance and losses, with load measurements as close as possible to nominal values. Machine efficiency tests and full tests of the generator guaranteed values can last 2 months.

11.2 LAYOUT OF THE GENERATING UNIT

The arrangement of the turbine generator set can be horizontal or vertical, depending on the type and size of the machine. Pelton turbines can have both horizontal and vertical arrangement. In general, the horizontal arrangement is only used for medium and small plants, with a maximum of two jets per rotor and a maximum of two rotors.

For high-powered machines, the vertical arrangement is allowed almost exclusively. Turbine speed can be increased by using a larger number of injectors, or by building the rotor with two or four injectors. Machines with up to six injectors have already been built.

Physically, it is impossible to build a horizontal-axis Pelton turbine with more than two injectors, because the arrangement of the water pipe and injectors presents almost insurmountable difficulties. The common trend of manufacturers is to offer rotors with vertical axis and a large number of injectors.

Figures 11.17 and 11.18 show sketches of the Naturno Plant (Trentino, Italy), which houses horizontal-axis Pelton turbines. The open-air plant has two 55 MW wheels, that went into operation in 1983, with the following characteristics: $H = 1,021$ m; $Q = 5.77$ m^3/s; $D_1 = 2.26$ m; $D_0 = 0.22$ m; $D_1 = 0.21$ m. The ball valve is 700 mm in diameter. In 1986, another 110 MW machine went into operation (Figure 11.19).

Figures 11.20–11.22 show a vie and sketches of the San Carlos Plant, owned by Interconexion Electrica, 50 km from Medelin, Colombia. The plant is underground

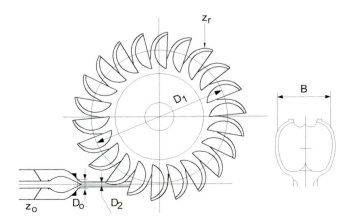

Figure 11.17 Naturno HPP. Pelton wheel dimensions. $D_1 = 2.26$ m; $D_0 = 0.21$ m; $D_2 = 0.22$ m; $B = 0.53$ m (Henry, 1992).

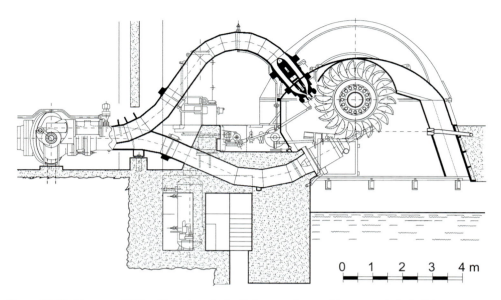

Figure 11.18 Naturno HPP – Trentino, Italy. Pelton turbine – Horizontal shaft (Henry, 1992).

Figure 11.19 Naturno HPP. Turbine generator set in progress assembly (Henry, 1992).

and houses 8 vertical-axis Pelton turbines of 170 MW each. It uses water from Nare and Guatape rivers that join the Magdalena river, which flows north into the Caribbean Sea.

The machines have the following characteristics: $H = 578$ m; $Q = 33.4$ m^3/s; $D_1 = 3.25$ m; $D_0 = 0.335$ m; $D_2 = 0.264$ m. The ball valve is 1,900 mm in diameter. The hydraulic and

Electrical equipment: operation and maintenance 381

(a)

Figure 11.20A San Carlos HPP. Intake towers.

(b)

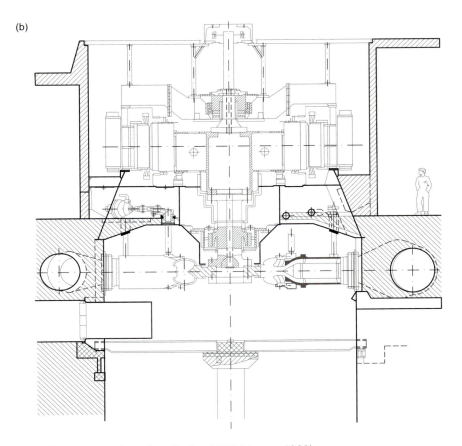

Figure 11.20B Pelton turbine San Carlos HPP (Henry, 1992).

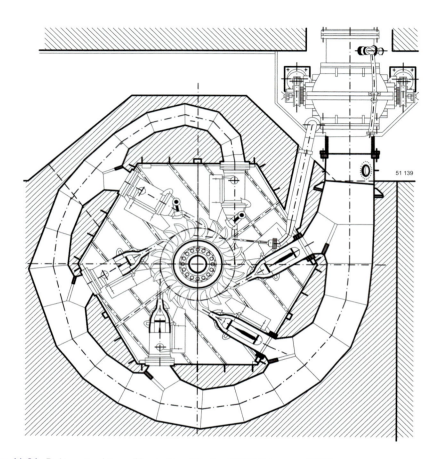

Figure 11.21 Pelton turbine. Plant. San Carlos HPP (Henry, 1992).

mechanical designs along with manufacturing was elaborated by Sulzer-Escher Wyss S.A., Zurich. The machines went into operation between 1984 and 1987 (Figures 11.20B, 11.21 and 11.22).

For the Francis and Kaplan turbines, different reasons determine the fixation of their position. In plant, the vertical arrangement takes up less space. The width of the powerhouse, in both cases, is determined by the spiral case. However, in the longitudinal direction, the length of the horizontal-axis unit is longer due to the draft tube elbow.

The elevation of the turbine is fixed by the permissible suction height as a function of cavitation and in reference to the rotor blade exit edge at the highest position.

The excavation depth for both turbines is about the same. As the length of the horizontal-axis powerhouse is longer, consequently, the excavation volume is larger.

For these reasons, the vertical axis is used almost exclusively for large units.

Turbine rotor mounting and dismounting solutions, with or without generator disassembly, should be considered in the design definition.

Electrical equipment: operation and maintenance 383

Figure 11.22 Pelton stainless steel wheel. Diameter = 4,100 mm. San Carlos HPP, Colombia (Henry, 1992).

The typical section of the Tucuruí intake/powerhouse is shown in Figure 9.2. Francis machines have an external diameter of 8.975 m (Figure 11.23) for a nominal drop of 60.8 m and a nominal flow of 574 m³/s. Other images referring to the Tucuruí HPP can be seen in Figures 11.24–11.28.

It also presents a sketch of the Taquaruçu HPP Kaplan turbine (Paranapanema river, São Paulo).

Figures 11.29 and 11.30. Figure 11.31 shows the Estreito HPP Kaplan turbine (Tocantins river, Pará).

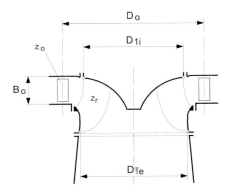

Figure 11.23 Tucuruí HPP. Main dimensions of distributor and Francis wheel. D_{1e} = 8.15 m; D_{1i} = 5.48 m; D_0 = 8,975 m; B_0 = 2,475 m (Henry, 1992).

Figure 11.24 Tucuruí HPP. Unit 14. Sroll case. Inlet diameter = 9.85 m. May 2002.

Figure 11.25 Tucuruí HPP. Unit 13 pre-distributor assembly. July 2001.

Figure 11.26 Tucuruí HPP. Turbine rotor. External diameter = 8,975 m; Weight = 255 t. June 28, 2002.

Figure 11.27 Tucuruí HPP. Stator shaft (20/09/2002).

Figure 11.28 Tucuruí HPP. Unit 13 generator rotor descent. (salient pole rotor). Weight = 918 t. 2nd Stage of the Plant (20/09/2002).

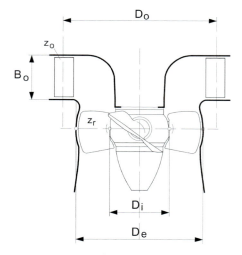

Figure 11.29 Taquaruçu HPP (Henry, 1992). Main dimensions of distributor and the Kaplan wheel. $H = 17.7$ m; $Q = 490$ m^3/s; 5 Kaplan 103 MW each. $D_e = 7.70$ m; $D_i = 3.30$ m; $D_o = 9.31$ m; $B_o = 2.86$ m.

Figure 11.30 Kaplan turbine, Taquaruçu HPP (Henry, 1992).

Figure 11.31 Kaplan turbine. Estreito HPP. 14 m high, 9.5 m diameter, 470 t. (It is the largest piece of its kind ever installed in Brazil).

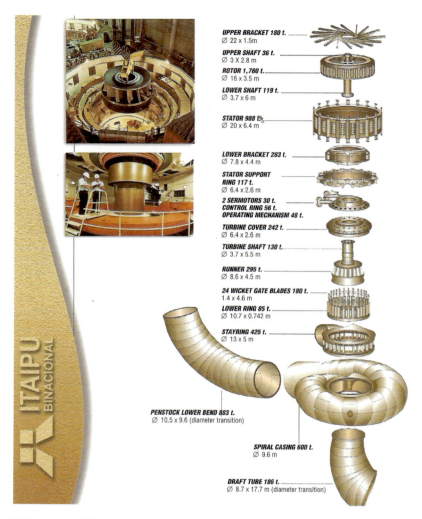

Figure 11.32 Itaipu HPP generating unit.

Figure 11.32 shows an assembly sequence of the Itaipu HPP generating unit (Paraná river, Brazil-Paraguay).

11.3 MAIN TRANSFORMERS

Theoretically, the transformers belong to the transmission system and not to the generation system. However, above 20 MVA, in general, the generator is connected directly to the transformer through the isolated phase bus forming a block called unit system. It is noteworthy that the isolated phase bus is very expensive and, therefore, the distance between the generator and the transformer should therefore be as short as possible, which also reduces the possibility of failures (Figure 11.33).

Electrical equipment: operation and maintenance 389

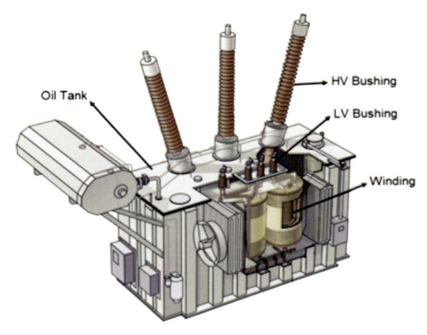

Figure 11.33 Transformer.

Transformers can be three-phase (joining the three-phase windings in one unit) or single-phase, with a separate transformer for each phase.

A three-phase transformer is cheaper than three single-phase transformers. However, transport may determine the choice of type of single-phase transformers of smaller dimensions. Highways and railways capacities may limit the dimensions and weight of the parts to be transported.

The type of transformer and cooling mode are determined by the requirements of the installation (weather, cave interior), duty cycle, dimensions and weight, and other considerations not directly linked to the plant design itself. Transformers are placed over shafts that collect any oil that leaks from them and pipe it to a shaft with an oil separator system that allows rainwater to flow directly into the river, retaining only the oil to avoid river pollution. They are equipped with wheels, with variable position of up to 90°, and are moved on rails to the maintenance area. There are abacuses for determining the size and weight of transformers for Basic Design purposes, but the ideal solution involves consulting with manufacturers.

11.4 AUXILIARY ELECTRICAL SYSTEMS

Generating units (turbines and generators), elevator transformers, hydromechanical equipment (gates), cargo-handling equipment (overhead cranes, gantries, elevators), mechanical auxiliary systems (drainage, emptying, compressed air, sewage), service water, cooling, fire protection, ventilation, air conditioning, etc.) and plant substation operate on motors having as their source the three-phase low-voltage AC system with the most commonly used rated voltage of 380 or 460 V.

Lighting systems, sockets, telecommunications, as well as the supply of other equipment, devices and electrical loads, have as their source the AC system operating with reduced voltage to 220 V/127 V.

The DC auxiliary service system, powered by accumulator batteries, is used to give the plant greater reliability and operational safety. This system provides power supply to the protect equipment control and control system of the generation equipment and its most used voltages are 125 and 48 V, respectively. This source is independent of the alternating current source (AC) which, due to failure or shutdown of the generating system, may not be available to supply the equipment protection and control system supply in an emergency situation. The same concept applies to the plant's telecommunications system.

11.4.1 Alternating current system (AC)

The AC system has as sources of energy supply:

- the generators of the units, by shunt to an auxiliary service transformer, which reduces the average generation voltage to that of use (380 or 460 V);
- the transmission system to which the plant is connected via an elevator transformer, or an auxiliary services transformer installed in the plant substation;
- the local distribution network near the plant;
- the installation of small power turbine-generator units for medium and high-drop plants (high speed and, consequently, small size/low cost);
- the emergency diesel generator used in the absence of other sources.

The main source used is usually from the generators, where when one of the generators is out of operation the other, or the others, replace it as a source of supply. The design of the AC system, according to the level of flexibility and reliability to be adopted for the operation of the plant, takes into account the number of sources of supply, the arrangement defined in the single-line diagram of the distribution of the load supply boards, the level of automation for the automatic transfer of sources and the redundancy of the power supply circuits. In large hydroelectric power plants, where distances between supply sources and loads are high, the use of medium voltage for power distribution should be evaluated (13.8 kV is the most used) when it is reduced to 380 or 460 V at points near their use. This solution lowers the voltage drop values and reduces the conductor cross-section of the feeders and may be more economical than the low-voltage distribution (380 or 460 V).

11.4.2 Direct current system (DC)

The direct current system uses the batteries of accumulators the most. These batteries can be of the lead-acid or alkaline type. These batteries are sized to meet a particular duty cycle, which is considered to be the most severe operating occurrence of the 125 V DC powered load, which is the most common voltage. Battery chargers are equipment

Electrical equipment: operation and maintenance **391**

designed to keep them capable of supplying a certain defined cycle. Operate connected to batteries meeting system load requests. In the event of a lack of alternating current in the system, the batteries supply the charges. Battery chargers are powered by the 380 or 460 V AC auxiliary service system, depending on the system adopted for those services and, through a set of controlled rectifiers, provide the 125V voltage at the output of equipment connected to batteries and direct current system.

11.5 PROTECTION SYSTEMS

Protection systems for generating plants, switchyards and transmission lines are intended to prevent a failure in any part of the system from spreading. In other words, the system is selective, minimizing the damage caused to the part where the fault occurred and to the system as a whole. In this way, we try to maintain the stability of the rest of the system and minimize downtime. Protection system means the set of protective relays, auxiliary relays, teleprotection equipment and accessories intended to protect various components of electrical power system in the event of electrical failure (such as short circuits) and other abnormal operating conditions.

11.5.1 Protective relays

Protection relays are the main components of the protection systems, since they are the ones that continuously measure the main electrical quantities that characterize the behavior of the electrical system, detect the abnormalities in the system and, due to their severity, initiate the shutdown of the defective section. It should be dedicated to each component of the electrical system – generator, transformer, isolated phase bus and transmission lines – and, in the current state of the art, use digital numerical technology, which may be multifunctional.

11.5.2 Current protection criteria

Each main equipment shall be protected by two completely independent systems, in addition to the inherent or intrinsic protection of each equipment. These systems shall be functionally similar and identified as primary protection and alternating protection. Major equipment, with the exception of substation isolated phase buses, shall not rely on remote rear guard. For substation isolated phase buses, remote rear protection shall be provided to cover the eventual unavailability of their only protection.

All relays that make up the protection systems shall have oscillography facilities. Protection systems shall be integrated at the installation level, allowing, through a dedicated station, local and remote access to adjustments, event logs, oscillography, etc. Analysis software should be included with this dedicated station. The architecture and protocols used should not impose restrictions on the integration of new equipment or the operation of the facility.

To avoid unprotected zones, current transformers supplying protection systems should be installed in such a way as to override the restricted protection zones of adjacent equipment. Restricted protections are those that detect and selectively eliminate faults that occur only in the protected equipment, without any time delay. Examples: differential protection, relay direct communication protection, teleprotection schemes, phase comparison schemes, etc.

The currents and voltages for the supply of each protection system shall be obtained from independent main transformers of current and secondary transformers, different from potential transformers. Each protection system shall have its independent direct current supply, which shall come from separate battery banks and rectifiers.

Digital relays should be able to allow the interaction of Boolean logic elaborated from the "status" of some binary relay inputs with the protection functions, so as to enable their operation only when the logic conditions allow it. The protection system shall provide a 100% degree of redundancy for the various types of electrical faults and, in the case of high-speed protections, allow a total fault clearing time, which is a function of the system voltage level – the higher the system voltage level, the shorter the time. For example, for systems with 230 kV and above, this minimum time is in the order of 100 milliseconds.

Protective relays shall be interconnected with the Digital Supervisory and Control System of the plant or substation in order to comply with ONS Network Procedures through dry contact or serial communication. If serial communication is adopted, IEC 870-5-103, 870-5-101 or DNP 3.0 protocol should be used. The relays shall be insensitive to the saturation of the current transformers to which they connect. In addition, they must have the necessary resources to enable the connection of current transformers with different linearity and nonlinearity ratios and characteristics to perform the same protection function.

11.5.3   Protection of generating nits

The generating units are composed of the turbine/generator set, which is responsible for the electromechanical energy conversion. Because they are made up of rotating machines, the generating units are subject to electrical failure and mechanical failure.

11.5.3.1   Electrical faults

As mentioned above, the electrical faults are related to the generator and its excitation system and are detected by the numerical digital relays and to the protection devices specific to the excitation system. For each protection system of each generating unit (Main Protection and Alternate Protection), the protection functions listed below are recommended.

- Generator percent differential protection (87G).
- Stator winding earth fault protection 95% (64GA).
- Stator winding earth fault protection 100% (64GB).
- Negative sequence current protection (46).
- Over-excitation protection (24 V/Hz).
- Over-voltage protection with instantaneous and timed units (59).
- Overload protection (49).
- Voltage/sub impedance dependent overcurrent protection (51V/21).
- Instantaneous and time phase overcurrent protection (50/51).

- Instantaneous and timed earth overcurrent protection (50/51N).
- Reverse power protection (32).
- Protection against loss of excitation (40): Function (40) should be able to detect the field breaker opening to characterize a loss of excitation of the equipment, in addition to monitoring the underexcitation limits,
- Frequency relay (81).
- Protection against improper energization (50/27).
- Secondary voltage supervision of generator potential transformers (60).
- Loss of synchronism protection (78).
- Breaker failure protection (50FD/62FD).
- Field Protection Relay (64R). The relay with function 64R may be part of the excitation system supply.
- Short circuit protection between turns (61). If the stator winding is constructed with Roebel bars, then short-circuit protection between turns is not required. Otherwise, it is recommended to add this function (61).

Detection of any electrical failure in the generating unit shall cause it to be immediately disconnected from the electrical system by instantaneous (unintentionally delayed) tripping of the circuit breaker connecting it to the electrical system and shutdown of its excitation system.

This actuation will cause load rejection and, consequently, overvoltage and overspeed of the generating unit, which will be controlled by its voltage and speed regulators. For details, it is recommended to consult "IEEE Tutorial on the Protection of Synchronous Generators" (IEEE Catalog Number 95 TP 102).

11.5.3.2 Mechanical faults

The mechanical failures that can occur in the generating units are mainly related to their bearings, their cooling system and their speed regulation hydraulic system. These failures are associated with abnormal temperature elevations, abnormal pressure losses, overspeed, etc. These faults are detected by devices installed in the mechanical systems associated with the generating unit.

In case of actuation of any of these devices, the system generating unit shutdown should only occur after the active power generated has been set to zero to avoid overspeed and increased damage to the failing element.

In the event of a speed regulation hydraulic system failure, the generating unit can reach the runaway speed, which is catastrophic and can even destroy the unit. In this condition, the generating unit shall be disconnected from the system in a manner analogous to the power failure shutdown and shall also initiate emergency closing of the water supply in the case of hydroelectric plants and fuel in the case of thermal plants.

11.5.4 Protection of elevator transformers

The generation voltage is defined by the generator design, taking into account, mainly, its power and its dimensions. This voltage is usually lower than that of the electrical system to which the generator will be interconnected and, therefore, to enable this interconnection, transformers are used that raise the generation voltage to the voltage of the electrical system to which the plant will be connected.

The protection criteria for each step-up transformer are the same as described in Section 11.5.3. It is recommended to also apply the following protection functions:

- Percent differential protection of the transformer elevator with restriction to the second and fifth harmonics (87 TE). Depending on the distance between the powerhouse and the substation, instead of short line differential protection, the percent differential protection of the step-up transformer may be extended to the substation or, in the case of unit arrangement, a differential protection incorporating the generator, the elevator transformer and their interconnection with the plant substation shall be added;
- Overcurrent and neutral current protection (50/51N);
- Overcurrent protection in the elevator transformer neutral (50/51N). Reverse time non-directional overcurrent relay connected to the TC installed on the transformer neutral;
- Short line differential protection (87L) to protect the interconnection between the HV of each elevator transformer and the plant's substation;
- Breaker failure protection (50BF/62BF);

As a transformer generally immersed in oil, in addition to these functions, the transformer also has its intrinsic protections related to internal overpressure. Detection of any fault in the elevator transformer shall cause it to be immediately disconnected from the electrical system together with the generating unit by instantaneous shoot (without intentional delay) of the high- and low-voltage circuit breakers (if any).

This actuation will cause load rejection and, consequently, over-voltage and over-speed in both the elevator transformer and the generating unit, which will be controlled by the voltage and speed regulators of the generating unit.

11.5.5 Transmission line protection

The protection criteria for the transmission line are the same as described in Section 11.5.3 and the protection functions for lines with voltage equal to or greater than 138 kV described below are as recommended.

- Distance protection for phase to phase and phase to ground fault detection with zone independent timers (21/21N);
- Synchronism check function (25SC);
- Protection against undervoltage (27L).
- Overcurrent protection against defects during transmission line energization (50LP);
- Overcurrent protection against transmission line gap defects with open line isolating switch (50SB);
- Overvoltage protection (59);
- Neutral directional overcurrent protection (67N);

Electrical equipment: operation and maintenance   395

- Distance protection blockade function, function of power swing (68);
- Protection against loss of synchronism (78);
- Automatic reclosing function (79).

The protection sets shall operate with independent telecommunications equipment. It is recommended to use optical fibers, installed in the transmission line protection cables, as physical means for the traffic of the communication signals between the protections located in their terminals. Each protection set must be able to operate correctly with either, or a combination of the following teleprotection schemes:

- Directional lock type comparison;
- Directional unlock type comparison;
- Direct, underreach trigger transfer (DUTT);
- Allowable trigger transfer with underreach (PUTT);
- Permissible, overreaching trigger transfer (POTT).

Protection systems shall be multi-processed, with independent algorithms for each function (21, 67N, 68, 25, 27, 59, 78, 79, fault locators, etc.).

Each set shall be provided with facilities for performing single- and three-pole breaker shutdowns and reclosures following the occurrence of the internal short circuit. Recloser type selection must be done by the operator.

11.5.6   Breaker failure protection

All circuit breakers shall be protected against failure. For each breaker, the protection shall consist of a current supervision function, an adjustable timer, a high-speed trip relay (with the necessary contacts to energize the adjacent breaker trip coils) and a lock relay (capable of, from trip transfer schemes, to open remote line terminals and to block circuit breaker closing and/or reclosing).

This protection must be activated by the protection functions (of the generating units, or of the transmission lines, of the transformers or of the bars) that give command of opening of the respective circuit breakers. Current detectors should be insensitive to the "post fault" transients that appear on the secondary of current transformers. The protection must also act correctly in the event of a "no current" fault, for example when receiving trip transfer signals or the transformer buchholz relay actuation.

For plant substation circuit breakers, breaker failure protections may be integrated into the busbar protection system. For these circuit breakers, the total fault clearing time by the circuit breaker failure scheme, including the auxiliary relay protection relay operating time and the breaker opening time, shall not exceed 250 m/s.

11.5.7   Substation bar protection

The busbar protection shall be performed by digital relay, numeric, with differential principle, by percentage differential overcurrent. An independent set of protection shall be provided for each substation busbar. The protections shall be selective,

396 Design of Hydroelectric Power Plants – Step by Step

turning off only the circuit breakers connected to the faulty busbar and having the ability to detect ground faults and phase to phase faults.

Each protection set must be provided for each substation span with a stub-bus function to cover the connections between the current transformers and the circuit breakers. The operating time shall be less than or equal to 20 m/s and be independent of the number of connected spans and protective functions used in the system.

Moreover, each protection set must have immunity to different current transformer saturation levels, with stability for external faults and sensitivity for internal faults. It should also allow connection to a current transformer with different magnetic characteristics and transformation ratios.

There should be no limitations on the resistive load of the secondary windings, i.e., it should be possible to connect via long cables. There should be supervision for the secondary windings of the current transformers within their area of operation, with actuation lock and alarm in case of secondary circuit opening.

Each protection assembly must shut down and lock all protected busbar circuit breakers, and each assembly must have a self-diagnostic and self-monitoring system that supervises its hardware and software and identifies the defect found.

11.6 SUBSTATION INTERCONNECTION OF THE PLANT TO THE SYSTEM

11.6.1 Switchyard, or substation, equipment

The plant switchyard, or substation, connects the power plant to the associated transmission system, raising the value of the generation voltage to the value of the transmission voltage through the elevator transformers. It is composed of the following equipment:

- elevator transformers;
- switchgear: circuit breakers and disconnecting switches;
- instrument transformers required for protection and measurement systems;
- surge arresters;
- control, protection, measurement and supervision system;
- ancillary service systems.

Circuit breakers have the function of maneuvering and shutting down the circuits when the system is experiencing an overload or a fault which may be a short circuit.

Disconnecting switches are equipment that can only be operated without load and have the function of isolating the equipment in operating maneuvers and for the maintenance of the substation.

Instrument transformers consist of current transformers and potential transformers.

Current transformers reduce the primary system current values to values suitable for protection and measuring instruments (below 5 A).

Potential transformers reduce the value of the primary high voltage to values suitable for protection and measuring instruments (below 115 V).

Surge arresters have the function of limiting overvoltage caused by electrical system voltage surges or lightning strikes to values that can be supported by substation equipment, protecting them from damage.

11.6.2 Other components and installations

In addition to high-voltage equipment, switchyards have the following facilities:

- metal or concrete structures, with the purpose of anchoring the transmission lines and supporting the internal isolated phase buses of the substation and its high-voltage equipment;
- control room where the control, protection and measurement panels, telecommunications equipment, and alternating and direct current auxiliary switchboards are located;
- stormwater drainage network, channels and cable ducts, grounding grid (for protection of equipment and persons against over-voltages), protection system against atmospheric discharges, lighting and water and sewage systems.

11.6.3 Switchyard types

Switchyards can be installed over time, with equipment being placed in a courtyard, or they can be indoor, protected from the weather by a ceiling. In the 1960s, with the need to reduce space, the SF6-sulfur hexafluoride isolated substation was designed, which brought about a huge reduction in the installation area, reaching up to 80% of the space of a conventional substation.

11.6.4 Equipment arrangements

Elevator transformers can be installed near the powerhouse or in the switchyard area. For small plants, including SHPs, transformers are usually installed in the switchyard. For medium and large plants, the transformers are usually installed near the powerhouse, due to the high values of the generation current, which leads to the need to use isolated phase buses that cannot be extended to the switchyard site due to cost reasons and difficulties in the route. It does not occur in the case of SHPs which can be connected by insulated medium voltage cables installed in duct.

11.6.5 Maneuvering schemes

A high-voltage switchyard (AT) will have an electrical scheme and equipment arrangement depending on the importance of the system to which it belongs. As the importance increases, there should be more alternatives for its operation in order to satisfy the highest demands for continuity of service provision.

The degree of response to contingencies that may occur depends, therefore, on the design of its electrical circuits and the distribution of equipment, which is commonly referred to as its maneuver scheme. Alongside the factors that contribute to the selection of the maneuver scheme, it is observed that the experience of the operation of existing facilities plays a major role.

11.6.5.1 Simple bar

This is a maneuver scheme widely used in SHP (Figure 11.34). The initial investment is minimal, but there is maximum simplicity and good defect identification, although there is in general, low reliability, low flexibility, and it requires full shutdown of boom defects for extension or maintenance. From the standpoint of arrangement, it requires minimal area and allows exits in any direction without intersections (provided expansions have been anticipated).

11.6.5.2 Main transfer bar, single breaker

This simple maneuvering scheme is widely used in small and medium sized plants (Figure 11.35). In it, all circuits are connected to the same bar. With the circuit loaded, it allows the possibility of maintaining or replacing a circuit breaker using the transfer circuit breaker. Regarding this replacement of a circuit breaker, the switching maneuvers require the transfer of the respective closing operation protection.

11.6.5.3 Double bar, single breaker

This maneuvering scheme is used in medium and large plants (Figure 11.36). For all circuits, this scheme offers the possibility of connection to any of the isolated phase buses by choosing the disconnecting switch of the selected busbar.

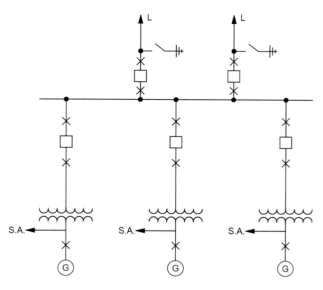

Figure 11.34 Simple bar – single-line diagram. X = Disconnecting switches, □ = Breakers, S.A. = Auxiliary services, G = Generator.

Electrical equipment: operation and maintenance 399

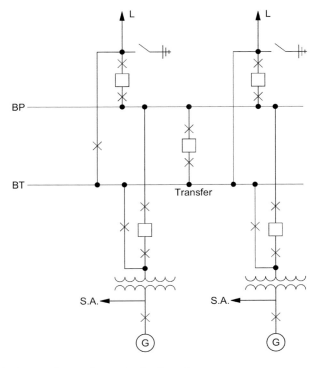

Figure 11.35 Main bar and transfer – single-line diagram.

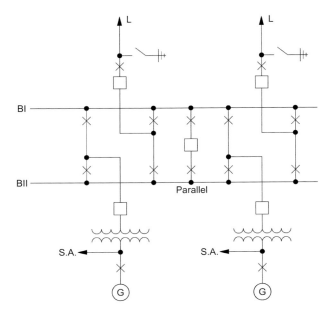

Figure 11.36 Double bar, single breaker – single-line diagram.

The change of connection of a circuit from one bar to another is made possible by the existence of a parallel circuit breaker, which allows this operation without having to disconnect the corresponding circuit.

This maneuvering scheme does not allow the circuit breaker to be maintained with the corresponding circuit under load.

11.6.5.4 Double bar, single circuit breaker with bypass disconnecting switches

This arrangement allows the selection of any of the bars as a transfer bar. As a result, it entails additional complications in blocking schemes and operating procedures.

It is essential to perform the protection transfer maneuver before closing the by-pass disconnect switch, otherwise, if a short circuit occurs in the transferring bay during the closing operation, it will open circuit breaker and will allow the short-circuit current to pass through the closing disconnecting switch and damage it.

Its reliability is far less than that of the double breaker, ½ breaker, single ring and multiple ring. It may require closed circuit change maneuvers by actuating disconnecting switches, which makes it not advisable for voltage levels above 345 kV.

The command, control and protection of circuits is relatively complex due to the large number of interlocks and the mentioned need for protection transfer. The enlargements do not present great difficulties and the physical arrangement is simple (Figure 11.37).

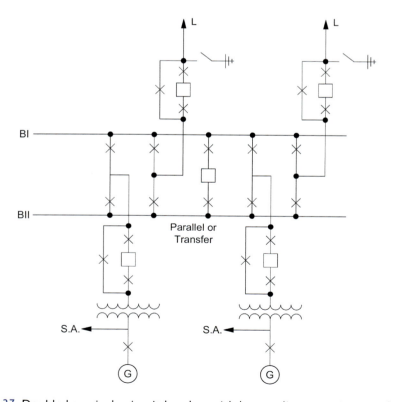

Figure 11.37 Double bar, single circuit breaker with bypass disconnecting switches.

This diagram is rarely used in hydroelectric plants, but is often used in switchyards of the power transmission system.

11.6.5.5 Double bar and transfer bar

This scheme has the same inconveniences as the double bar, single circuit breaker with bypass disconnecting switches type for voltage levels greater than 345 kV.

Its merits are significantly superior in terms of the greater simplicity of command, control and protection and the fact that the maneuvers of changing circuit conditions require the use of fewer number of disconnecting switches (Figure 11.38).

However, it is inferior in the ease of arrangement aspect when there are outlets on both sides of the busbars. The occupied area is also larger than the previous design.

This diagram is rarely used in hydroelectric plants, but is often used in switchyards of the power transmission system.

11.6.5.6 Double bar, one breaker and a half

This scheme has lower reliability than the double bar/double circuit breaker scheme that will be presented below. Operating flexibility has the disadvantage that circuit

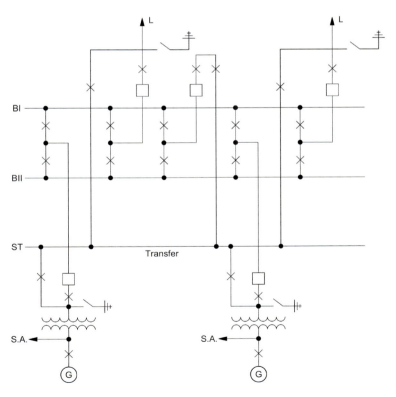

Figure 11.38 Double bar and transfer bar – single-line diagram.

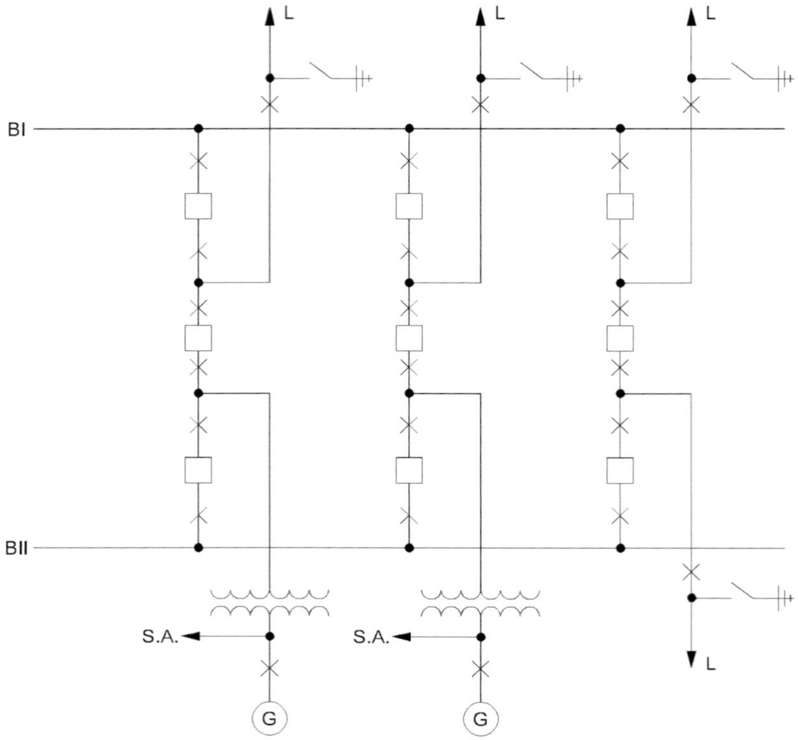

402 Design of Hydroelectric Power Plants – Step by Step

Figure 11.39 Double bar, one breaker and a half – single-line diagram.

breakers are required by more than one circuit. This effect is mitigated in relation to conjugated bays that can be kept in service even with defective bars.

The command, control and protection system are quite complex as the middle circuit breaker serves two circuits. However, it requires fewer switchgears than the double bar/double circuit breaker scheme. In maintenance, two or three voltage levels make work difficult, but it is easy to isolate parts of the circuit for these services.

Expansions must be provided for and arrangements should be made for them in order to avoid disturbances to the installations in service. The physical arrangement is simple, especially when conjugate bays are placed in opposite directions.

The following diagram is rarely used in hydroelectric plants, but is often used in switchyards of the power transmission system (Figure 11.39).

11.6.5.7 Double bar, double breaker

The following scheme (Figure 11.40) has the best rates for all factors. However, it is the most expensive design scheme, and for this reason, it is rarely used in hydroelectric plants, but it is used in switchyards of the power transmission system.

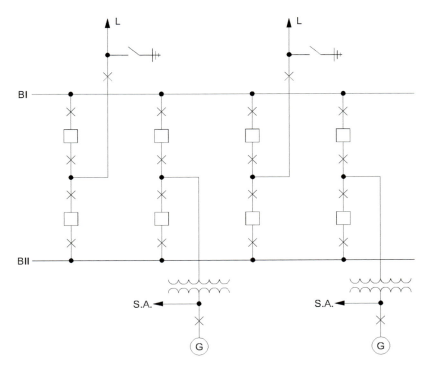

Figure 11.40 Double bar, double breaker– single-line diagram.

11.6.6 Maneuvering scheme selection criteria

In choosing the most suitable type for substation, it will be necessary to list the influencing factors, to select them as best as possible by their independent characteristics and to quantify them. The sum of the quantification of each alternative will be the indicator that will support the decision. The following factors are often taken into consideration: cost, reliability, operating flexibility, simplicity of the control system, control and protection, ease of maintenance, simplicity of physical arrangement and standardization.

The difficulty of ordering the various factors by their importance and defining them by exact values is well known. Notice, for example, how difficult it is to value the standardization factor of one in relation to the others. In a superficial assessment, it can be concluded that the cost factor will be decisive. However, its weight will tend to decrease as this assessment attempts to weigh the damage caused to industrialized centers by power outages, even for a short time. The reliability of the substation that serves these centers is a factor inversely proportional to the cost and may derive its apparently preponderant value.

Also the interdependence of the factors indicated should be noted. In fact, it is not easy to establish the boundaries of analysis between reliability and operational flexibility, simplicity of the control, control and protection system, and ease of maintenance, as all others compete for the first factor.

In addition to the complexity of the characterization of each of these factors, motivated by the multiplicity of influence, they have an interrelationship that exists between some of them. It must also be considered that it is difficult to give certain evaluations a rigorous value, given that the definition of relative weights between factors usually appears with a subjective influence.

Note that in evaluating the merits of each scheme type for the substation, one must not forget that it will be an integral part of a network. There will be no objective sense in assigning, for example, a degree of reliability to a substation that is much higher than a single line that feeds it. On the other hand, it may happen that a network is supplied in whole or in part by either of two sufficiently close switchyards. This fact will eventually advise choosing a less reliable but more economical type of scheme for switchyards, without diminishing the reliability of the assembly as the switchyards complement each other.

With the aim of a more accurate ordering of merits, the solution is to study and evaluate circumstantially the behavioral hypotheses of each switchyard element, within each type of scheme and arrangement, keeping in view the various situations that may arise. A study in this direction will hardly be free from an analysis of subjective characteristics, if not supported by sufficiently numerous and precise experimental statistical data.

Additional consideration is that the frequency of maneuvering of the generating unit circuits varies with the nature of the plant, being minimal in base plants, higher in peaking power plants and maximum in reversible plants. Although the reliability of the generating units and their auxiliaries is actually lower than the associated high-voltage equipment, basic plants should be prevented from contributing to unnecessary shutdowns that waste energy or shorten the life of the generating unit. Hence, if these plants require maneuvering equipment whose maintenance interval is quite long, or even redundant.

In peaking power plants, however, this equipment, although much more requested, can be checked and maintained frequently during scheduled generation interruptions and an unforeseen shutdown does not harm them in the same way. In addition, Brazilian plants are rarely "base power plants" after the project is fully installed, and there may always be some units available for short shutdowns daily.

When choosing the circuits of the generating units, the unit power of these units relative to the output of the plant and the system must also be considered. Indeed, the higher that power relative to the capacity of the plant, the more harmful the loss of a unit on the stability of long substation-linked transmission lines, and the higher its value relative to the total system capacity, the more severe the effects of unit loss over system frequency and voltage. Thus, for the same output, in the first approximation, simpler the circuits of the units, higher their number and the lower their unit power.

In the vast majority of plants, the single-circuit, single-breaker scheme on the high-voltage side seems most appropriate. Only in plants with a small number of units and which represent an appreciable share of the generating capacity of the system, a more complex scheme can be justified, either in these circuits or in the bus itself (for example, with the use of a bypass disconnect switch in the unit circuit breaker).

11.6.7 Powerplant connection to electrical system

In design planning, one of the items considered in the attractiveness of the project is the connection of the powerplant to the electrical system. In Brazil, interested parties should consult the system operator (ONS) for details. It is noteworthy that the Brazilian system has size and characteristics that allow it to be considered unique worldwide.

The generation voltage (medium voltage) is raised through the elevator transformer to the transmission voltage (high voltage), appropriate to the power to be transmitted.

Power generated at the plant is transferred to the electrical system through the powerplant substation that will be connected to the electrical system through an existing system substation (called the receiving substation), or by sectioning a transmission line at the appropriate voltage.

11.6.7.1 Receiving substation

The cost of the line gaps to be added to the existing receiving substation (Basic Grid) and all necessary adaptations in this substation to integrate these plant gaps into the interconnected system is also included in the investment of the power plant's connection to the electrical system.

11.6.7.2 Transmission line

The design of the stretch of the transmission line that connects the powerhouse (elevator transformer) to the plant's substation, as well as that of its transmission line to the receiving substation, must be prepared by the investor and will be included in total investment. The civil design will involve surveying the implementation range and preservation areas, as well as conducting surveys at the towers' base locations.

The electrical design will involve the definition: the power to be transmitted (MW); the transmission voltage (kV); line structures (concrete or metal structures); the conductor cables; arrester cables; fiber optic cable; protection against atmospheric discharges and over-voltages; and wind stress in the region and seismic effects, if applicable.

Typically, power companies have standard designs for line and substation civil works as well as electrical designs.

11.7 OPERATION AND MAINTENANCE

This subject is not within the scope of this book, but a brief mention of the importance of this complex subject is necessary. The operation and maintenance of the powerplants over their useful life must be done by the operation team in accordance with the rules and routines defined in specific manuals prepared for this purpose, both for civil structures and for electromechanical equipment, in order to ensure energy production. Equipment stops must be reduced to those required for maintenance. If both the plant and equipment have been designed, built and manufactured to internationally recognized technical specifications, the life expectancy of parts is long. Experience has shown that units have been operating under a full range of conditions with generators requiring no spare parts for 30–40 years. It has hydraulic turbines that continue to operate efficiently even after 50 years (see Chapter 10; Gulliver and Arndt (1991) and USBR (1965)). Regarding the checklists for inspection and maintenance work the main topics are:

406   Design of Hydroelectric Power Plants – Step by Step

- dams and canals;
- powerplant buildings;
- penstocks, gates and valve;
- cranes, hoists and elevators;
- miscellaneous station auxiliaries;
- hydraulic turbines, governors, and large pumps;
- generators, motors, and synchronous condensers;
- low-voltage switchgears, buses, and cables;
- oil and air circuit breakers;
- transformers and regulators;
- disconnecting switches and fuses;
- lightning arresters;
- switchboards and control equipment;
- communications equipment;
- switchyards and substations.

Each of these topics is subdivided into several subtopics with their respective inspection intervals: daily, weekly, monthly, quarterly, semiannual and annual. Prudence, knowledge and diligent attention to detail are necessary characteristics of the team involved in keeping the hydroelectric power station running, producing electricity. The appreciation and preservation of the heritage represented by hydroelectric facilities during their lifetime involves a number of multiple issues, which come up to ensure both a high level of safety and a high level of performance (response to facility requirements and availability).

Asset management is carried out within a context of strong demands, particularly within the areas of improved economic, social and environmental performance. The aging of the hydroelectric park and its increased level of demand are also two important elements. The multiplicity of these diverse requirements has a direct impact on the governance of proprietary companies. It requires managing risks globally, coherently, and as per the appropriate level of organization.

Chapter 12 presents an approach to control those requirements that are inherent to hydroelectric generation activities in order to develop and perpetuate the practice of risk analysis and management.

Chapter 12

Construction planning

12.1 CONSTRUCTION PHASES

The phases of construction depend on the types of arrangement of hydroelectric plants, which in turn depend on the topographic/geological aspects of each site. In conventional designs, two diversion phases are usually provided. In some projects, however, the diversion may have to be done in three phases, as the Tucuruí HPP. The specifications for the civil, mechanical and electrical aspects of the project, common around the world, are not part of the scope of this book.

12.1.1 First phase diversion

In the case of dams in open valleys, usually in the first phase of diversion, a cofferdam is built that strangles part of the river bed and keeps the river flowing through the lateral channel.

In the case of dams in narrow valleys (V) the river is usually diverted by a tunnel or side diversion gallery on one of the banks.

The area is depleted to allow for partial construction of the plant's permanent structures, dams, water intake, powerhouse, spillway, including the culverts through which the river will flow in the subsequent diversion stage.

12.1.2 Second phase diversion

In the second diversion phase, the side channel is closed and the river is diverted to other conduits which may be: the culverts under the spillway; lowered spillway spans; another tunnel; or diversion by intake - powerhouse. As an example of diversion, we present the case of the Tucuruí HPP in the lower section of the Tocantins River (Northern Region). The diversion was expected to be done in two phases. During the course of the works, the discovery of the deep channel in the riverbed caused the change to occur in three phases of diversion. The first phase remained as originally planned, with strangling part of the river chute. A new second phase was introduced by building a cofferdam to protect the deep channel in the middle of the river and enable its treatment. The river flowed through a 40 m wide channel near the right bank (Figure 12.1). In the third phase, this channel was closed (Figures 12.2) and the river was diverted by 40 culverts under the spillway (Figure 12.3).

DOI: 10.1201/9781003161325-12

408 Design of Hydroelectric Power Plants – Step by Step

Figure 12.1 Tucuruí HPP. Third phase of river diversion. Foreground – river flowing down the canal near the right bank. In the middle of the river is the "new second phase cofferdam" protecting the deep channel area. In the background, there is also the first phase cofferdam isolating the spillway and intake/powerhouse area; you can also observe in the left bank the facilities of the construction site and part of the residential village.

Figure 12.2 Tucuruí HPP. Third phase of river diversion. River closure (final breach). Observe the gap (Δh) at the tip of the landfill and the acceleration of the flow. Also observe three people positioned at risk condition in the foreground.

Construction planning 409

Figure 12.3 Tucuruí HPP. Third diversion phase: river through the culverts.

The diversion design flow was changed from 51,000 to 56,000 m³/s (recurrence time = 25 years) as a result of the review of flood studies following the exceptional flood in March 1980 – 68,400 m³/s.

Before entering the text about the "river diversion design", special mention should be made of the block slip that occurred in the Tucuruí HPP on the slope of cofferdam B in the embrace of the cell wall. This solution, a "cylinder" shaped wall of cells (metal planks) filled with sand, was adopted to move the cofferdam slope away from the turbulent flow zone during the 3rd phase of diversion by the culverts at the end of January 1982 (see the sketch and pictures in Figures 12.4–12.10). In this period the flow reached 41,000 m³/s.

12.2 RIVER DIVERSION DESIGN

The river diversion design is directly linked to the type of layout foreseen for the hydroelectric plant works, which in turn is conditioned by the topographic and geological-geotechnical characteristics of the site (as mentioned in Chapter 3). For each type of arrangement, the river diversion should be thoroughly analyzed.

12.2.1 Discharges and risks

The river diversion shall be designed for the discharges determined according to the methodology presented in Chapter 4. Depending on each entrepreneur/owner and considering the types of contracts to perform the works, the risks to the diversion phases have varied greatly.

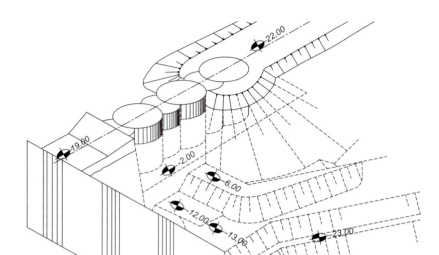

Figure 12.4 Tucuruí HPP – Third phase of river diversion. Perspective of cofferdam B hugging the cell wall.

Figure 12.5 Tucuruí HPP. Cofferdam B hugging the cell wall, concrete slab and rock interface.

Figure 12.6 Tucuruí HPP – Third phase of river diversion through the culverts. Left concrete guide wall, cells and cofferdam B. Turbulent flow – $Q = 41,000 \, m^3/s$.

If the cofferdam encompasses the powerhouse construction area that, if flooded, produces high damage, the project has been designed for the flow with a recurrence time (TR) of 100 years, even to meet the requirement of insurance companies. Naturally, it results in a higher cofferdam with higher risk of overtopping, but the cost of the work is much higher.

If the cofferdam only encompasses the region of the dam and spillway works, with excavation, concreting and embankment services, which if flooded does not cause high losses, the project has been designed for flow rates with lower TRs, thus admitting increased risk of overtopping. These projects have been made for 25 years TR. The Brazilian experience records cases of projects that aimed to reduce the cost of the diversion works, and contemplated the overtopping of the first phase of cofferdam (CBDB, 2009- examples of the designs of Corumbá I HPP and Serra da Mesa HPP).

It is known that the risk or probability r of a flow with "TR" recurrence years, considered here as the river diversion design flow, to be equaled or exceeded at least once in n years is given by the equation:

$$r = 1 - (1 - 1/TR)^n \tag{12.1}$$

where:

$1/TR$ = probability of occurrence of design flow in any given year; and
n = diversion phase duration time.

Figure 12.7 Tucuruí HPP – Third phase of river diversion through the culverts - cofferdam B, cells interface, broken concrete slab after sliding rockfill blocks.

In Brazil, the design must consider the risk values that are presented in Frame 12.1.

Substituting these values in the risk formula (Equation 12.1) for various times of diversion phases gives the respective probabilities. The higher the "TR" the lower the risk r for a given n (Table 12.1). The higher the TR the greater the design flow of the river diversion. The cofferdam quotas will be higher and the risk of being hit and climbed will be lower. Of course, the cost of diverting the river and the enterprise as a whole will be higher.

Construction planning 413

Figure 12.8 Tucuruí HPP – Third phase of river diversion through the culverts - cofferdam B. Repair work begins by laying the rock block rosaries.

Figure 12.9 Tucuruí HPP – Third phase of river diversion through the culverts - cofferdam B, Bar of rock blocks in formation.

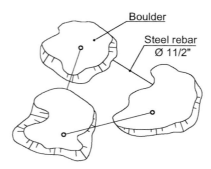

Figure 12.10 Tucuruí HPP. Third phase of river diversion. Rock blocks rosary. The rock blocks were at least 1.5 m^3 and should be 50 cm spaced. In addition, the rosary should weigh a maximum of 20 tons.

Frame 12.1 Risk Criteria (CBDB/Eletrobras, 2003)

Damage Category	Annual Risk
There is no danger of loss of human life, nor is it anticipated that major damage will occur to the work and its progress.	5.0%–20%
There is no danger of loss of human life, but major damage to the work and its progress is already expected.	2.0%– 5%
There is some danger of loss of human life and major damage to the work and its progress is predicted.	1.0%–2.0%
There is a real danger of loss of human life and major damage to the work and its progress is predicted.	<1.0%

Table 12.1 Risk Variation for Arbitrary Values of "TR" and n

TR (years)	r (%)		
	n = 3	n = 2	n = 1
2	87.50	75.00	50.00
5	48.80	36.00	20.00
10	27.10	19.00	10.00
25	11.53	7.84	4.00
50	5.88	3.96	2.00
100	2.97	1.99	1.00
200	1.49	1.00	0.50
500	0.60	0.40	0.20

CBDB (2009) provides statistics on river diversion types in Brazil, which relate design flows and their respective "TRs" to the diversion stages in case of a large number of projects. This statistic is summarized in Figure 12.11. The most frequent occurrences are the 25 and 50 years, but there are some cases of 100 years TRs, as well as, TRs for the time range between 2 and 5 years for SHPs.

Construction planning

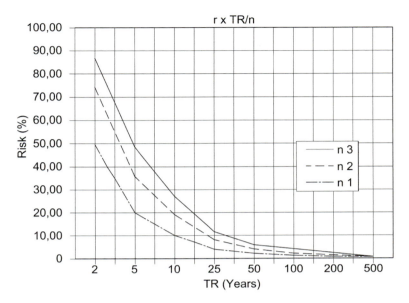

Figure 12.11 Risk variation (r) for arbitrary values of "TR" and n.

12.2.2 Phases of river diversion

For open valleys plants designs with direct development with no permanent diversion usually the river diversion is conceived in two phases:

- in the first phase the river is diverted to a channel dug in one of the banks; in this phase, in the dry area, the structures of the spillway/culverts, intake and powerhouse and closing dams are built;
- in the second phase the river is diverted by the culverts under the spillway and the construction of the generation circuit and of the dam is completed; at this stage, the intake and the powerhouse are protected either by a cofferdam or by the upstream cofferdam gates and downstream of the powerhouse; there are several possible cases.

The river diversion design for closed-valley plants, V, with direct developments without diversion, usually considers the river diversion in two phases:

- in the first phase, with the river flowing into its natural channel, the tunnels, or diversion galleries, are built on one of the banks; concomitantly with the execution of the tunnels, or galleries, the construction of the structures of the hydraulic adduction and generation circuit begins; if necessary, depending on the schedule and arrangement of the works, a cofferdam can be built in the powerhouse region;
- in the second phase, the river is diverted by the tunnels, or galleries, and the construction of the generation circuit and of the main dam on the riverbed is completed.

416 Design of Hydroelectric Power Plants – Step by Step

The river diversion design for the closed valley works with diversion development. It considers river diversion in two phases:

- in the first phase, with the river flowing into its natural channel, the adduction tunnel is built on one of the banks; concurrently with the execution of the tunnel, the construction of the spillway, dam and powerhouse structures begins on the dry stretch of river or banks;
- in the second phase, the river is diverted by the spillway galleries, or the lowered spillway blocks, and the construction of the generation circuit and the dam is completed.

12.2.3 River diversion dimensioning

In the hydraulic dimensioning of the partial river choke and in the estimation of the waterline profile along the cofferdam, the HEC-RAS program is used, which incorporates the SSM – Standard Step Method. The design of the rock blocks for the construction of river closure dikes is based on the Isbash formula, as defined in:

- Isbash, S.V. Construction of dams by depositing rock in running water, Communication nº 3, 2º ICOLD, Washington, DC (1936).
- USACE, HDC (1973), Chart 712-1, Stone Stability x Stone Diameter;

The phenomenon was thoroughly stdied by Isbash at the Scientific Research Institute of Hydrotechnics (SRIH), Leningrad (now Saint Petersburg) from 1930 to 1935. It used the data from river diversion works of powerplants under construction in the Soviet Union, and also from extensive laboratory testing campaigns in which the process of launching the rockfill coffer was simulated.

Applying Wilfred Airy's Law to the movement of a block of rock under the action of runoff, Isbash defined his equation as follows:

$$V = C \left[2g \left(\frac{\gamma_r - \gamma_a}{\gamma_a} \right) \right]^{0,5} D^{0,5} \tag{12.2}$$

where:

V = flow velocity (m/s);
C = Isbash stability coefficient (0.86–1.20);
g = gravity acceleration (m/s^2);
γ_r = rock specific weight (tf/m^3);
γ_a = water specific weight (tf/m^3);
D = rock blocks diameter (m).

For further use, it is reported that the diameter (D) of a spherical block of rock as a function of its weight (W) is given by formula 12.3:

$$D = \left(\frac{6W}{\pi \gamma_r} \right)^{1/3} \tag{12.3}$$

12.2.4 River diversion – execution

At the beginning of the closure, the tractors push large amounts of loose rock blocks into the river without overlapping. The HDC (Chart 712-1) records that in this case the movement is sliding.

The speed that will remove these loose blocks at the top of the landfill has been set to a coefficient $C = 0.86$ in laboratory tests, confirmed by field measurements.

The construction process is intense. Tractors continue to throw blocks into the river in large quantities, an always well-planned operation. The landfill evolves and begins to take shape in a trapezoidal shape, functioning hydraulically as a broad crest weir submerged low monolith (HDC Chart 711).

The blocks at the top of the landfill are now "protected" juxtaposed with each other, already showing some imbrication. Under these conditions, the blocks can withstand a maximum speed for $C = 1.20$. The HDC (Chart 712-1) records that in this case the movement is rolling or overturning.

The embankment proceeds and exceeds the water level and already has a characteristic track until the river closes. Head and energy losses are estimated by classical hydraulics and are also measured in the reduced model.

Table 12.2 lists the block diameters required for river closure as a function of the flow velocity at the tip of the embankment, considering the coefficient C values already mentioned and the specific rock weight equal to 2,700 kgf/m^3.

In the author's opinion, whenever possible, the gap of the flow at the tip of the landfill should be limited to 1.30 m, which would imply a flow velocity of 5.0 m/s and would require blocks with a diameter ranging from 0.50 to 1.00 m (see Figure 12.2).

A larger difference, 1.8 m, for example, would increase the speed to 6.0 m/s and the diameter of the blocks would increase to the range of 0.75 m–1.46 m, which are already large, special blocks.

Table 12.2 River Closure – Diameter and Weight of Blocks

Velocity (m/s)	Loose Block C = 0.86	Protected Block C = 1.20	Loose Block C = 0.86	Protected Block C = 1.20
	$D = 0.0405\ V^2$ (m)	$D = 0.0208\ V^2$ (m)	$P = 1,413.72\ D^3$ (kgf)	
2.0	0.162	0.083	6.0	0.8
2.2	0.196	0.101	10.6	1.5
2.4	0.233	0.120	17.9	2.4
2.6	0.274	0.141	29.1	4.0
2.8	0.318	0.163	45.5	6.1
3.0	0.365	0.187	68.7	9.2
3.2	0.415	0.213	101	13.7
3.4	0.469	0.241	145.8	19.8
3.6	0.525	0.270	204.6	27.8
3.8	0.585	0.301	283	38.6
4.0	0.649	0.333	386.4	52.2
4.2	0.715	0.367	517	70
4.4	0.785	0.403	683	93
4.6	0.858	0.441	892	121
4.8	0.934	0.480	1.152	156
5.0	**1.013**	**0.521**	**1.471**	**199**
6.0	1.46	0.75	4.369	594

418 Design of Hydroelectric Power Plants – Step by Step

But if it is indeed mandatory, these special blocks should be selected and used but the operation has a higher cost.

It should be noted that the author witnessed the launching of large rock block "rosaries" in the Tucuruí HPP in January 1982, to protect the cofferdam B hugging the cell wall (shown earlier in Figures 12.5–12.9).

12.2.5 Hydraulic models

In addition to theoretical dimensioning, the diversion design of a large river is tested and optimized in tests related to reduced hydraulic models. In these models, extensive testing campaigns are programmed for the full range of flows that may occur during the construction phases, in which:

- accommodating the flow at the inlet of the diversion structures for the full range of predicted flows to achieve a smooth flow without localized turbulence;
- the flow capacity of the channel, tunnel or diversion gallery, as well as the cofferdam elevations;
- the protective materials of cofferdam and of the diversion channel slopes;
- the dimensioning of the rock blocks required for the construction phases of the cofferdams, to protect the slopes exposed to contact with the flow and to close the final breach;
- the overall stability of the cofferdam and erosion trends; eventual deflecting structures to detach the flow of the slope in longitudinal sections of the cofferdams, etc.;
- behavior with lower downstream water levels, simulating flow-rating curve inaccuracies, that worsen hydrodynamic conditions along the cofferdam slopes and, as a result, can lead to problems with uncontrolled erosion and even rupture situations.

12.3 CONSTRUCTION PLANNING

The planning of the construction is done considering the layout and the steps involved in planned construction. Compliance with the project's implementation schedule, and even an anticipation of the commercial generation deadline, is a goal that has been rigorously pursued because it positively affects the internal rate of return of the invested capital.

In Brazil the planning became common after the "partial" privatization of the electricity sector. It is noteworthy that the implementation times after the changes were drastically reduced. Construction methods and techniques have successfully evolved to meet new contractual challenges and milestones.

The success, in addition to changes in constructive technologies, however, could only be achieved on the basis of experience gained over more than 40 years of state works carried out between 1960 and 2000.

In the past, in works without concrete cooling, the height of the concreting layers was limited to 1.5 m and the recovery time between layers was 76 hours. Nowadays, sliding forms have been routinely used even in massive, precast structures, which provided a substantial reduction in the time of implantation of these structures.

Table 12.3 shows the average progress of concreting in the works of the 1970–1980s. The average advance in relation to the dam height was 3.33 m/month.

Construction planning 419

Table 12.3 Concrete Structures – Advances (Works of the 1970s–1980s)

Structure	Powerplant	Height (m)	Time (month)	Advance (m/month)
Spillway	Itaúba HPP	31.2	9	3.46
	Palmar HPP	48	14	3.43
	Salto Santiago HPP	38.3	8	4.79
	Foz do Areia HPP	40	19	2.11
	Avanço médio	-	-	3.45
Intake	Itaúba HPP	33.4	10	3.3
	Palmar HPP	53.5	21	2.55
	Salto Santiago HPP	57	22	2.6
	Foz do Areia HPP	40	18	2.22
	Avanço médio	-	-	2.67
Powerhouse	Itaúba HPP	40	15	2.7
	Palmar HPP	75	22	3.4
	Salto Santiago HPP	47	12	3.9
	Foz do Areia HPP	43.5	13	3.3
	Avanço médio	-	-	3.33

It is noted that the work of the 74 MW Monjolinho powerplant was carried out in 2.5 years, between July 2007 and December 2009 in the Passo Fundo river, Rio Grande do Sul. The 33 m high powerhouse was built in 5 months with full use of sliding forms. For the purposes of the schedule, if we had considered the advance of 3.33 m/month, the construction period would have been double.

Roller-compacted concrete dams, as well as concrete-faced rockfill dams have also had shorter construction times. Production of the main civil works services were all reduced, whether excavations of open-pit foundations or underground excavations of the many powerhouses and tunnels executed during this period (Figure 12.12).

Following are some self-explanatory photos of the construction of the spillway of the Tucuruí HPP spillway construction, from October 1981 (Figure 12.13).

It is also worth mentioning the construction of tunnels and shafts of hydroelectric plants in rock masses, which in Brazil are mainly performed by the traditional method (drilling, loading and blasting). The most recent TBM (Tunnel Boring Machine) method was little used in hydroelectric works in Brazil (Figure 12.14).

For illustration purposes only, it should be noted that, with the tunnel section defined, as set out in Section 9.4.2, the construction of the tunnel should be planned, which includes the excavation front geometry, feed length and curb shape; the need for mass treatment; the need for ancillary services such as lowering, drainage, compressed air, etc. partitioning of the excavation section, such as temporary invert, side drift, pilot tunnel, etc.; and sizing of the supports. These items are interdependent and should have their design integrated.

Tunnels in soft rock masses or on ground are usually constructed by partial construction methods. They are usually lined with sprayed concrete. Considering that geology is variable from place to place, it is noteworthy that the hydraulic equivalence between the two processes with respect to the roughness of the walls was discussed earlier in Section 9.4.2.

In terms of productivity, it has been possible in Brazil to make 10–12 m tunnel/day on two conventional excavation fronts, for a 5 m diameter SHP tunnel (20 m² area) in good quality rock working 20 hours a day with hydraulic jumps, efficient ventilation

Figure 12.12 Tucuruí HPP. Spillway construction. Preparation for concreting of layers 20 and 21 downstream in the bucket.

Figure 12.13 Tucuruí HPP Spillway construction. Concreting of downstream layer 21 using sliding form.

Figure 12.14 (a–b) Tucuruí HPP – Spillway construction. Temporarily fixed shape example (Layers 24–28). Panel Removal. (Note aspect of concrete surface.)

and optimum utilization of equipment. For larger tunnels Nielsen and Thidemann (1993) presents the data in Figure 12.15, considering 100 working hours per week.

The productivity of the excavation process using the TBM is a function of several parameters of the rock massif, the DRI (Drilling Rate Index), and the TBM machine (Cutter Life Index-CLI). The most important parameter is the joint spacing of the rock mass. It is much more important for the TBM process than for the conventional process.

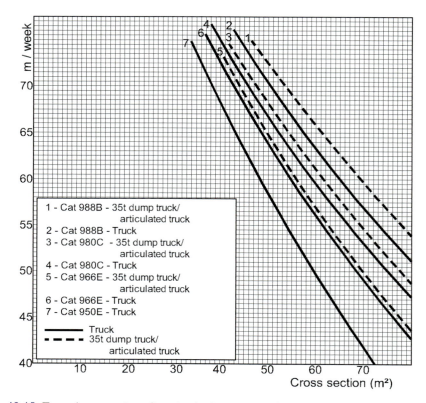

Figure 12.15 Tunnel excavation. Standard advance rate (Nielsen and Thidemann, 1993).

According to Nielsen and Thidemann (1993), the productivity prognosis for TBM excavation of a 4.5 m diameter and 6 km long tunnel in a Class I gneiss granite (10 cm joint distance) and a gneiss Class II (5 cm distance between joints), ranges from 110 to 155 m per week, considering a worked week of 108 hours. As stated earlier, the process is very specific and has many details.

Optimizing the construction time from the perspective of private initiative has immediate return and substantially improves the attractiveness of the business.

12.4 ASSEMBLY OR ERECTION PLANNING

Equipment manufacturing and assembly or erection times were also reduced. In medium-sized plants it is possible to assemble spiral box in the assembly area and launch with cranes even during the construction of the powerhouse or immediately after the release of the rolling path, with significant time savings in the assembly of the units (examples in Brazil: Monjolinho HPP mentioned above and the Antas River Energy Complex cited in Chapter 9).

Construction planning   423

12.5   ACCESSES TO THE CONSTRUCTION SITE

Site logistics, which involves access and preparation, mounting of the job site – offices and accommodations, water supply, drainage and sewerage, access to borrow areas, including sand deposits and quarries, crushing and concrete plant, power supply, compressed air, etc., varies from site to site, depending on the experience and culture of each contractor. These aspects will not be covered in this book.

12.6   CONTRACTING PROCEDURES

To contract the works of the infrastructure designs, the following contracting modalities have been used: "Classic", "EPC – Engineering procurement and construction", "Turn-key", "Alliance" and "Guaranteed maximum price", as summarized below. The main reference used to prepare this text was McNair (2001).

The execution of public and private works has been done directly by the owner based on a logical and traditional sequence of steps listed below.

- elaboration of technical, economic-financial and socio-environmental feasibility studies;
- elaboration of basic designs;
- elaboration of executive designs;
- acquisition of materials and equipment;
- execution of civil works and erections;
- financial Engineering;
- management or quality control;
- training of operation and maintenance teams, etc.

These designs and services are normally contracted to different companies or organizations. The following diagrams illustrate the basic contractual structure of a financed energy sector design, using an EPC contract. The detailed contractual structure varies from project to project.

In general, there is an energy company that has received a government concession contract (usually 30 years). This company (SPE – Specific Purpose Company) is responsible for the tripartite agreement involving the sponsor and the bank for structuring equity, borrowing, financing (based on PPA - Power Purchase Agreement) and insurance; along with EPC contract (Engineering, Procurement and Construction) and O&M contract (Operation and Maintenance).

12.6.1   Classical modality

In the traditional (classical) modality, the owner maintains a staff of professionals prepared to analyze the elements for decision and hiring, selection of suppliers of goods and services, contract administration, supervision of execution of designs and works and other activities for the purpose implementation of the project (Figure 12.16).

Execution oversight is usually assigned to consultancy firms, which may be the same as the ones who did their engineering designs (Figure 12.17).

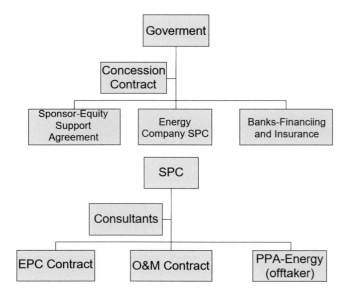

Figure 12.16 Organizational charts of a financed project.

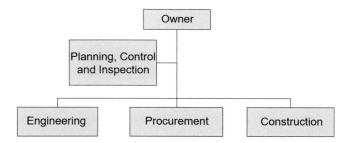

Figure 12.17 Classic organization chart.

More recently, it became common to hire an engineering consulting firm for the full management of the venture's construction, which was previously done by the owner's technicians and administrative staff. In this case, the manager uses its professional staff for all activities previously listed, and the owner has the decision-making power in each event based on the reports submitted by the management company. The advantages of this model are as follows:

- the owner may have the qualified teams of the manager, who is able to allocate specialized professionals for each specific activity or stage, and exempt the maintenance of own technical staff for tasks of limited duration, which do not constitute their core activity;
- limit the interfaces with multiple contractors, focusing their attention on monitoring the manager's performance; the owner participates in meetings with the various actors involved in the venture to make decisions more secure.

At the same time, the owner should avoid the risk of staying away from everyday events due to the manager's good performance as that may result in transferring too much power to the manager for decisions that go beyond what can and should be delegated.

12.6.2 Turn-key

Currently, new types of hiring are frequently taking place with the new objective of hiring on project basis, also called "packages", such as "turn-key". From the feasibility studies, with detailed terms of reference and functional specifications, the owner contracts the venture with a single organization, capable of developing the executive projects, supplying the materials and equipment, performing the works and assemblies, starting up the venture undertaken, preparing personnel for operation and other possible tasks that may include their own operation and maintenance.

For this purpose, the owner hires an EPC. At this moment, Consortia are constituted to meet this type of demand. Typically, Consortia consist of engineering consulting firms (studies, projects and management), construction and industrial assembly, equipment manufacturers and other suppliers of other goods and services. Some of these partners may be subcontracted, and not participating in the Consortium.

Consortia thus formed may be organized into SPCs (or SPEs) composed of companies formally constituted specifically for the execution of that enterprise (Figure 12.18).

The SPCs have their own legal personality, with the partners being their partners or shareholders, or stable Consortia, lasting alliances, forming a consolidated partnership for usual joint action in all or in certain sectors or regions. Other partners may participate as partners in SPC, such as the insurer, the financial company, as well as others interested in the business itself.

In all these alternatives, the contract between the partners for the formation of SPC, or Consortium, is extremely complex and will have to be carefully negotiated due to some peculiarities of the association, including:

- the unequal economic size of the partners (builder/consultant, for example);
- the co-responsibility assumed by partners of unequal capacity in offering guarantees;
- the professional responsibility of the designer in constant confrontation with the consortium's interest in optimized financial results;

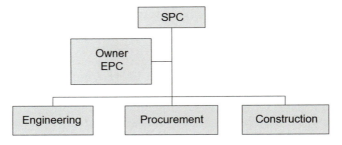

Figure 12.18 Owner hires an EPC and the Consortium organizes an SPC.

- the generally reduced ownership interest of the design company in the consortium or SPC, disproportionate to its non-transferable technical responsibility for any design errors;
- the overload of unforeseen tasks on the designer, in the interest of the SPC or consortium to study or adopt frequent design modifications, carry out technical and economic studies of successive design alternatives or construction systems, placing a burden on the less able partner.

All of these difficulties can certainly be overcome with the help of well-designed contracts, that foresee all possible occurrences and problem-solving procedures, taking into account the disproportion and asymmetries between partners mentioned here.

As usual, the contract will try to reconcile divergent interests to transform them into convergent interests, theoretically enabling everyone to win. One party pursues the lowest price, highest quality and shortest construction time, and expects the targets to be met without increasing costs. The other party intends to impose the highest admissible price under market conditions and comfortable deadlines, being willing to fulfill only strict quality and deadline obligations, only exceeding the conditions initially proposed by appropriate additional remuneration.

In hiring a turn-key, currently the most usual model, the following steps are usually fulfilled:

- the owner obtains financing under conditions and grace periods compatible with the enterprise;
- based on technical specifications, terms of reference, descriptive memoranda and other defining documents of the enterprise prepared by the owner, the invited bidders prepare their proposals with detailed budgets and other conditions required at the time of the call;
- after analyzing the proposals, the owner, according to his criteria and interests, selects the bidder and hires him, and the contractor is responsible for planning and executive designs, supply of materials and equipment, construction and assembly, management and other services provided in the scope.

Generally, in addition to a financial company, an insurance company intervenes in the hiring, which will guarantee the performance of the contractor through a performance bond, which may be partial or full, at the owner's criteria. After the works are completed, amortization of the financing begins.

The great advantage for the owner of this type of contracting is to concentrate on a single entity (SPC or the Consortium) the responsibility for the accomplishment of the project, thus restricting the management of the deployment to a single interface.

The disadvantages, or risks, are:

- less control of the owner over the work, less likely to include changes that are convenient for him;
- the contractor has no interest in seeking innovative solutions that are beneficial to the owner, as it may represent an increase in study and project costs without the corresponding financial compensation; may also lead to longer implementation time;

- possibility of using equipment that meets specifications, but of non-traditional origin; a vendor list previously defined by the owner can mitigate this risk;
- non-compliance with contractual clauses such as delays in works may not be compensated for by applicable penalties;
- the financial guarantees in the financing operation are exclusive of the owner, who assumes full responsibility for the business.

It is also possible to differentiate the most relevant aspects of the two most commonly used ways for hiring an engineering enterprise:

- turn-key associated with an EPC type scope;
- turn-key based on the minimum overall performance requirements set for the project over its entire lifetime.

In the first, "turn-key" associated with an EPC-type scope, the owner provides the contracted organization (SPC or consortium) with a detailed technical specification of the enterprise (for example, an Advanced Basic Design, or a FEED – "Front-End Engineering Design").

During the completion of the stages of the project's implementation, the owner is responsible for the management and supervision of the contractor's activities (the owner may use specialized management and oversight companies to supplement his own ability to perform these duties), in order to ensure compliance with all items of the technical specification (basic design or FEED) of the enterprise.

This first form of engineering contracting is well suited for large and highly complex works, which include a number of technological options and unusual or even innovative building standards. Its success depends primarily on the quality of its technical specification (basic design or FEED) and the owner's demanding management and enforcement activities.

The second form, "turn-key" based on minimum performance requirements, is characterized by the lowest possible interference of the owner in the implementation of the enterprise. The technical specification, the main part of the contracting, has a functional character, based on the minimum parameters required for the operational performance of the main equipment and systems and the expected overall performance of the project throughout its useful life.

In this second form of hiring, the contracted organization (SPC or Consortium) may offer the most convenient technological option, provided that this option meets the requirements of the functional specification, and, by adopting this technological option, make the implementation of the enterprise cheaper, or carry it out in a shorter period.

Therefore, turn-key hiring based on the minimum performance requirements of an engineering enterprise is adequate in principle for two situations which, oddly enough, are diametrically opposed:

- when the object (works and other supplies, such as systems and equipment) presents low complexity, coupled with a high standardization rate;
- when it comes to proprietary technologies that may necessarily give rise to considerably different engineering solutions for the enterprise.

428 Design of Hydroelectric Power Plants – Step by Step

In both forms of hiring, success will depend on the accuracy of the functional specification. It must absolutely set all minimum parameters required for the operational performance of main equipment and systems. The overall performance of the project over its entire lifetime should also be accurately determined.

The SPC formation helps to make the design viable through project finance, in which the plant itself is the guarantee of the financing. It is noteworthy that, as in SPC all parties involved are partners (owner, financial, insurer, designer and contractor), sharing the benefits and risks of the business, costs can be minimized, including financial ones.

12.6.3 Alliance

Another type of contracting is the alliance, very similar to the previous one. The group is constituted with greater advancement, integrating the teams of all partners from the preliminary stages of project definition and budgeting. Throughout the execution, all gains from good negotiations and design modifications that result in financial or quality benefits are shared between all partners in the previously established proportions. The same happens with any losses or reductions in profits. In this mode the interests are effectively convergent, all interested in reducing costs, improving quality and shortening lead times. It requires a change of culture, as it does not correspond to the usual practices that put the parties in permanent conflict of interest. The formalization of this type of alliance is made through a Head Alliance and a Main Agreement comes into existence governing the performance of the parties throughout the execution of the enterprise.

12.6.4 Guaranteed maximum price

There is even more to the type of contracting called "guaranteed maximum price". In this mode, the contracted organization (SPC or Consortium) guarantees a maximum value for the implementation of the project.

It also resembles the latter as regards to the search for more appropriate technologies that can minimize the implementation costs of the enterprise, thus resulting in financial benefits that will be shared between the owner and the contracted organization, according to agreed percentages. and contained in the contract.

On the other hand, if the maximum guaranteed price is exceeded, the contracted organization will have to bear its additional costs – which may (and should) be mitigated by the coverage of appropriate insurance.

As with the other contractual models, it also requires the management teams involved – both the owner and the contracted organization – to have their objectives convergent, non-conflicting, and to work in a partnership so that due benefits can be achieved.

In particular, in this mode, it is up to the designer, as a participant of the contracted organization, to play a prominent role regarding the search for well-timed technological options (possibly even innovative), which may bring about cost reductions and implementation times. It is precisely the design company that should print increasingly modern engineering solutions, adapting them to the particular project needs of the enterprise.

Of greatest relevance is the role of the engineering consulting firm hired by the owner for this purpose - owner engineering. In fact, in this role, it is up to this company to safeguard the interests of the owner, having strong performance in aspects that refer to the safety and operation of the enterprise. It is to be hoped that in the pursuit of lower costs and shorter lead times, one may risk compromising safety and sacrificing the operational facilities of the enterprise.

As part of its owner in engineering role, this engineering firm should then set minimum safety limits and arbitrate technology options so that, in the end, it has a venture within the required standards of quality, safety and operational ease.

12.6.5   Final considerations

The changes in hiring practices for implementation of infrastructure projects were profound and radical in many ways. It was a real change in the engineering market that also required a profound shift in entrepreneurial mindset and new company profile.

Changes of this size do not consolidate quickly. It will take some time for the new practices to be incorporated into the culture of people and organizations and to become effectively constructive, beneficial to all who participate in the business, with positive consequences for society and the country.

Chapter 13

Risks and management of patrimony

13.1 INTRODUCTION

A dam throughout its useful life can be affected by various phenomena, natural or otherwise, such as; climate of the region; foundations problems; deterioration of the landfill; exceptional floods; overtopping; earthquakes; slope slips; and building materials. The aging of dams can lead to accidents that normally result in intangible losses of human life and significant economic, social and environmental losses that exceed the capacity of affected communities to recompose themselves. For all these reasons, every dam should be systematically well constructed, monitored by trained and qualified teams.

The development of the theme "Dams Breaks" has been approached considering its main aspects:

* all geological and geotechnical issues;
* hydrological aspects concerning maximum design flow estimates;
* all the aspects of the construction, from excavation, the treatment of foundations to the top of the dam.

Hydroelectric layouts include a spillway to overflow the floods to keep the reservoirs water levels, to ensure the freeboard and provide protection against overtopping. It is noteworthy that the arrangements often do not include a bottom outlet to allow emptying the reservoir. In the author's opinion, the inclusion of this device should be mandatory in all hydraulic works that create a reservoir regardless of the cost. Some of the accidents that occurred could have been prevented if such a device existed and, where it existed, had to be unobstructed in order to function freely. All these topics are very complex and hence the concern to emphasize the need that the teams that participate in the design and construction must necessarily have proven experience in the subject, to reduce and minimize risks.

13.2 DAM BREAKS CAUSES STATISTICS

After some catastrophic dam breaks, from the 1950s onwards several countries around the world began to study this complex matter and to take measures to increase the safety of their dams. This process was slow and time consuming. In the United States

DOI: 10.1201/9781003161325-13

432 Design of Hydroelectric Power Plants – Step by Step

Table 13.1 Dam Failures Causes Statistics (USBR, 1983)

Cause of Failure	Percent (%)
Foundation failure	40
Inadequate spillway	23
Poor construction	12
Uneven settlement	10
High pore pressure	5
Acts of war	3
Embankment slips	2
Defective materials	2
Incorrect operation	2
Earthquakes	1

of America, it was not until after the Teton Dam ruptured in 1976 that the government launched an intense study program to support the planning and implementation of its dam safety system. These studies were published by USBR (1983) from which Table 13.1 is extracted. This table summarizes the statistics of the most frequent causes of dam ruptures.

It is noteworthy that only three causes total for 94% of accidents, which coincidentally are associated with the main and most complex disciplines of the project: 59% of accidents/disasters are caused by geological/geotechnical causes, including varying foundation problems, settlement, high neutral pressures, slope slippage and poor materials; 23% are caused by overflows caused by floods in excess of design floods that were underestimated; and 12% to various construction problems, either in treating the foundations of structures, in compacting landfills or in concreting various structures.

13.3 MAIN ACCIDENTS IN THE WORLD

The following is a summary of some notable accidents involving dams in the world. The author had no intention of discussing in detail the causes of accidents:

- Malpasset (1959, France): foundation deformability;
- Vajont (1963, Italy): sliding left abutment into reservoir;
- Teton (1976, USA): piping;
- El Guapo (Venezuela): insufficient flow capacity of the spillway;
- Lower San Fernando (1971, USA): slope liquefaction;
- Sayano – Susnhensk (2009, Russia): rupture of several machines of the plant by cavitation, among other causes, as presented later in this chapter;
- Bieudron (2000, Switzerland): penstock rupture by cavitation.

It is also worth mentioning the Brazilian accidents:

- Pampulha dam in Belo Horizonte (1954) and with SHPs: Camará (2004, Paraiba), Apertadinho (2008, Rondônia), Espora (2008, Goiás), and Algodões (2009, Piauí). Also, there were accidents with tailings dams Fernandinho (1985), Pico São Luiz (1985) and Cataguazes (2003), all in Minas Gerais State. Some of these accidents have already been mentioned in Chapter 6;

- Orós in the Jaguaribe river, Ceará, in 1960, which made 100 thousand people homeless (CBDB, 2000);
- Euclides da Cunha (CBDB, 1982), in São José do Rio Pardo (São Paulo), and Limoeiro, in Mococa (São Paulo), in 1977 (Oliveira and Leme, 1985).

13.3.1 Malpasset dam (Southeast France)

The accident occurred on the Reyran river, 12 km from the city of Fréjus (southeastern France) on 02/12/1959, destroying the small villages of Malpasset and Bozon and causing the deaths of 423 people (Jansen and USBR, 1983; Veiga, 2008; Ginocchio and Viollet, 2012). The dam was built between 1952 and 1954. The section was concrete with double arch curvature with a height of 65 m and a length of 222.66 m (Figures 13.1–13.7). The objectives of the reservoir were irrigation and water supply.

Geological and hydrological studies carried out in 1946 found the site appropriate, although with opposition from some consultants. Investigations were limited due to scarce resources. The foundation, metamorphic-gneiss rock, seemed to be waterproof. The right abutment was made up of healthy gneiss; already on the left abutment the gneiss was heterogeneous with schist, and it was necessary to build a concrete wing wall for better interaction of the foundation with the dam. Cracks were observed at the base of the downstream dam, but these were not investigated. Two weeks after the cracks were identified, the dam collapsed in a catastrophic manner causing the death of 500 people.

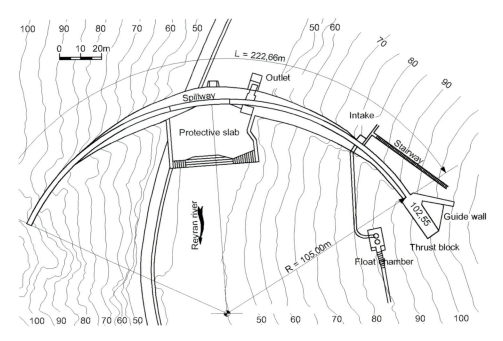

Figure 13.1 Malpasset dam. Non-scale plan.

Figure 13.2 Malpasset dam. Downstream view of the dam before and after rupture.

Figure 13.3 Malpasset dam. Downstream view of the dam after rupture.

No such dam had previously been breached, leading to numerous studies. According to Jansen (1980) and Veiga (2008), the investigations concluded that:

- the dam was built on top of two geological faults – F and F' (Figure 13.6); the F', a 45° inclined vertical fault, had not been recognized at the design stage;

Risks and management of patrimony 435

Figure 13.4 Malpasset dam. Downstream view of the broken dam.

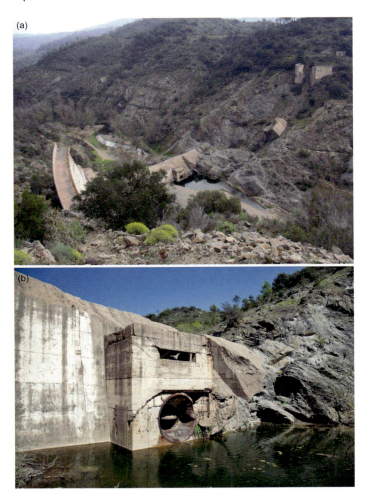

Figure 13.5 (a–b) Malpasset dam. Observe that the dam was totally destroyed on the left bank. Observe the bottom outlet.

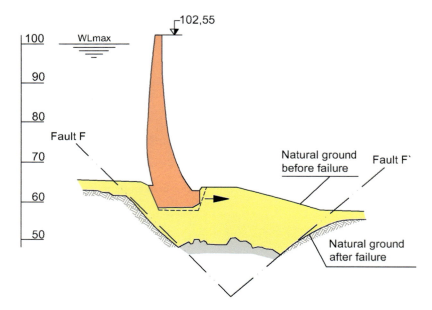

Figure 13.6 Malpasset dam. Cross section – non-scale.

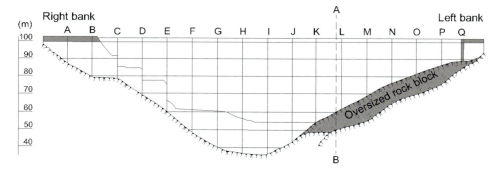

Figure 13.7 Malpasset dam. Longitudinal section – non-scale.

- heavy rain had rapidly raised the water level of the reservoir by more than 5 m, which contributed to increased pressure under the dam in the first reservoir filling;
- the rupture mechanism was triggered on the left abutment; the increase in under-pressure caused the rock dihedral to slip through the low resistance surface of the F′ fault (Figure 13.7), with the consequent opening of the foundation mass, leading to the sudden flow of water from the reservoir and the dam rupture.

According to Jansen and USBR (1983) and Veiga (2008), much emphasis was placed on the structural calculation of the dam, optimizing its shape and stress distribution, but neglecting the recognition and geological-geotechnical characterization of the foundation and the mechanisms deterioration that could lead to the instability of the structure.

Veiga also pointed out that the project did not contemplate the drainage of the foundation because the designer considered that the under-pressure would not have any consequence on the dam's stability, even in the case of a very slender arch dam.

13.3.2 Vajont dam (Italy)

The Vajont dam is located in the Italian Alps in Longarone, about 100 km north of Venice. The area is characterized by a thick section of sedimentary rocks, dominantly limestone with frequent clayey interbeds and a series of alternating limey and marl layers (Figure 13.8).

The dam is a concrete arch with a crest length of 1,850 m and a height of 266 m. The reservoir has a volume of $115 \times 10^6 \, m^3$. At the time of the accident, it was the tallest dam in the world.

During the reservoir filling, in 9/10/1963, a gigantic mass of soil and rock from the left abutment, about 270 million m^3, slipped into the reservoir at a speed of 30 m/s.

The generated wave, over 100 m high, climbed the dam and swept the valley downstream, destroying the city of Longarone and causing the deaths of over 2,000 people. Remarkably, the dam weathered the crash without collapsing (Figure 13.9).

Figure 13.8 Vajont dam during filling. Observe, on the left margin, an anterior sliding scar (Hoek, 2007).

Figure 13.9 Vajont dam after Mount Toc landslide (Hoek, 2007).

Prior to the completion of the dam, the site director was concerned about the stability of the left encounter. Recognitions made in 1958 and 1959 identified scars from old landslides on the abutment (Figure 13.10).

The first filling, even before the dam was closed (completed in September 1960), began in February 1960. In March, designers recognized that a large mass of terrain was unstable on the left bank. It was decided to maneuver with the reservoir fill level, while draining galleries were performed on the unstable slope. The problem seemed to be being resolved, but between April and May 1963, the reservoir level rose rapidly. It was decided to empty the reservoir, but in October, the slope slid (Figure 13.11).

13.3.3 Teton dam (USA)

The Teton earth dam, 100 m high, was built by the USBR on the Teton river, Rexburg, Idaho. It ruptured during the first fill in 05/06/1976 and cost 11 people their lives and a loss of about $ 1 billion. The analysis of the accident by Peck (1980) is presented in Chapter 6. Figures 13.12 and 13.13 have an overall downstream view of the dam before and after the accident. Figure 13.14 shows the accident sequence.

On the morning of 06/05/1976, a water leak appeared in the downstream slope of the dam near the right jamb. The alarm rang but there was no time for anything because the breaking process was already well under way. Figure 13.14 shows the sequence of the dam break. In the first figure you can see two tractors that sought to

Figure 13.10 (a–b) Longarone before and after the accident (Hoek, 2007).

bridge the gap and were caught by the wave of mud. Near noon, the dam broke. At night, the whole reservoir was emptied. The cities of Idaho Falls and American Falls suffered considerable losses.

13.3.3.1 US dam safety

From this accident, by order of then President Jimmy Carter, USBR designed, implemented and began to comply with a strict dam safety program. Each structure was inspected periodically for stability, internal faults and physical deterioration.

Figure 13.11 Longarone 40 years later (2003). Note disused dam in the background.

13.3.4 El Guapo dam (Venezuela)

Following are some photos of the rupture in 16/12/1999 of the El Guapo dam in Barlovento, Miranda State, 150 km from Caracas, built for flood control, water supply (415,000 people) and irrigation. The rupture was due to insufficient flow capacity of the spillway. Data: height: 60 m; length: 524 m; original spillway capacity: $102 \, m^3/s$; new spillway capacity: $2,700 \, m^3/s$ (new design flood is 27 times larger than original) (Figures 13.15–13.17).

Volumes of reconstruction services: concrete = $30,000 \, m^3$; CCR = $350,000 \, m^3$; compacted fill = $350,000 \, m^3$.

13.3.5 Lower San Fernando dam (USA)

Figure 13.21 shows the lower and upper San Fernando dams of the water supply system near Los Angeles, California. Figure 13.22 shows the upstream slope of the lower dam broken by liquefaction after an earthquake in 09/02/1971. The 44 m high dam was built on a landfill on a 5 m thick alluvial layer. The upper dam suffered minor damage (Figures 13.18–13.20).

Risks and management of patrimony 441

Figure 13.12 Downstream view before the rupture.

Figure 13.13 Downstream view after the rupture.

Figure 13.14 Teton dam break sequence photos (Höeg, 1992).

Figure 13.15 El Guapo dam (Prego Ing. Geotecnica).

13.3.6 Sayano-Shushensk accident (Russia)

The Sayano – Shushensk HPP, Republic of Khakassia, Siberia, was presented in Chapter 3. The following is a summary of the accident in August 2009. In this accident, 75 people lost their lives. At the time, the 6,500 MW plant was the sixth largest in the world (see Table 1.1).

Research of articles available on the Internet revealed the flaws in the design of civil structures and equipment, as well as in the operation of the plant (see: Reflections on Russian Accident on Sayano-Shushensk HPP, by Eugenio Kolesnikov, Miami, 10/12/2009). According to him, the plant often operated outside the range specified for the equipment. Excessive vibration problems have occurred since 1982. Maintenance was poor. The plant was flooded twice in 1979 and 1985. There were many problems reported.

Possible causes include the heavy water hammer following the sudden closure of unit 2, which gave rise to an upward force that destroyed the civil structure above the spiral box and the forced duct, as illustrated in the following figures.

Recovery work from an accident of this size has been long. In 2010, machines 3, 4, 5 and 6 were repaired. In 2011, new machine 1 was assembled and connected. In 2012, new machines 7, 8 and 9 were assembled and commissioned. In March 2013, the new machines went into operation. Machines 5, 6, and 10 were new machines that were previously repaired in 2010.. Throughout 2014, new machines 2, 3 and 4 were commissioned.

It is noteworthy that a new, uncontrolled, complementary surface spillway was designed and built on the right abutment, which went into operation in October 2011, to increase flow capacity and optimize the operation of the discharge organs. This made

Figure 13.16 (a–b) El Guapo dam. Flow capacity exceeded. Spillway overflowed. Beginning of dam break.

Figure 13.17 (a–b) Reconstructed dam and spillway (Prego Ing. Geotecnica).

it possible to repair the main spillway dissipation basin after the erosion of the bottom slabs due to flood traffic from 1985 onward. This 200 m wide spillway was designed in steps (five) with a dissipation basin whose end-sill had downstream labyrinth teeth, as shown in Figures 13.23 and 13.24.

13.3.7 Bieudron plant - breakdown of the penstock (Switzerland)

The Grande Dixence dam reservoir supplies, through penstocks, four hydroelectric plants: Chandoline commissioned in 1934 ($H = 1{,}748$ m, 120 MW), Fionnay ($H = 874$ m, 290 MW), Nendaz ($H = 1{,}008$ m, 390 MW) and Bieudron ($H = 1{,}883$ m, 1,269 MW). It is opportune to represent in Figure 3.6A. (Figure 3.6A). Bieudron HPP. Localization map.

Figure 13.18 Dams of San Fernando water supply system.

The characteristics were previously listed in Table 1.6. The effluent flow from the Nendaz, Bieudron and Chandoline plants was discharged into the Rhone river as mentioned in Chapter 3. Chandoline had been out of service since July 2013. Fionnay and Nendaz plants operated normally. Bieudron started up was in 1988. Figure 13.25A presents the geological profile to the stretch Lake Dix-Bieudron.

Figure 13.25B shows the orographic aspects of the stretch Lake Dix, Cleuson dam and Dent of Nendaz, consisting of the surge tank. After the surge tank, the pressure tunnel dips towards the powerhouse of Bieudron.

Figure 13.25B Aspects of the stretch Lake Dix, Cleuson lake, Dent de Nendaz, Bieudron HPP.

By the way, the Bieudron's penstock exploded on December 12, 2000, under more than 1,000 m of head, at elevation 1,234 m. On this subject, see Ginocchio and Viollet (2012). The rupture was credited to several factors, including the low strength of the rock mass surrounding the penstock at the site of rupture. A 9.0 m long gap was opened at the site. Flow was estimated at 150 m^3/s. Three people died. An area of 1.0 km^2 of forest, pastures and orchards along with several chalets and stables was destroyed. Legal actions are still going on, and there is still no clear public information available on the causes of the break (Figure 13.25C).

The new project included improvements in the penstock layout, and the injections of consolidation and waterproofing of the rock mass aimed to reduce the percolation between the penstock and the massif. When the penstock works, a crack appears in the massif depending on the operating conditions: the penstock expands when under load and contracts when in the position of unload. Penstock alignment has been relocated

Risks and management of patrimony 447

Figure 13.19 (a–b) Lower dam. Upstream slope break.

Figure 13.20 Sayano – Shushensk HPP – Destroyed stretch.

to a nearby region where no damage has occurred and where the massif is of the highest quality. Reconstruction ended in December 2009, and the plant went into operation in January 2010 (Figure 13.25D).

Figure 13.25C Bieudron HPP reconstructed.

Figure 13.25D Bieudron HPP. Dent de Nendaz where is the surge tank.

13.4 RISKS ASSOCIATED WITH HYDROELECTRIC PLANTS

13.4.1 Risks of dam breaks – submersion waves

High and medium head plants generally have reservoirs with large volumes of water (see Section 3.2.3). The main risk for the riverside population, as well as the surrounding property is the accidental and instantaneous release of this water, implying the formation of a submersion wave that will propagate through downstream.

The main phenomenon that causes a submersion wave is the rupture of the dam itself. This rupture may be progressive or instantaneous, partial or total, depending on the type of dam. A submergence wave can also result from a large landslide into the reservoir - large proportions with respect to the water depth of the reservoir.

Although rare, this is what happened in Vajont (Italy) on October 9, 1963. A huge volume of material from Monte Toc, $250 \times 10^6 m^3$, slid into the reservoir, which was $168 \times 10^6 m^3$, and generated a wave that overtopped the dam and submerged the city of Longarone downstream causing the death of 2,000 people. This accident is described

Figure 13.21 (a–b) In the foreground the crosshead. Unit 2 totally destroyed.

in the previous item. It is noteworthy that in the design phase no detailed geological-geotechnical studies were made of the stability of the reservoir slopes (Figure 13.26). In ICOLD Dam Failures Statistical Analysis Bulletin No. 99, you will find detailed statistics on dam failure types up to 1995.

Table 13.2 provides a partial list of accidents that occurred after 1900, with more than 50 victims. A detailed list is presented by USBR (1980). Examination of this list shows that flooding, submergence and piping ruptures of landfill dams are more frequent.

(a)

(b)

Figure 13.22 (a–b) – Sayano – Shushensk HPP. Recovering powerhouse.

1. Rupture circumstance: FF = first filling; PI = piping; FL = flood; W = war; SU = submersion; SC = submersion during construction; FO = foundation; UB = upstream break; GF = gate failure; T = typhoon.
2. Type: CGD = concrete gravity dam; MGD = masonry gravity dam; BA = vaulted dam; MVD = multiple vaulted dam; BD = buttresses dam; ED = earth dam; RD = rockfill dam.
3. The Malpasset accident, concrete arch dam, is described in the previous item.

Risks and management of patrimony 451

Figure 13.23 Sayano – Shushensk HPP. (Note new complementary spillway on the left abutment.)

Figure 13.24 Sayano – Shushensk HPP. New complementary spillway operating.

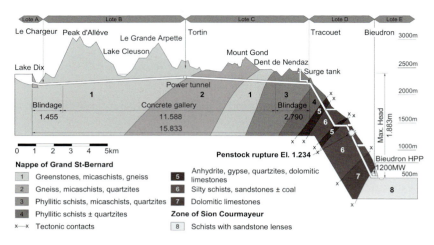

Figure 13.25A Bieudron HPP. Profile (sciencedirect.com).

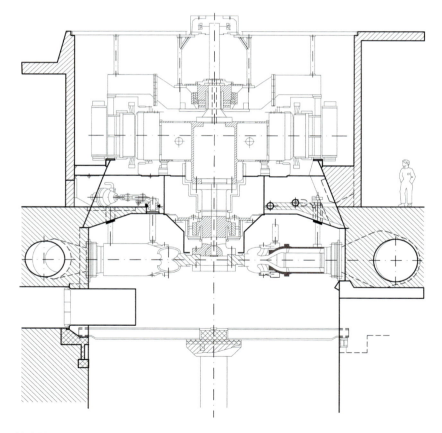

Figure 13.25B Aspects of the stretch Lake Dix, Cleuson lake, Dent de Nendaz, Bieudron HPP.

Risks and management of patrimony 453

Figure 13.25C Bieudron HPP reconstructed.

Figure 13.25D Bieudron HPP. Dent de Nendaz where is the surge tank.

454 Design of Hydroelectric Power Plants – Step by Step

Figure 13.26 Vajont. Sliding of the left abutment (environmentandsociety.org).

Table 13.2 List of Some of the Accidents with More Than 50 Victims (Ginnochio and Viollet, 2012)

Name (Country)	Year/Circ. Break (1)	Construction	Height (m)	Length (m)	Res Vol ($10^6 m^3$)	Type (2)	Victims
Bayless (USA)	1911/FF	1909	16	160	1.3	CGD	80
Bila Desna (Czech Rep.)	1916/PI/FF	1915	18	240	0.4	ED	65
Tigra (India)	1917/FL	1917	25	1,340	124	MGD	1,000
Gleno (Italy)	1923/FF	1923	35	225	5	MVD	600
S. Francis (USA)	1928/FF	1926	62	213	47	CGD	450
Sella Zerbino (Italy)	1935/FL	1924	16	70	10	CGD	100
Möhne (Germany)	1943/W	1913	40	-	134	MGD	1,200
Edersee (Germany)	1943/W	1914	48	400	200	MGD	100?
Heiwaike (Japan)	1951/SU	1949	22	82	0.2	ED	100
Malpasset (France) (3)	1959/FO	1954	60	222	48	MVD	423
Vega de Tera (Spain)	1959/FF	1955	33	270	7.3	BD	14
Khadakwasla (India)	1961/UB	1879	33	1,400	137	MGD	1,000
Panshet (India)	1961/SC	-	49	740	214	ED	1,000
Hyokiri (Korea)	1961/PI	1940	15	110	0.2	ED	139
Sempor (Indonesia)	1967/SC	-	60	228	56	RD	200
Nanak Sagar (India)	1967/PI	1962	16	19,300	210	ED	100
Banqiao (China)	1975/SU/T	1952	24.5	-	492	ED	>20,000
Machchhu (India)	1979/GF/SU	1972	26	3,900	101	ED	2,000
Gotvan (Iran)	1980/SU	1977	22	710	-	-	200
Kantale (Sri Lanka)	1986/PI	1869	27	2,500	135	ED	127
Shadi Kaur (Pakistan)	2005/FL	2003	-	148	-	-	70
Situ Gintung (Indonesia)	2009/SU	1933	16	-	2	ED	100

13.4.2 Dam breaks risk prevention – regulatory and Legal Aspects

The risk of dam ruptures is prevented by detailed control of the quality of studies and projects in their various phases. These phases include hydrological, geological-geotechnical studies of the foundation and massifs, the stability of all structures, electro-mechanical design, as well as the efficient supervision of the construction and commissioning of the works is necessary before putting the dam into operation.

All member countries of the ICOLD have a Basic Dam Safety Guide (2001). The owner must be responsible for the safety of the dam and should be responsible for the consequences of any failure.

All dams are classified according to the consequences of a potential rupture, where the following factors must be considered: downstream populations; materials damage; damage to the environment; and damage to infrastructure. This guide, which is available on the Internet, prescribes routine, periodic and formal inspections.

It establishes that all dams should be periodically subjected to a reassessment of their safety conditions, according to their classification. In addition, operation and maintenance personnel must have procedures and training to respond to emergency situations such as during exceptional flooding and instrumentation alerts. It also establishes that the dams must be provided with an emergency plan, aiming at the preservation of downstream residents in the event of an accident.

Regarding legislation, in Brazil Law No. 12,334 of September 20, 2010, establishes the National Dam Safety Policy intended for the accumulation of water for any use, the final or temporary disposal of tailings and the accumulation of waste, along with creation of National Dam Safety Information System.

In summary, this law defines the policy instruments that consider, among others, the dam classification system by risk category and associated potential damage and the Dam Safety Plan as detailed in Section II of the law. In Art. 7, the law defines that the dams will be classified by the inspection agents, by risk categories, associated potential damage and their volume, based on general criteria established by the National Water Resources Council:

- § 1 defines that the classification by risk category as high, medium or low will be based on the technical characteristics, the state of conservation of the enterprise and compliance with the Dam Safety Plan;
- § 2 defines that the classification by category of potential damage associated with the dam at high, medium or low based on potential for loss of human life and the economic, social and environmental impacts resulting from the dam failure.

The National Water Resources Council:

- through Resolution No. 143, of July 10, 2012, it established general criteria for the classification of dams by risk category and associated potential damage and by reservoir volume;
- through Resolution No. 144, of July 10, 2012, it established the guidelines for the implementation of the National Dam Safety Policy and the application of its instruments and the performance of the National Dam Safety Information System.

456 Design of Hydroelectric Power Plants – Step by Step

13.4.3 Flood risks

Hydrological risks are exposed in Sections 4.1.5 and 12.2. As discussed earlier, 23% of accidents result from insufficient spillway flow capacity, which means that the floods have been undersized (see example of the El Guapo dam accident). These structures are designed for floods with TR = 10,000 years.

The risk or probability "r" of a flow with "TR" recurrence years, considered as the design flow of a given structure, to be equaled or exceeded at least once in "n" years, is given by the equation:

$$r = 1 - (1 - 1/\text{TR})^n \qquad (13.1)$$

where:

1/TR = probability of occurrence of the design flow in any given year;
n = duration, or useful life, of the work (or event – example: first phase of river diversion).

Considering $n = 50$ years and TR = 10,000 years, there is $r = 0.5\%$ over the lifetime of the work (50 years), i.e., a very low risk.

The study should consider the values recorded in Table 13.3 (which was already presented in Chapter 12).

13.4.4 Geological and geotechnical risks

As already mentioned, Malpasset and Teton dam breaks were caused by problems in the foundation. Table 13.2 shows other cases of accidents due to problems with the foundations, including piping.

It is evident that the programming of a detailed program of foundation investigations, laboratory tests, as well as basic geological-geotechnical studies must be done by experienced staff, observing the worldwide consolidated knowledge in such works as mentioned in Chapter 5, it is important to minimize these risks.

Table 13.3 Risk Criteria (CBDB/Eletrobras, 2003)

Damage Category	Annual Risk (%)
There is no danger of loss of human life, nor is it anticipated that major damage will occur to the work and its progress.	5.0– 20
There is no danger of loss of human life, but major damage to the work and its progress is already expected.	2.0– 5
There is some danger of loss of human life and major damage to the work and its progress is predicted.	1.0–2.0
There is a real danger of loss of human life and major damage to the work and its progress is predicted.	<1.0

13.4.5 Risks related to the constructive aspects

To minimize the risks, the construction should be done by a company with proven experience in works of this nature and strictly observing the technical specifications that consider, in general, the state of the art in the subject.

Special care throughout the process is fundamental, from excavation and treatment of foundations, to the execution of earthworks and rockfill, as well as the execution of concrete structures and equipment assembly.

13.4.6 Risks related to penstocks

In hydroelectric plants, the penstocks are subject to high hydrodynamic pressures, as well as variations of these pressures due to sudden closure of a turbine protection valve, with high risks.

This issue was previously presented in Chapter 9. Section 13.1.7 provides a summary of the accident with the Bieudron plant penstock in Switzerland in December 2000.

13.4.7 Risks related to turbine start-up

Quick start-up of a turbine to meet a system's energy demand discharges into the leakage channel a flow that can cause a rapid variation in the downstream water level as well. These maneuvers can cause accidents with people indulging into leisure activities downstream, such as people doing fishing or boating on small boats. Prevention involves informing the population about this risk and placing appropriate warning signs, alerting through possible siren ringing.

13.4.8 Risks during operation and maintenance

Considering that the projects have been properly done, the maintenance of the equipment and its operation within the specified ranges must be strictly observed.

A summary of the Sayano-Shushensk (2009) plant accident in which 75 people lost their lives was presented earlier. Research has revealed a number of defects in the design of civil structures and equipment as well as problems in the operation of the plant. To read more about it, the article "Reflections on Russian Accident on Sayano-Shushensk HPP", by Eugenio Kolesnikov, Miami, 10/12/2009 is especially recommended. According to internationally accumulated knowledge, it is permissible to admit that these errors can be avoided.

13.5 MANAGEMENT OF HYDROELECTRIC PATRIMONY

13.5.1 Context evolution

Appreciation and preservation of the patrimony represented by hydroelectric facilities during their lifetime involves a number of multiple issues that come up to ensure both a high level of safety and a high level of performance (response to requests, facility

458 Design of Hydroelectric Power Plants – Step by Step

availability). This patrimony management is carried out within a context of strong demands, particularly within the areas of improved economic, social and environmental performance. The aging of the hydroelectric park and its increased level of solicitation are also two important elements.

The multiplicity of these diverse requirements has a direct impact on the governance of proprietary companies. It requires managing risks globally, consistently, and ensuring an appropriate level of organization. The following is an approach to control those requirements that are inherent in HPP activities, to develop and perpetuate the practice of risk analysis and management. This text is based on Ginnochio and Viollet (2012).

13.5.2 The three Issues of asset management in hydraulic production

Approaches to these three issues are put into practice within a framework of requirements and performance structured around three key issues: Hydraulic Safety; Economic Performance; and the Ability to Produce Respecting Regulation.

13.5.3 Risk management: key issues

13.5.3.1 The technical questions

Hydroelectric patrimony management is divided into several key risk management issues:

- Where are the risks and how important are they?
- What precaution should be taken given the importance of the risk? Defenses consist, as appropriate, of exploration pressures (operation), surveillance or control, modernization and maintenance pressures.
- When to act? Immediately, in the short, medium and long term? What about the effectiveness of preventive actions?

The scope of risk analysis is wide and not part of this book. They concern a set of operation and maintenance activities of hydroelectric plants: operation of works in normal situation, during floods, management of current maintenance, and maintenance of facilities.

13.5.3.2 The coordination of actions

Maintaining the assets represented by hydroelectric installations involves a number of actors with varying roles and prerogatives: central management, local management, specialists, chiefs, operators, and asset managers. These actors necessarily carry different objectives and pressures (technical, financial, human resources, etc.). In addition, there is a need to understand the risks in the short, medium and long term, without meeting one imperative to the other.

Statistics show that the various actors in a company, in playing their multiple roles, end up with different individual views that form a very diverse and sometimes divergent set of views. Risk management, through more or less formalized decision-making techniques, achieves coherent things and contributes to the coordination and organization of interactions between multiple stakeholders.

13.5.3.3 Decision support for measurement of issues posed

Maintaining assets raises technical and organizational issues that can be resolved with the help of decision support systems. The level of formalization is the measure of importance of the problems. There are two decision techniques: decision system for triggering maintenance and renovation/modernization actions, based on a risk hierarchy; and multi-criteria decision support system to choose between maintaining different scenarios and costly renovation/modernization.

13.5.3.4 Principles governing the development of decision approaches

The two decision approaches are designed and implemented respecting fundamental principles as:

- separate the qualification of the problem (diagnosis), the maintenance operation and the renovation / modernization (solution);
- separate the role of expertise that brings the elements of knowledge about dam materials and services from the decision that is based on the choice of priorities, and arbitrage between interest and context-specific appreciation;
- have a legible and traceable decision support system; this requirement goes through a shared process, avoiding the "black box" effect on the one hand, and allowing the introduction of easily interpretable indicators on the other;
- have a live decision-making system, i.e., building on the experience gained and capable of being updated on the basis of new information.

13.5.4 Risk hierarchy

The first approach is to rank the risks to prioritize and trigger maintenance and renovation/modernization works. Each year several thousand maintenance and renovation/modernization operations are initiated on several hundred works. Within this context, the role of the decision system is to provide a risk position according to company policy, discriminating and capable of handling a large number of materials and structures in a necessarily limited time.

The risk hierarchy allows, among others, to prioritize the operations to be undertaken over time. It also allows for temporary measures to be taken to limit the risk to an acceptable value pending permanent maintenance treatment.

To conduct this analysis of the hydroelectric park, all structures and materials are grouped into families: dams, water intakes, forced ducts, mechanical installations, etc.

13.5.4.1 Operations prioritization process

Hierarchy is part of a general process in which the objective is to specify the timing of operations to be performed on materials and structures. The objective has two parts:

- the first part of the process is to classify the materials or structures of a family according to unwanted events that may occur;
- the second part is, in particular, the responsibility of the asset manager: it consists defining a maintenance operation schedule according to the classification results, and taking into account the management of unavailability, the consolidation of field operations, and the profitability of the HPPs.

Hierarchy of hydraulic patrimony risks to a rule-based system. The first question, to be underlined, is where are the risks and what is their importance. More formally, the triggered risk hierarchy step responds to a screening problem called $P\beta$: it has the purpose of assigning x elements (of risk) into categories (criticisms) according to defined rules based on the intrinsic information of element x.

The steps of risk hierarchy can be organized into five steps, from defining unwanted events to global criticality, as summarized below. For process details, it is recommended to refer to cited reference (Figure 13.27).

13.5.4.2 Define unwanted events

This preliminary step consists of an analysis of the functions associated with the family, and the loss of function to consider in relation to the maintenance of patrimony.

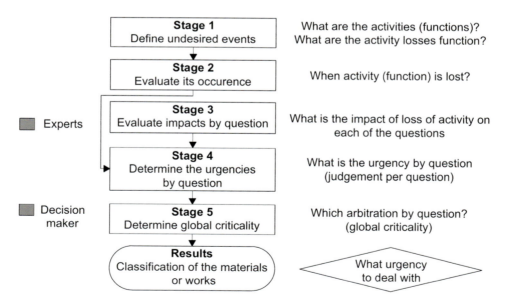

Figure 13.27 Steps of hierarchization. Part one of the process (Ginnochio and Viollet, 2012).

It allows to define the link between the material cut (in the family) and the functional ratio; It is the effect of loss of function that has an impact on the issues.

All function losses have no impact on the maintenance of assets: they are certainly relevant to the ongoing maintenance, or corresponding to those of non-critical functions. This step then aims, in terms of functional analysis, to define the generic function losses that should be considered in terms of heritage maintenance, these function losses are considered as unwanted (feared) events.

The definition of the unwanted event must meet several requirements:

- analyze to the appropriate level of detail: if unwanted events are well defined, an overview of materials or structures is no longer available, and a risk of "combinatorial explosion" is in process. However, if unwanted events are not defined enough, one can overlook a few flaws that may be important;
- define unwanted events to take into account a common horizon for all materials and structures, and consistent with budget planning (0–15 years) (see next step). Unwanted events are of course family specific. If failures are limited, the various families have very different time constants (a few years for an automatism card, several decades for the dam).

For example, an unwanted failure for a forced duct is ruptured. The maintenance policy of penstocks seeks to prevent this risk well in advance: it is the presence of corrosion protection that prevents metal wear and tear. Preventing this wearing off prevents breakage and also extends the life of the penstock. Thus, within a perspective of "patrimony maintenance", there is an "urgency to be addressed", not when the penstock is close to rupture, but when the painting no longer assures its steel protection function. If the unwanted event relies on the painting (there is a strong impact here), what drives the case profile, among other factors, is the absence of painting that has to fulfill the function of protecting the steel.

Function losses are thus defined as a function of the problematic "maintenance of patrimony".

13.5.4.3 Evaluate occurrences

The occurrence indicates a degree of technical urgency to intervene, that is, a period determined for the "level of impact"; the second information, the impact of loss of function, then modulates the degree of technical urgency. Within this framework of process, the notion of occurrence is no longer a statistical notion, and its assessment "at a given moment" is insufficient if it is not accompanied by an estimate of its future evolution, considering aging. Thus, it encompasses the kinetics of the evolution of a degradation.

The criteria for assessing the occurrence are unique to each family, as they depend on physical and functioning parameters of materials and structures. That is, the occurrence classification categories must be general, that is, transversal families, precisely in order to define the overall prioritization of operations, all families combined. Four occurrences' categories are defined in Frame 13.1.

The proposed indicator is simple to quote. It is very common to the set of materials and structures and relevant to each family. Finally, it allows to provide temporal

Frame 13.1 Occurrence Categories of a Threatening Event

Occurrence of a Threatening Event	Abbreviation
Short term	CT
Mid-term	MT
Long term	LT
Very long term	TLT

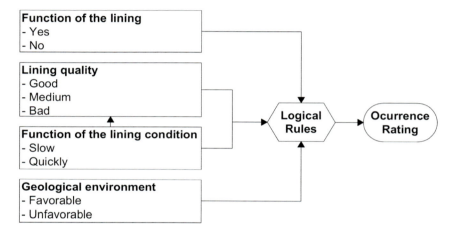

Figure 13.28 Establishing an occurrence of unwanted events to the Family "Gallery" (Ginnochio and Viollet, 2012).

information on the degree of "technical" urgency to intervene, here on an absolute time scale.

Example: for the "Gallery" family (see example in the following figure), the occurrence establishment is defined by the combination of different technical elements: the role of the coating, the state of the coating, and the geological environment. Each element is defined by an expert and its combination, within a logical rule, allows to define the occurrence (Figure 13.28).

13.5.4.4 The impacts per question

For each question, four categories of "impact" are defined: 0 – no impact; 1 – weak impact; 2 – strong impact; and 3 – very strong impact. As for their occurrence, these categories are defined to cross families according to the impact of loss of function on safety, performance and regulation.

There is an ordinal scale, that allows to establish only an order between different quotations. This type of scale does not allow arithmetic manipulations, but only logical and rational manipulations.

Risks and management of patrimony   463

Frame 13.2 Urgency Level Related to Notes
and Colors

Urgency	Note	Associated Color
Without urgency	0	White
Little Urgent	I	Light blue
So Urgent	2	Light violet
Urgent	3	Dark blue

Frame 13.3 Level of Urgency Related to Impact and Occurrence

Occurrence Impact	?	Occurrence Impact	CT	MT		LT	TLT
0	Without urgency	0		Without urgency			
I	Very urgent	I		Very urgent			
2	Urgent	2		Urgent			
3	Urgent	3		Urgent			

*   Urgency Levels per Question

    For an unwanted event on a given material, crossing the impact level for a given issue and the occurrence level allows you to define a priority level (relative to a question) of handling this event, determining a level of urgency (Frame 13.2).

    The following tables allow you to construct the urgency per question from impact (on a scale of 0-1-2-3) and occurrence (on a CT-MT-LT-TLT scale) (Frame 13.3).

*   Global Criticality

    The aggregation of three urgency quotes per question at a global criticality (global priority level) is defined by the hydraulics manager. This way of building global criticality ensures good readability of strategic intentions. It is recalled that the occurrence indicator, as well as the urgency per issue indicator, was designed to be the "common family measures" of occurrence and urgency per issue.

Similarly, the overall criticality constructed from these indicators defines a common measure for all families of the priority of treating the unwanted event, without foreseeing the choice of protection (defense), the mode of treatment of inter-family problems (mutualization of operations, profitability or value).

From three urgency levels per question (*), eight criticality levels were defined: A^+ (maximum criticality), A, A^-, B^+, B, C^+ and C (minimal criticality), as per the following aggregation rules:

Logic rule i	If $U_s = x$ (logical operator) $U_p = y$ with $(x, y) \in \{0,1,2,3\}^2$
	then score $= z$ with $z \in \{A^+, A, A^-, B^+, B, B^-, C^+, C^-\}$

(*) U_s, designating security urgency, U_p the urgency performance, U_r the regulatory urgency.

The logical operator may take the form of an ET, an inclusive or exclusive OU, whether or not followed by a negation. It is also an example of a logical rule that can take other, more general forms such as predicates (Azibi et al., 2002; Azibi and Vanderpooten, 2003).

The system of logical rules built to elaborate the global criticality score is based on a body of ten logical rules. This system complies with certain commonly accepted principles:

- all quotation settings must be covered by the rules;
- two different rules do not affect the same urgency quotation setting at a different overall criticality level;
- two different rules do not affect settings close to the global criticality level away;
- If one setting is worse than another for the urgency level set, it is not affected by a better overall criticality.

Use of risk hierarchization on structures and materials by family.

The process of risk hierarchy leads to classifying materials and works by increasing criticality within a family. This criticality determines the date of execution of interventions: the materials deemed most critical are the subject of earlier monitoring and maintenance interventions. This process thus allows to systematically learn the risks of failure of the overall structure and materials of the hydraulic park families, and provides a macroscopic view of the risks in relation to the three patrimony issues.

This systematic approach of risk helps asset management define a tailored maintenance strategy per family and anticipate early failures to reduce suffering to help achieve the goal of maintenance over the life of current park hydraulic equipment.

13.5.5 A multicriteria decision support

A second approach is multicriteria decision support that allows you to determine maintenance and renewal/modernization choices. Once the risks to materials and structures are identified, defenses must be precisely determined and dimensioned, i.e., in terms of maintenance and renovation/modernization actions. In certain cases where the risks are high costs, and the level of investment is undoubtedly significant, the choice of detailing these actions proves to be complex (it is difficult to set the boundary of the analysis, views of different experts, etc.). To scale out these renovation/modernization investments at best, hydraulic engineering uses a multicriteria approach based on a probabilistic risk assessment.

In terms of investment choice, economic theory essentially offers three alternatives for analysis: cost-benefit, cost-effectiveness or multicriteria approximations. In the present case, support should be given to the multi-criteria decision essentially due to:

- absence of a monetary equivalent to assess an improvement in the level of security;
- the nature of the information to be processed: the consequences of decisions are not deterministic but allegedly probabilistic;

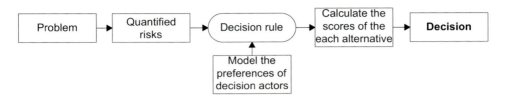

Figure 13.29 Multicriteria decision steps (Ginnochio and Viollet, 2012).

- the presence of multiple issues (safety issues, economic performance, above mentioned regulations);
- the multiplicity of actors: multi-criteria decision support allows a finely described viewpoint, a construct of a discussed value system to prepare the final decision.

The multicriteria decision system, based on the principal quantities, must meet the following essential conditions:

- the first condition stems directly from the imperative of economic efficiency. The goal is to scale the expenses and therefore better allocate them according to various issues. This goal no longer simply needs to prioritize operations, but to be able to compare them from performance gaps between them. The punctuation is no longer ordinal as above, but cardinal;
- as a result, uncertainty must be estimated by a probability measure; which, in other words, means that risk is modeled in this context as a random variable, which should estimate the probability density.

The set of approaches underway is therefore a generalization of the MAUT (Multi-Attribute Utility Theory) approach to the American school of decision. The following are the main steps.

The fundamental idea is to consider renovation/modernization investments as a cover (i.e., a defense) against an unwanted event. The cost that the company commits to cover depends not only on a technical data, the risk, but also on subjective data of the decision maker, which reflects their preferences in strategic intentions.

For details on this subject, readers are advised to consult Ginnochio and Viollet (2012) (Figure 13.29).

13.6 CONCLUSION

The multicriteria decision was successfully developed for the choice of investments in the French hydraulic park (EDF). Each time, it is considered by different actors as a help to make decisions converge, by objectifying the risks and clarifying the preference criteria of different actors. The main difficulties are linked to the risk analysis carried out previously, always difficult to establish when the frequencies are low and the stakes are high. But these difficulties are compensated by the strong expertise available, the ability to model numerous physical phenomena, the triggering of decision-making processes, and for the associated traceability requirement. The obtained results allow to recommend the maintenance of the multicriteria approach. They must be interpreted with caution, then validated by the decision maker who remains the master of the final decision. If the quantitative approach provides the necessary references to size investments, the decision maker always takes into account the specific elements of the problem posed that remain untreatable to modeling.

References

Abecasis, F. M. M. Spillways: Some special problems. LNEC. Memory n. 175. Lisbon. 1961.

ABGE. Associação Brasileira de Geologia de Engenharia e Ambiental. Manual de Sondagens (Brazilian Association of Engineering and Environmental Geology. Sounding Manual. Bulletin n. 3). São Paulo. Brazil, 2013.

Aksoy, S., et al. Cavitation damage at the discharge channels of Keban dam. In: *International Conference on Large Dams*, 1979, New Delhi. Proc. ICOLD, 1979.

Alla, A. A. The role of fusegates in dam safety. Hydropower&Dams, Wallington, n. 6, 1996.

Almeida and Carvalho. Effects of Reservoirs Silting on Electricity Generation. The Case of Mascarenhas Hydroelectric Power Plant. Brazil. In: Brazilian Water Resources Association, Tenth Symposum. Recife. State of Pernambuco. Brazil. 1993.

Anderson, D., et al. Zambesi hydro-electric development at Kariba, first stage. *ICE - Institution of Civil Engineers Proceedings, London*, v. 17, n. 1, 39–60, 1960.

Azibi, R. & Vanderpooten, D. Aggregation of dispersed consequences for construction criteria: The evaluation of flood risk reduction strategies. *European Journal of Operational Research*, Elsevier, v. 144, n. 2, 16, Jan. 2003.

Azibi, R., et al. Construction of rule-based assignment models. *European Journal of Operational Research*, v. 138, n. 2, 274–293, April 2002.

Baecher, G. B. & Christian, J. T. *Reliability and Statistics in Geotechnical Engineering*. Chichester: John Wiley & Sons Ltd, 2003.

Bakhmeteff, B. A. *The Mechanics of Turbulence Flow*. Princeton, NJ: Princeton University Press, 1936.

Ball, J. W. Cavitation from surface irregularities in high-velocity flow. *Journal of the Hydraulic Division*, v.102, n. 9, 1283, ASCE, HY9, Set. 1976.

Barton, N. R., et al. Engineering classification of rock masses for the design of tunnel support. *Rock Mechanics*, Berlin, v. 6, n. 4, 189, 1974.

Barton, N. R., et al. Updating the Q-System for NMT. International symposium of sprayed concrete - modern use of wet mix sprayed concrete for underground support, 1993, Fagernes. Proc. Oslo: Norwegian Concrete Association, 1993.

Bieniawski, Z. T. *Rock mechanics design in mining and tunneling*. Roterdam: A.A. Balkema, 1984.

Brekke, T. L.; Ripley, B. D. Design guidelines for pressure tunnels and shafts. Berkeley: University of California, Electric Power Research Inst. Final Report, 1987.

Brito, R. J. R. Analysis of aeration in high-speed flows in spillways. MSc Thesis. Engineering School of São Carlos. University of São Paulo, 2011.

468 References

Broch, E. Unlined pressure tunnels in areas of complex topography. *WP&DC*, v. 36, n. 11, 21, 1984.

Carvalho, N. O. Practical hydrosedimentology. Eletrobras, Rio de Janeiro. 1994.

CBDB. The history of dams in Brazil, Centuries XIX, XX and XXI: 50 Years of CBDB. RJ, 2011.

CBDB. Diversion of large Brazilian rivers. Rio de Janeiro, 2009.

CBDB. Main Brazilian Dams. Design, construction and performance. Rio de Janeiro, 2009.

CBDB. Large Brazilian spillways: An overview of Brazilian practice and experience in designing and building spillways for large dams. Rio de Janeiro, 2002.

CBDB. Main Brazilian Dams. Design, Construction and Performance. Rio de Janeiro, 2000.

CBDB. The Trans. International Symposium on Layout of Dams in Narrow Gorges. Rio de Janeiro, 1983.

CBDB. *Main Brazilian Dams: Design, Construction and Performance*. Rio de Janeiro: Brazilian Committee on Large Dams, 1982.

CBDB. Brazilian Committee on DAms. Topmost Dams of Brazil. Rio de Janeiro, 1978.

Cedergren, H. R. *Seepage, Drainage and Flow Nets*. New York: John Willey & Sons, 1967.

Chaudhry, M. H. Applied hydraulic transients. British Columbia Hydropower Authority, 1979.

Chavarri, G., et al. Spillway and tailrace design for raising of Guri dam using large scale hydraulic model. In: *ICOLD*, n. 13, New Delhi. Proc. 1979.

Chow, V. T. *Handbook of Applied Hydrology*. New York: McGraw-Hill Co., 1959.

Cooke, J. B. Develop in high concrete face rockfill dams. *Hydropower & Dams*, Wallington, V. 4, n. 4, 1997.

Cooke, J. B. The developments of today's concrete face rockfill dams. In: International symposium on concrete face rockfill dams, 2., Florianópolis, 1999. Proc. Paris: ICOLD, 1999.

Cooke, J. B. & Sherard, J. L. Concrete face rockfill dams - design, construction and performance. Symposium on the concrete face rockfill dam, 1985, Detroit. Proc. NY: ASCE, 1985.

Cruz, P. T. 100 Brazilian dams (100 barragens brasileiras). Oficina de Textos. São Paulo, 1996.

Cruz, P. T., Materón, B. & Freitas, M. Concrete face rockfill dams (Barragens de enrocamento com face de concreto). Oficina de Textos. São Paulo. 2009.

Costa, W. D. *Geologia de Barragens (Geology of Dams)*. São Paulo: Oficina de Textos, 2012.

Davis, C. V. *Handbook of Applied Hydraulics*. New York: McGraw-Hill Books, 1952.

Deppo, L. & Datei, C. Optimal diameters for pressure pipes of hydro plants. WP&DC, abr. 1984.

Design of Small Dams. United States Department of the Interior. Bureau of Reclamation. A Water Resources Technical Publication, 1973.

Edvardsson, S. & Broch, E. Underground powerhouse and high-pressure tunnels. Hydropower Development, n. 14, Dept. Hydraulic and Environmental Engineering, Norwegian University of Science and Technology, Trondheim, 2002.

Elevatorski, E. A. *Hydraulic Energy Dissipators*. New York: McGraw-Hill, 1959.

Erbisti, P. C. E. *Design of Hydraulic Gates*. The Netherlands: A. A. Balkema Publishers, 2004.

Falvey, H. T. Cavitation in chutes and spillways. Engineering Monograph, Denver, n. 42. USBR, 1990.

Feldman, G. M. Stabilization of spillway scour hole at grand rapids. Canadian Electrical Association. Hydraulic Power Section, 1970.

Fell, R., et al. *Geotechnical Engineering of Dams*. 2nd edition. School of Civil and Environmental Engineering, University of New South Wales, Sidney, Australia. Boca Raton, FL: CRC Press, 2017.

Ginocchio, G. & Viollet, P. L'énergie hydraulique. Collection EDF R&D. Editions TEC&DOC. Paris, 2012.

Gomide, F. L. S. Dimensioning of Reservoirs to Regulate Flow Rates. *Brazilian Water Resources Association*, V. 4, 1981.

Gomide, F. L. S. Water storage for sustainable development and poverty eradication. Curitiba, Brazil, 2012.

Gomide, F. L. S. Contribution to the Study of Drought Periods. *International Association of Hydro-Environment Engineering and Research*, Oaxtepec. Mexico, V. 2, 25–36, 1970.

Gordon, J. L. Vortices at intakes. WP&DC, abr. 1970.

Grishin, M. M. *Hydraulic Structures*. v. 1–2. Moscow: Mir Publishers, 1982.

Gulliver, J. S. & Arndt, R. E. A. *Hydropower Engineering Handbook*. New York: McGraw-Hill, 1991.

Hartung, F. & Hausler, E. Scours, stilling basins and downstream protection under free fall overfall jets at dams. In: *International Conference on Large Dams*, n. 11, Madrid. Proc. 1973.

HDC. U.S. Army Corps of Engineers. Hydraulic Design Criteria. Vicksburg, 1977.

Henry, P. Turbomachines hydrauliques. Choix illustré de réalisations marquantes. Presses Polytechniques et Universitaires Romandes. CH-1015 Lausanne. 1992.

Höeg, K. *Asphaltic Concrete Cores for Embankment Dams - Experience and Practice*. Oslo: NGI-Norwegian Geotechnical Institute, 1993.

Hoek, E. Practical rock engineering, 2007.

ICOLD. Embankment dams with bituminous concrete facing. Review/. Bulletin 114, 1999.

ICOLD. Dam failures-statistical analysis. Bulletin 99, 1995.

IS Code 5330. Criteria for Design of Anchor Blocks for Penstocks with Expansion Joint, 1984.

Jabara, M. A., et al. Selection of spillways, plunge pools and stilling basins for earth and concrete dams. ICOLD, 11, Madrid. Proceedings, 1986.

Jaeger, C. *Hydraulique Technique*. Paris: Dunod, 1954.

Jansen, R. B. USBR - U.S. Bureau of Reclamation. Dams and public safety: A water resources technical publication. Denver: USBR, 1983.

JNCLD - Japanese National Committee on Large Dams. Dams in Japan, No. 11, 1988.

Kjaernsli, B., Valstad, T. & Höeg, K. *Rockfill Dams: Design and Construction*. Oslo: NIT-Norwegian Institute of Technology, 1992.

Kohler, W. H. Selection of outlet works gates and valves. ASCE Annual Meeting, Chicago, 1969.

Kolesnikov, E. Reflections on Russian accident on Sayano-Shushenskaya. Miami, 2009.

Kollgaard, E. C. & Chadwick, W. L. Development of dam engineering in the USA. USCOLD, 16º ICOLD, 1988.

Kramer, K. Development of aerated chute flow. PhD Thesis. Swiss Institute of Technology, Zurich, 2004.

Lambe, T. W. & Whitman, R. V. *Soil Mechanics, SI Version*. Massachusetts Institute of Technology. New York: John Wiley & Sons, 1969.

Lambe, T. W. & Whitman, R. V. *Soil Testing for Engineers*. New York: John Willey & Sons, 1979.

Lana, A. E. *Statistical and Probability Elements*. Brazil: Federal University of the State of Rio Grande do Sul., 1993.

Lemos, F. O. Behavior of structures of some dams built in narrow valleys. *The Transactions of the International Symposium on Layout of Dams in Narrow Valleys*, CBDB. Rio de Janeiro, 1982.

Lencastre, A. *General Hydraulics Manual*. São Paulo: University of São Paulo, 1966.

Linsley, R. K., et al. Hydrology for engineers. New York: McGraw-Hill, 1959.

Lowe III, J., et al. Tarbela service spillway plunge pool development. WP&DC, Nov. 1979.

Lysne, D. K., et al. Norwegian University of Science and Technology. Hydropower Development, v. 8. Hydraulic Design, 2003.

Magela, G. M. and Brito, S. N. *Erosion in Jet Diving Basins: Hydraulic and Geotechnical Aspects*. Rio de Janeiro: CBDB Magazine, 1996.

Magela, G. M. and Brito, S. N. The use of ski jump spillway in highly fractured rock masses. The case of Jaguara Hydroelectric Powerplant. In: Brazilian Seminar on Large Dams. 19. 1991. Aracaju, Brazil. Proceedings. CBDB, 1991.

Mahmood, K. Reservoir sedimentation, impact, extent and mitigation. World Bank Tec. Paper, n. 71, Sep 1987.

Marsal, R. J. & Nuñez, D. R. *Earth and Rockfill Dams*. Mexico City: Limusa, 1975.

Martins, R. Kinematics of the free jet in the context of hydraulic structures. LNEC. Memory n. 424. 1977.

Mason, P. J. & Arumugan, K. A review of 20 years of scour development at Kariba Dam. In: *Conference of Flood and Flood Control*, v. 2, 1986. Cambridge.

Mayo Jr. H. A. Low-head hydroelectric unit fundamental. *Engineering Foundation Conference*. Hydropower: A national resource. Department of the Army. USACE, 1979.

Mays, L. W. *Hydraulic Design Handbook*. New York: McGraw-Hill Handbooks, 1999.

McNair, D. EPC contracts in the power sector. DLA Piper, 2011.

Miller, D. S. Internal flow systems. Cranfield: British Hydromechanics Research Assoc., 1978.

Miranda, J.C.M. Hydraulic Turbines. Engevix Engineering. Technical Information Group, November, 1982.

Moody, L. F. Friction factors for pipe flow. In: *Semi-Annual Meeting of the ASME*, 1944, Pittsburg.

National Environmental Council, Resolution 237, December, 19, 1997.

National Environmental Council, Resolution 01, January, 23, 1986.

Neidert, S. H. Performance of spillways. Energy dissipation. CEHPAR. Pub. 37. Brazil. 1980.

Nielsen, B. & Thidemann, A. Rock engineering. Hydropower development, v. 9, Division of Hydraulic Engineering, NIT – Norwegian Institute of Technology, Trondheim, 1993.

Nikuradse, J. Laws of flows in rough pipes. National Advisory Committee for Aeronautics Technical Memorandum 1292, 1933.

Oliveira, A. R. & Leme, C. R. Adding $1,000\,m^3/s$ to Euclides da Cunha spillway. ICOLD, 15, 1985, Lausanne. Proc. Paris, 1985.

Oliveira, A. R., et al. Case histories of repairs of concrete surfaces subjected to water erosion. ICOLD, 15, 1985, Lausanne. Proceedings. Paris, 1985.

Palmström, A. & Broch, E. Use and misuse of rock mass classification system with particular reference to the q-system. *Tunnelling/Underground Space Technology*, Amsterdam, v. 21, n. 6, 575–593, 2006.

Parmakian, J. Minimum thickness for handling steel pipes. WP&DC, Jun. 1983.

Peck, R. Where has all the judgment gone? The fifth Laurits Bjerrum memorial lecture, 1980.

Peterka, A. J. Hydraulic design of stilling basin and energy dissipators. EM n. 25, Denver, 1983.

Pigott, R. J. S. The flow of fluids in closed conduits. National process meeting of the American Society of Mechanical Engineers, Buffalo, 1932.

Pinto, N. L., et al. *Basic Hydrology*. São Paulo: Edgard Blücher, 1976.

Pinto, N. L., et al. *Surface Hydrology*. São Paulo: Edgard Blücher, 1973.

Pinto, N. L. Prototype aerator measurements. In: Wood, I. R. (Ed.) *Air Entrainment in free-surface flow*. New Zealand: A. A. Balkema, 1991.

Pinto, N. L. Cavitation and aeration in high velocity flows. Federal Univ. of Paraná. Pub. 35. Brazil. 1979.

Pinto, N. L. Energy dissipation and erosion downstream of dams. In: Symposium on the Upper Paraná river Geotechnics. Proceedings. ABGE. 1983.

Pinto, N. L. S. Designing aerators for high velocity flow. WP&DC, July 1989.

Power, G. Siltation is threat to whole world's storage dams. World Water, June 1988.

Quintela, A. C. & Ramos, C. M. Protection against cavitation erosion. LNEC. Memory, n. 539. 1980.

Reinius, E. Rock erosion. WP&DC, Jun. 1986.

Roig, H. L., et al. Determination of Reservoir Silting Using Geoprocessing Techniques. The Case of Funil Hydroelectric Power Plant. In: Brazilian Symposium on Remote Sensing, November 2003. Belo Horizonte. Brazil, 2003.

Rudavsky, A. B. Selection of spillways and energy dissipators in preliminary planning of dam developments. ICOLD, 12, 1976, México. Proc. Paris, 1976.

Sarkaria, G. S. Economic penstock diameters: A 20-year review. WP&DC, Nov. 1979.

Saville Jr. T., et al. Freeboard allowances for waves in inland reservoirs. *Journal of the Waterways and Harbors Division. Transactions of the ASCE*, v. 88, n. 4, 195–226, 1962.

Schreiber, G. P. *Hydroelectric Powerplants*. Rio de Janeiro: Edgard Blücher, 1977.

Sherard, J. L., et al. *Earth and Earth-Rock Dams*. New York: John Wiley & Sons, 1963.

Silveira, J. F. A. *Instrumentação e Comportamento de Fundações de Barragens de Concreto (Instrumentation and Behavior of Concrete Dama Foundations)*. São Paulo: Oficina de Textos, 2003.

Souza, et al. *Hydroelectrics Stations. Implementation and Commissioning*. Rio de Janeiro, Brazil: Interciência, 2009.

Spurr, K. J. W. Energy approach to estimating scour. WP&DC, jul. 1985.

Stucky, A. Chambres d'équilibre. École Polytechnique de L'Université de Lausanne. 1958.

Taylor, D. W. *Fundamentals of Soil Mechanics*. New York: John Wiley & Sons, 1948.

USBR. Hydraulic design of stilling basin and energy dissipators. EM, n. 25, Denver, 1983.

USBR. Freeboard criteria, guidelines. Denver, 1981.

USBR. Design of small dams, Denver, 1974.

USBR. Power Operations-Maintenance Bulletin n. 19, "Maintenance Schedules and Records", Oct. 1965.

Veiga, A. P. Dam risk and safety management. *Symposium on Dams and Associated Risks*. LNEC, 2008.

Vinogg, L., et al. Mechanical equipment. Hydropower Development, NIT - Norwegian Institute of Technology, Trondheim, v. 9, 2003.

Warnick, C. C., et al. *Hydropower Engineering*. Hoboken, NJ: Prentice-Hall. Inc., 1984.

Watermark Engineering. Basic Design Conde d'Eu HPP. Rio de Janeiro, Brazil, 2007.

Wengler, R. P. The layout of Mossyrock arch dam in a narrow canyon. *International Symposium on Layout of Dams in Narrow Gorges*, v. I, Rio de Janeiro: CBDB, 1982.

Westgaard, A. K., et al. Electrical equipment, hydropower development, NIT-Norwegian Institute of Technology, Trondheim, v. 13, 1994.

White, W. R. World water: Resources, usage and the role of man-made reservoirs. FR/R0012. Marlow: Foundation for Water Research, 2010.

Whittaker, J. G. & Schleiss, A. Scour related to energy dissipators for high head structures. ETH - Zürich, 1984.

Wilson, D. Many and varied uses of asphalt. Hydraulic asphalt eng. Stafford. Walo, U. Kingdom Ltd., 2013.

Xavier, L. V. Unconventional solution for the Spillway at the Itapebi Powerplant. In: Brazilian Seminar on Large Dams. 25. 2003. Salvador, Brazil. Proceedings. CBDB, 2003.

Yuditskii, G. A. Hydrodynamic action of the overfall nappe spilled on fragments of a rocky bed and conditions of rupture of this. LNEC. Translation n. 442, 1963.

Glossary

To prepare this Glossary, the author used information available on the internet and information extracted from books in his private library, whose credits are given below:

- Dams and Public Safety-USBR, Jansen R. B. (1983);
- Glossary of Engineering Geology Technical Terms. Tognom, A. A. ABGE, São Paulo, 1985;
- Hydropower Engineering Handbook, Gulliver and Arndt (1991); and
- Water Words Dictionary, Nevada Division of Water Planning (2000);
- A very complete dictionary available on the internet;
- The Hydraulics of Open Channel Flow: An Introduction, Chanson H. (2004).

The author tried not to limit the content to the book and the resulting Glossary was quite complete and may be useful as such for hydraulic works in general. For more details, it is recommended to consult the references.

Abrasion   Mechanical process of wear on rock surfaces caused by solid material carried by currents and waves, rivers, glaciers and wind.

Abutment   The part of the valley side against which the dam is constructed. Artificial abutments are sometimes constructed to take the thrust of an arch where there is no suitable natural abutment; Right abutment; Left abutment.

Accretion   Increase of channel bed elevation resulting from the accumulation of sediment deposits. See Sanding up.

Adhesion   Shear resistance between a soil and any other material without external pressure acting.

Aerator   Device used to introduce artificially air within a liquid. Spillway aeration devices are designed to introduce air into high-velocity flows. Such aerators include basically a deflector and air is supplied beneath the deflected waters. Downstream of the aerator, the entrained air can reduce or prevent cavitation erosion.

Afflux   Rise of water level above normal level (i.e., natural flood level) on the upstream side of a culvert or of an obstruction in a channel. In the US, it is commonly referred to as maximum backwater.

474 Glossary

Afterbay dam A dam constructed to regulate the discharges from an upstream powerplant.

Aggradation Rise in channel caused by deposition of sediment material. Another term is accretion. See Sanding. In Brazil it's common to use the term "sedimentation".

Air concentration Concentration of undissolved air defined as the volume of air per unit volume of air and water. It is also called void fraction.

Alluvium Generic name that includes deposits of cuttings, sands, silts and clays, of current deposition, of fluvial, lacustrine, lacustrine-fluvial, marine, fluvial-marine origin, and can be formed in the floodplains, in the alluvial fans at the foot of mountains and escarpments, as well as on coastal plains.

Alternating current An electric current changing regularly from one direction to the opposite direction.

Ampere The common unit of measure of electrical current.

Anchor block A concrete block on a hillside for supporting and fixing the penstock.

Angle of internal friction Angle corresponding to the slope of the tangent to the Mohr curve that establishes the relationship between the shear resistance of soil and the corresponding normal stress.

Anisotropy Variation of a physical property of a given rock as a function of the direction along which this property is measured.

Anomaly Deviation from uniformity, in terms of physical properties, of exploratory interest.

Aquifer Geological formation capable of storing and transmitting water in appreciable quantities. See Water-bearing stratum.

Arc-gravity dam a dam which is only slightly thinner than a gravity dam.

Arch dam A concrete or masonry dam which is curved in plan so as to transmit the major part of the water load to the abutments.

Arch buttress dam or Curved buttress dam a buttress dam which is curved in plan.

Artesian aquifer See Confined aquifer.

Assembly floor (erection bay) A separated area inside or next to the powerhouse, having a platform of adequate load-bearing capacity for mounting one, exceptionally two machine units.

Auxiliary spillway A dam spillway built to carry runoff in excess of that carried by the principal spillway; a secondary spillway designed to operate only during exceptionally large floods. Also referred to as Emergency Spillway. See Spillway.

Average annual runoff (Yield) The average of water-year runoff; or the supply of water produced by a given stream; or water development project for a total period of record.

Axial-flow turbine A collective term for turbines with axial flow through the runner blades connected axially to the turbine shaft. Both propeller turbines and Kaplan turbines are axial-flow turbines.

Backwater A small, generally shallow body of water attached to the main channel, with little or no current of its own; (2) water backed up or retarded in its course as compared with its normal or natural condition of flow. In Stream Gaging, a rise in Stage produced by a temporary obstruction such as ice or weeds, or by flooding

Glossary 475

the stream below. The difference between the observed stage and the indicated by the Stage-Disc rage Relation, is reported as backwater.

In a tranquil flow motion (i.e., subcritical flow) the longitudinal flow profile is controlled by the downstream flow conditions: e.g., an obstacle, a structure, a change of cross-section. Any downstream control structure (e.g., bridge piers, weirs) induces a backwater effect. More generally the term backwater calculations or backwater profile refer to the calculation of the longitudinal flow profile. The term is commonly used for both supercritical and subcritical flow motions. Backwater calculation: the first successful calculations were developed by J. B. Bélanger who used a finite difference step method (MEF) for integrating the equations (Bélanger, 1828).

Bank   The strip of land adjoining a body of water, especially a river.

Barrage   A barrier built across a river that even contain a spillway where the floods pass - excess water.

Baseload   Typically, the minimum load over a given period of time.

Base flood (100-year flood)   The flood having a 1% average probability of being equaled or exceeded in a given year at a designated location. It may occur in any year or even in successive years if the hydrologic conditions are convenient for flooding.

Base flow   The sustained flow of streams; that part of stream discharge not attributable to direct runoff from precipitation, snowmelt, or a spring. Discharge entering streams channels as effluent from groundwater reservoir. Also referred to as Groundwater Flow.

Base level   The lower level to which land surface can be reduced by the action of running water.

Basin   A single area drained by a single major stream; consists of a drainage system comprised of streams and often natural or man-made lakes. Also referred to as Drainage Basin, Watershed, or Hydrographic Region.

Bedding   Lodging, Stratification.

Bed-load   Sediment material transported by rolling, sliding and saltation motion along the bed.

Berm   A horizontal step in the sloping profile of an embankment dam.

Blade servomotor   The hydraulic cylinder actuated by governor oil pressure which supplies the force necessary to adjust the runner blades.

Blanket   See Upstream blanket.

Bottom outlet   Opening near the bottom of a dam for draining the reservoir and eventually flushing out reservoir sediments.

Boundary layer   Flow region next to a solid boundary where the flow field is affected by the presence of the boundary and where friction plays an essential part. A boundary layer is characterized by a range of velocities from zero at the boundary to the free-stream velocity at the outer edge.

Brake jet   The water jet that provides the counterrotational force used to decelerate an impulse runner.

Broad-crested weir   A weir with a flat long crest weir when the crest length over the upstream head is greater than 1.5–3. If the crest is long enough, pressure distribution along the crest is hydrostatic; the flow depth equals the critical flow depth d_c: $(q^2/g)^{1/3}$ and the weir can be used as a critical depth meter.

476   Glossary

Bulb   The streamlined watertight housing for the bulb turbine generators.

Buoyancy   Tendency of a body to float, to rise or to drop when submerged in a fluid at rest. The physical law of buoyancy (or Archimedes' principle) was discovered by the Greek mathematician Archimedes. It states that a body submerged in a fluid at rest is subjected to a vertical (or buoyant) force of magnitude equal to the weight of the fluid displaced by the body.

Butterfly valve   A disk-shape closing body rotating around a shaft perpendicular to the penstock axis.

Buttress dam   A dam consisting of a watertight part supported at intervals on the downstream side by a series of buttresses; can take many forms:

Capacity   The greatest load which a piece of equipment can safely serve.

Capillarity   Property by which the interstitial water of a soil reaches points above the water table. This phenomenon is related to capillary tensions and its intensity increases due to decrease in voids index.

Cascade   (1) A steep stream intermediate between a rapid and a waterfall. The slope is steep enough to allow a succession of small drops but not sufficient to cause the water to drop vertically (i.e., waterfall).

(2) A man-made channel consisting of a series of steps: e.g., a stepped cascade, a staircase chute, a stepped sewer.

Casing   Coating, covering, encasement, jacket, lining, revetment.

Cataract   A series of rapids or waterfalls.

Cavitation   Formations of vapor bubbles and vapor pockets within a homogenous liquid in low-pressure regions where the liquid has been accelerated. Cavitation is characterized by damaging erosion, additional noise, vibrations and energy dissipation.

Cellular gravity dam   See Hollow gravity dam.

Clay   Earthy material that is plastic when moist and that become hard when dry.

Clearance loss   The discharge which escapes through the gaps between the casing and the runner and through the sealings of the runner of the turbine, and therefore is lost for power production. Also known as leakage.

Cofferdam   A temporary structure enclosing all or part of the construction area so that construction can proceed in the dry. A diversion cofferdam diverts a river into a pipe, channel or tunnel.

Cohesive sediment   Sediment material of very small sizes (i.e., less than $50\,\mu m$) for which cohesive bonds between particles (e.g., intermolecular forces) are significant and affect the material properties.

Collapsing pressure   See Uplift.

Concrete lift   In concrete work, the vertical distance between successive horizontal construction joints.

Conduit/penstock valve   See control valve.

Confined aquifer   Geological formation completely saturated with water, limited at the top and bottom by an impermeable layer or formation. The water stored in it is subjected to a pressure higher than the atmospheric level.

Conjugated depth   See Sequent depth. In open channel flow, another name for Sequent depth.

Connecting rods   Elements connecting the servomotor piston rod to the gate operating ring.

Consolidation Progressive reduction of the volume of a soil mass under the effect of its own weight or the increase of external pressure, in three successive stages of densification: initial, primary and secondary. The reduction is made at the expense of the volume of voids with the consequent expulsion of air and water.

Consolidation grouting Strengthening an area of ground by injecting grout.

Construction joint The interface between two successive placings or pours of concrete where bond, and not permanent separation, is intended.

Contact grouting Filling, with cement grout, any voids existing at the contact of two zones of different materials, e.g., between a concrete tunnel lining and the surrounding rock. The grout operation is usually carried out at low pressure.

Control Considering an open channel, subcritical flows are controlled by the downstream conditions. This is called a "downstream flow control". Conversely. Supercritical flows are controlled only by the upstream conditions (i.e., "upstream flow conditions").

Control section In open channel, cross-section where critical flow conditions take place. The concept of "control" and "control section" are used with the same meaning.

Control valve Conduit/penstock valve: the valve installed between the surge tank or the head pond and the penstock.

Core, impervious core, or impervious zone A zone of material of low permeability in an embankment dam. Hence the expressions "central core", "inclined core", "puddle clay core", and "rolled clay core".

Core wall A wall of substantial thickness built of impervious material, usually of concrete or asphaltic concrete in the body of an embankment dam to prevent leakage. See also Membrane or Diaphragm.

Creager profile Spillway shape developed from a mathematical extension of the original data of Basin in 1886–1888 (Creager, 1917).

Crest of dam The upper part of an uncontrolled spillway. The term "Crest of Dam" should not be used when "Top of Dam" is intended.

Crest length The developed length of the top of the dam, including all structures: spillway, powerhouse, navigation lock, fish pass, etc., where these form part of the length of the dam.

Crest of spillway Upper part of a spillway. The term "crest of dam" refers to the upper part of an uncontrolled overflow.

Crib (1) Framework of bars or spars for strengthening; (2) frame of logs or beams to be filled with stones, rubble or filling material and sunk as a foundation or retaining wall.

Crib dam Gravity dam built up of boxes, cribs, crossed timbers of gabions, and filled with earth or rock.

Critical depth See Broad-crested weir.

Critical flow conditions In open channel flows, the flow conditions such as the specific energy (of the mean flow) is minimum are called the critical flow conditions. With commonly used the Froude number definitions occur for Fr: 1. If the flow is critical, small changes in specific energy cause large changes in flow depth. In practice, critical flow over a long reach of channel is unstable.

Cross section at crown Cross section at crown of an arch dam which generally corresponds with the point where the height of the dam is a maximum.

478 Glossary

Culvert Covered channel of relatively short length installed to drain water through an embankment (e.g., highway, railroad, dam).

Curved buttress dam See Buttress dam.

Curved gravity dam a gravity dam which is curve in plan.

Cutoff An impervious construction by means of which water is prevented from passing through foundation material.

Cutoff trench The excavation later to be filled with impervious material so as to form the cutoff. Sometimes, used incorrectly to describe the cutoff itself.

Cutoff wall A wall of impervious material (e.g., concrete, asphaltic concrete, steel sheet piling) built into the foundation to reduce seepage under the dam.

Cyclopean dam A gravity dam in which the mass masonry consists primarily of large one-man or derrick stone embedded in concrete.

Dam A massive wall or structure built across a valley or river for storing water.
  Top of dam – the upper part of a dam.

Darcy law Law of groundwater flow motion which states that the seepage flow rate is proportional to the ratio of the head loss over the length of the flow path.
  It was discovered by H. P. G. Darcy (1856) that for a flow of liquid through a porous medium, the flow rate is directly proportional to the pressure difference.

Darcy-Weisbach friction factor Dimensionless parameter characterizing the friction loss in a flow, named after the works of H. P. G. Darcy and the J. Weisbach. Darcy-Weisbach. Equation: h_f: f (l/D) $V^2/2g$, where,
  h_f, loss of head (m); f, friction factor; L, tube length (m); D, diameter (m), V, velocity (m/s), and g, gravity acceleration (m/s^2).

Debris Debris comprises mainly large boulders, rock fragments, gravel-sized to clay-sized material, tree and wood material that accumulate in creeks.

Degradation Lowering in channel bed elevation resulting from erosion.

Depth of cutoff The vertical distance that the cutoff penetrates into the foundation of dam.

Design head The head at which the turbine is designed to operate at a maximum efficiency; the head on the spillway crest to pass the design discharge.

Diaphragm See Membrane.

Dike, dyke, or levee A long low embankment. The height is usually less than four to five meters.

Dimensional analysis Technique used to reduce the complexity of a study, by expressing the relevant parameters in terms of numerical magnitude and associated units, and grouping them into dimensionless numbers. The use of dimensionless numbers increases the generality of the results.

Direct current The electric current going in one direction only.

Discharge Discharge means outflow; is used as a measure of the rate at which a volume of water passes a given point. Discharge, streamflow or runoff of drainage basins is distinguished as follows: (1) Yield – the total water runout or "water crop" and includes runoff plus underflow; (2) Runoff – that part of water yield that appears in streams; and (3) Streamflow – the actual flow in streams, whether or not subject to regulation or underflow.

Discharge, average The arithmetic average of the annual discharges for all complete water years of record whether or not they are consecutive. The term average is generally reserved for average of record and mean is used for averages of shorter periods; namely, daily mean discharge.

Discharge coefficient   The ratio of actual rate of flow to the theoretical rate of flow through orifices, weirs, or other hydraulic structures.

Discharge curve   A curve that express the relation between the discharge of a stream or open conduit at a given location and the stage or elevation or the liquid surface at or near that location. Also called Rating Curve and Discharge Rating Curve.

Discharge ring   The structural member of a Francis turbine that surrounds the runner band. On a propeller turbine, it surrounds the blades and forms a guide for the flow. It may be integral with the bottom ring. The draft tube liner is attached to the downstream and the discharge ring.

Diversion channel, canal, or tunnel   A waterway used to divert water from its natural course.

   The term is generally applied to a temporary arrangement, e.g., to bypass water around a damsite during construction. "Channel" is normally used instead of "canal" when the waterway is short. Occasionally the term is applied to a permanent arrangement (diversion canal, diversion tunnel, diversion aqueducts).

Diversion dam   Dam or weir built across a river to divert water into a canal. It raises the upstream water level of the river but does not provide any significant storage volume.

Diversion rate   A rate of water flow diverted into a canal or through a headgate.

Double curvature, arch dam   an arch dam which is curved vertically as well as horizontally.

Double curvature, arch dam   an arch dam which is curved vertically as well as horizontally.

Downstream   In the direction of the current.

Downstream slope (of a dam)   The slope or face of the dam away from the reservoir water, which, for Embankment Dams, requires some form of protection such as grass to protect it from erosive effects of rain and surface flows.

Downstream toe of dam   The junction of the downstream face of a dam with the ground surface.

Draft tube   The diffuser which regains the residual velocity energy of the water leaving the turbine runner.

Draft tube liner   The steel lining used in the draft tube to protect the concrete from the high velocity of the water.

Drainage basin   See Basin, Hydrographic region; Watershed.

Drainage blanket   A drainage layer placed directly over the foundation material.

Drainage curtain   A line of such wells forms a drainage curtain.

Drainage layer   A layer of pervious material in an earthfill dam to relief pore pressures or to facilitate drainage of the fill.

Drainage well or relief well   Vertical wells or boreholes downstream of an embankment dam, or in downstream shoulder, to collect and control seepage through or under the dam and to reduce water pressure.

Drop   (1) A rapid change of bed elevation called step; (2) volume of liquid surrounded by gas in a free-fall motion (i.e., dropping); (3) by extension, small volume of liquid in motion in a gas.

Droplet Small drop of liquid.

Drop structure Single step structure characterized by a sudden decrease of bed elevation.

Earth dam Earth fill dam: See Embankment dam.

Earth dam or earthfill dam an embankment dam in which more than 50% of the total volume is formed of compacted fine-grained material obtained from a borrow area.

Elbow draft tube A diffuser in the form of an elbow.

Embankment A fill constructed usually using earth or rock, placed with sloping sides and with a length greater than its height. An "embankment" is generally higher than a "dike".

Embankment dam Any dam constructed of excavated natural materials or of industrial waste materials.

Energy The power of doing work, for a given period. Usually measured in kWh.

Erection bay See Assembly floor.

External pressure See Uplift.

Face External surface which limits a structure, e.g., air face of dam (i.e., downstream face), water face of a weir (i.e., upstream face).

Facing With reference to a wall or concrete dam, a coating of a different material, masonry or brick, for architectural or protection purposes, e.g., stonework facing, brickwork facing. With reference to an embankment dam, an impervious coating or face on the upstream slope of the dam.

Filter or filter zone A band of granular material which is incorporated in an embankment dam and is graded (either naturally or by selection) so as to allow seepage to flow across or down the filter zone without causing migration of the material from zones adjacent to the filter.

Finger drains A series of parallel drains of narrow width (instead of a continuous drainage blanket) draining to the downstream toe of the embankment dam.

Fixed-bed channel The bed and sidewalls are non-erodible. Neither erosion nor accretion occurs.

Fixed-blade propeller type turbine An axial-flow reaction turbine with blades keyed to the hub, unlike those of the Kaplan turbine.

Flashboards A board or a series of boards placed on or at the side of a dam to increase the depth of water. Are usually of the lengths of timber, concrete or steel placed on the crest of a spillway to raise the upstream water level.

Flash flood Flood within short duration with a relatively high peak flow rate.

Flashy Term applied to rivers and streams whose discharge can rise and fall suddenly, and is often unpredictable.

Flat slab dam Ambursen dam, or Deck dam a buttress dam in which the upstream part is a relatively thin flat slab usually made of reinforced concrete.

Flip bucket A flip bucket or ski-jump is a concave curve at the downstream end of a spillway, to deflect the flow into an upward direction. Its purpose is to throw the water clear of the hydraulic structure and to induce the disintegration of the jet in air.

Forebay The upstream part of the reservoir extension of the river.

Foundation of dam The undisturbed material on which the dam is placed.

Glossary 481

Francis turbine A radial-inflow reaction turbine, where the flow through the runner is radial to the shaft.

Freeboard The vertical distance between a maximum water level and the top of the dam.

Free flow Flow through or over a structure not affected by submergence or backwater.

Free-flowing stream A stream that is unmodified by works of man or, if modified, still retains its natural scenic qualities and recreational opportunities.

Free-flowing weir A weir that in use has tailwater lower than the crest of weir.

Free surface Interface between a liquid and a gas.

Friction head Energy required to overcome friction due to fluid movement with respect to the walls of the conduit or containing medium.

Friction losses Total energy losses in the flow of water due to friction between the water and the walls of a conduit or channel.

Friction slope The energy loss per unit of length of open or closed conduit due to friction.

Froude number The Froude number is proportional to the square root of the ratio of the inertial forces over the weight of fluid. It is used generally for scaling free-surface flows, open channels and hydraulic structures. Several French researchers used it before William Froude.

Gabion dam A gabion consists of rockfill material enlaced by a basket or a mesh. The word "gabion" originates from the Italian "gabbia" that means "cage". Gabion dam is a crib dam built up of gabions.

Gage, or gauge (1) An instrument used to measure magnitude or position: WL elevation; the velocity; the pressure; the amount of precipitation; the depth of snowfall, etc.; (2) The act or operation of registering or measuring magnitude or position; (3) The operation, including both field and office work, of measuring discharge of a stream of water in a waterway.

Gage height The height of the water surface above the gage datum (reference level). Gage height is often used interchangeably with more general term, Stage, although Gage Height is more appropriate when used with a gage reading.

Gaging station A particular site on a stream, canal, lake, or reservoir where systematic observations of Gage Height or discharge are obtained.

Gallery (1) A passageway within the body of a dam or abutment, hence the terms "grouting gallery", "inspection gallery", and "drainage gallery; (2) a long and rather narrow hall; hence the following terms for a power plant: "valve gallery", "transformer gallery" and "busbar gallery".

Gate A device in which a leaf or member is moved across the waterway from an external position to control or stop the flow Port, equipment for controlling the passage of a fluid. It is of several types: bulkhead gate; crest gate; emergency gate; fixed wheel gate (fixed roller gate); flap gate; flood gate; guard gate (guard valve); outlet gate; radial gate: underflow gate for which the wetted surface has a cylindrical shape; regulating gate (or valve); slide gate (sluice gate).

Gate chamber The part of the power conduit where the gate is accommodated.

Gate house Located above the gate chamber, contains the hoist and pertaining equipment for operating the gate.

482 Glossary

Gate linkage Any linkage connecting the gate operating ring and the wicket gates.

Gate operating ring The ring rotated by servomotors which distributes the force from the servomotors to the individual wicker gate linkages to provide simultaneous movement of all wicket gates.

Gate servomotors The hydraulic cylinders actuated by oil pressure which supply the force necessary to operate the wicket gates through the gate operating ring.

Gate valve A leaflike closing gate, sliding in a plane perpendicular to the penstock.

Generator A machine powered by a turbine which produces electric current.

Generator brake A device for stopping the revolving part of the generating unit.

Geologic log A detailed description of all underground features (e.g., depth, thickness, type of formation, etc.) discovered during the drilling of a well.

Geologic time (History) Geologic history can be divided into five great Eras of recorded time. These Eras and approximate time periods include: (1) Archeozoic, 4,500 million years ago (MYA) to 3,500 MYA; (2) Proterozoic (or Prepaleozoic), 3,500 MYA to 570 MYA; (3) Paleozoic, 570 MYA to 230 MYA; (4) Mesozoic, 230 MYA to 66 MYA; and Cenozoic, 65 MYA to present. Each time Era (except the first) is divided into Periods (e.g., the Cenozoic into the Quaternary and the Tertiary) and Periods are further divided into Epochs (e.g., the Tertiary into the Pliocene, Miocene, Oligocene, Eocene, and the Paleocene). For each time period, whether an Era, Period, or Epoch, there is a corresponding rock formation by which the time period has been dated. Rock formations constituting a specific (time) Era form a Group of rocks; those rocks having been formed during a specific (time) Period constitute a rock System; and those rock formations originating during a specific (time) Epoch are said to belong to a particular Series of rocks. Series of rock formations are further subdivided into Formation, Stages, etc.

Geological age (Archeology) A period of time, earlier than the present postglacial period, which can only be effectively dated geologically, that is by its rock formations and fossilized matter within those rock formations.

Geological survey A systematic examination of an area to determine the character, relations, distribution and origin or mode formation, of its rock masses and other natural resources. United States Geological Survey (USGS) – an agency of the USBR, established in 1879, responsible for providing extensive earth-science studies of the Nation's land, water, and mineral resources.

Geology The science that studies the physical nature and history of the earth.

Geomorphology That branch of both physiography and geology that deals with the form of the earth, the general configuration of its surface, and its changes that take place in the evolution of land forms.

Gravel A mixture composed primarily of rock fragments 2 mm to 7.6 cm in diameter. Usually contains much sand.

Gravity dam A dam constructed of concrete and/or masonry which relies on its weight for stability.

Gross head The difference between the headwater level and the tailwater level.

Ground surface The original ground surface at a dam site prior to construction.

Grout blanket An area of the foundation systematically grouted to a uniform depth.

Glossary 483

Grout cap   A concrete pad or wall constructed to facilitate subsequent pressure grouting of the grout curtains beneath the grout cap.

Grout cutoff (Grout curtain)   A vertical zone, usually thin, in the foundation into which grout injected to reduce seepage under a dam.

Guide vanes   The streamlined movable blades regulating inflow to the turbine runner.

Head   See Gross head and net head.

Head cover   The axisymmetric member in vertical machines that spans the top of the distributor, provides separation between the watered runner chamber and the dry turbine pit, and supports the main shaft packing box and the main bearing.

Head gate   Built in the intake portion of the entrance flume of low-head and medium-head powerhouses, or in the head pond of a high-head power plant, to control inflow into the penstock.

Headrace   That portion of the power canal which extends from the intake works to the powerhouse.

Headwater   The water upstream from the powerhouse, or generally, the water upstream from any hydraulic structure creating a head.

Headwater elevation   The height of the headwater in the reservoir; the level.

Headworks   See Intake.

Heel of dam   The junction of the upstream face of a gravity dam with the ground surface. In the case of an embankment dam, the junction is referred to as the "upstream toe of dam".

Height of dam   See Hydraulic height.

Hollow gravity dam (Cellular gravity dam)   A dam which has the outward appearance of a gravity dam but is of hollow construction.

Homogeneous earthfill   See Embankment dam.

Homogeneous earthfill dam   an embankment type dam construction throughout of more or less uniform earth materials, except for possible inclusion of internal drains or blanket drains. Used to differentiate it from a zoned earthfill dam.

Housing   The enclosure, surrounding and impulse runner, which forms the aerated chamber in which the runner operates.

Hydraulic diameter   It is defined as the equivalent pipe diameter: i.e., four times the cross-section area divided by wetted perimeter (D_H: 4A/P).

Hydraulic efficiency   An efficiency parameter of the turbine, expressing exclusively the power decrease due to hydraulic losses (friction, separation, impact), including the losses in the scroll case and the draft tube.

Hydraulic fill dam   An embankment constructed of materials often dredged which are conveyed and placed by suspension in flowing water.

Hydraulic fill dam   Embankment dam constructed of materials which are conveyed and placed by suspension in flowing water. See Embankment dam.

Hydraulic head   See Gross head and net head.

Hydraulic height   Height to which the water rises behind the dam and is the difference between the lowest point in the original streambed at the axis of the dam and the maximum controllable water surface.

Hydraulic jump   Transition from a rapid (supercritical) to a slow flow motion (subcritical).

Hydroelectric power   The electric current produced from water power.

Hydroelectric powerplant   A building in which turbines are operated to drive generators, by the energy of natural or artificial waterfalls.

Hydrographic region   See Basin, Watershed.

Hydropower plant   The comprehensive term for all structures necessary for utilizing a selected power site.

Hydropower station   A term equivalent to the powerhouse, sometimes including the structures situated nearby.

Hydropower system   Two or more power plants (and therefore two or more powerhouses) which are cooperating electrically through a common network.

Ideal fluid   Frictionless and incompressible fluid. An ideal fluid has zero viscosity: i.e., it cannot sustain shear stress at any point.

Idle discharge   Old expression for spill or waste water flow.

Impeller   See Runner.

Impeller vanes   See Runner buckets.

Impervious core or zone   See core.

Inflow   (1) Incoming flow; (2) upstream flow.

Inlet   Upstream opening of a culvert, pipe or channel.

Inlet valve   The valve installed immediately ahead of the turbine, i.e., at the bottom of the penstock or the pressure shaft.

Intake   Any structure in a reservoir through which water can be drawn into a waterway or pipe. Intake tower: a pressure intake erected separately in the reservoir or dam for housing the flow control valves and gates.

Internal erosion   The formation of voids within soil, soft rock or fill caused by mechanical or chemical removal of material by seepage.

Intrinsic anisotropy   Anisotropy that the rock presents, being characterized by formation defects in the constituent minerals (microcracks, porosity, lamination) and by the presence of stratification and schistosity.

Invert   (1) Lowest portion of the internal cross-section of a conduit; (2) channel bed of a spillway; (3) bottom of a culvert barrel.

Inviscid flow   It is a non-viscous flow.

Jet nozzle   See Nozzle.

Kaplan turbine   An axial-flow reaction turbine with adjustable runner blades and adjustable guide vanes.

Kinetic energy   Energy which a moving body has because of its motion, dependent on the mass and the rate at which it is moving.

Laminar flow   Flow characterized by fluid particles moving along smooth paths in laminas or layers, with one gliding smoothly over the adjacent layer. Laminar flows are governed by Newton's Law of viscosity with relates the shear stress to the rate of angular deformations: $\tau : \mu \partial V / \partial y$.

Leakage   Free flow loss of water through a hole or crack (see water stop).

Lining   With reference to a canal, tunnel or shaft, a coating of asphaltic concrete, concrete, reinforced concrete, or shotcrete to provide watertightness, to prevent erosion, or to reduce friction.

Load   The amount of electric energy delivered at a given point.

Load demand   A sudden electrical load upon the generating units, inducing the rapid opening of the turbines.

Load factor   The ratio of the annually produced kilowatt-hours and of the energy theoretically producible at installed capacity during the whole year.

Load rejection   A sudden cessation of electrical load on the generating units, inducing the rapid closure of the turbines.

Lodging   See Bedding, Stratification.

Lowest point of foundation   The lowest point of dam foundation excluding cutoff trenches less than 10 meters wide and isolated pockets of excavation.

Magnitude   A rating of a given earthquake independent of the place of observation.

Main guide bearing   The bearing located nearest the runner.

Main shaft   The rotating element that transmits torque developed by the turbine runner to the generator.

Manifold (header)   The lowest portion of the penstock from which the unit penstocks bifurcate.

Masonry dam   Dam constructed mainly of stone, brick or concrete blocks jointed with mortar.

Meandering channel   Alluvial stream characterized by a series of alternating bends (i.e., meanders) as a result of alluvial processes.

Mechanical efficiency   An efficiency parameter of the turbine, expressing the power losses of the revolving parts, due to the mechanical friction.

MEL   Minimum energy loss.

Membrane or Diaphragm   A membrane or sheet or thin zone or facing, made of flexible impervious material such as asphaltic concrete, plastic concrete, steel, wood, copper, plastic, etc. A "cutoff wall" or a "core wall", if thin and flexible, is sometimes referred to as a "diaphragm wall" or "diaphragm".

Morning glory spillway   Circular or glory hole form of a drop inlet spillway. Usually free standing in the reservoir. Called because of its resemblance to the morning glory flower.

Mud   Slimy and sticky mixture of solid material and water.

Multiple arch dam   a buttress dam with its upstream part comprising of a series of arches.

Nappe flow   Flow regime on a stepped chute where the water bounces from one step to the next one as a sequence of free-fall jets (Chanson, 1999).

Needle valve   A streamlined regulating body moving like a piston in the enlarged housing of the valve.

Net head   The part of the gross head which is directly available for the turbines/generators to generate electricity.

Nozzle (jet nozzle)   A curved steel pipe supplied with a discharge-regulating device to direct the jet onto the buckets in impulse runners.

Obvert   Roof of the barrel of a culvert. Another name is soffit.

One-dimensional flow   Neglects the variation and changes in velocity and pressure transverse to the main flow direction. Example can be the flow through a pipe.

Outcrop   Any exposure of rocks or soils to the earth's surface. They can be natural (escarpments, flagstones) or artificial (in excavation surfaces).

Outflow   Downstream flow.

Outlet   An opening through which water can be freely discharged from a reservoir to the river for a particular purpose.

486 Glossary

Parapet wall   A solid wall built along the upstream or downstream edge of the top of dam for ornament or for the safety of vehicles and pedestrians.

Peak load   The greater amount of power given out or taken in by a machine or power distribution system in a given time.

Pelton turbine   The main type of turbine used under high heads. Impulse turbine with one to six circular nozzles that deliver high-speed water jets into air which then strike the rotor buckets. Lester Alan Pelton patented the actual double-bucket design in 1880.

Penstock   A pressurized pipeline conveying the water in high-head developments from the head pond or the surge tank to the powerhouse.

Penstock valve   See Control valve.

Pervious zone   A part of the cross section of an embankment dam comprising material of higher permeability.

Pier   The structural member used to support the upper surface of the horizontal portions of water passages such as the draft tube and the spiral case inlet; it's used to protect the pier nose with a steel lining.

Piping   The progressive development of internal erosion by seepage, appearing downstream as a hole discharging water.

Pit liner   The plate steel lining in the turbine pit. It serves as an internal form and as a protective liner for the surrounding concrete.

Pitting   Formation of small pits and holes on surfaces due to erosive or corrosive action (e.g., cavitation pitting).

Plant discharge capacity   The maximum discharge that can be utilized by the turbines with the full gate age. i.e., the entire discharging capacity of the turbines.

Plug valve   See Spherical valve.

Plunging jet   Liquid jet impacting (or impinging) into a receiving pool.

Poiseuille flow   Steady laminar flow in a circular tube of constant diameter.

Pondage   The rate of storage in run-of-river developments which can cover daily peaks.

Pore pressure   The interstitial pressure of fluid within a mass of soil, rock, or concrete.

Potential drop   Difference in total head between two equipotential lines.

Potential energy   The energy available function of a position (e.g., water held behind a dam). This energy can be converted to hydroelectric energy from falling water.

Potential flow   Ideal-fluid flow with irrotational motion.

Power   The rate at which energy is generated by an electric current or mechanical force, generally measured in Watts.

Powerhouse   The main structure of a water powerplant, housing the generating units and the pertaining installations.

Precipitation   Is the discharge of water, in liquid or solid state, from the atmosphere, generally onto a land or water surface.

Precipitation gage   A device used to collect and measure precipitation.

Precipitation intensity   The amount of precipitation in a unit of time interval.

Pressure pipe   See Penstock.

Pressure relief pipes   Pipes used to relief uplift or pore water pressure in the foundation/dam.

Glossary 487

Propeller-type turbine The collective term for axial-flow reaction turbines. In this terminology it denotes two types: fixed-blade propeller turbines and adjustable-blade propeller turbines, i.e., Kaplan turbines. (This is the original English terminology, whereas in continental practice, the fixed-blade type is also termed briefly as propeller-type turbine.)

Pumped-storage development A combined pumping and generating plant; hence it is not a primary producer of electrical power but, by means of a dual conversion, stores the additional power of the network and returns it in peak load periods as would a battery.

Radial gate See Gate.

Radial-inflow turbine A collective term for turbines in which the water enters radially into the runner and leaves it axially – Francis turbines.

Rating curve See Discharge Curve.

Reaction turbine A collective term for turbines in which the water enters the runner under a pressure exceeding the atmospheric value. The water flowing to the runner still has potential energy, in the form of pressure, which is converted into mechanical power along the runner blades.

Regulating dam A dam impounding a reservoir from which water is released to regulate the flow in a river.

Relief well See Drainage well.

Reservoir An artificial lake into which water flows and is stored for future use.

Reversible pump-turbine A hydraulic machine used in pumped-storage developments, suitable for operating both as a pump and as a turbine.

Reynolds number Dimensionless number proportional to the ratio of the inertial force over the viscous force.

Rheology Science describing the deformation of fluid and matter.

Riparian Pertaining to the banks of a river, stream, waterway, or water, typically, flowing body of water as well as to plant and animal communities along such bodies of water.

Riprap A layer of large uncoursed stones, broken rock or precast blocks placed in random fashion on the upstream slope of an embankment dam or on a reservoir shores or on the sides of a channel as a protection against wave and ice action. Very large riprap is sometimes referred to as "armoring".

River A natural stream of water of considerable volume, large than a creek. A river has its stages of development: youth, maturity, and old age. In its earliest stages a river system drains its basin imperfectly; as valleys are deepened, the drainage becomes more perfect, so that in maturity the total drainage area is large and the rate of erosion is high. The final stage is reached when wide flats have developed and the bordering lands have been brought low.

River basin A term used to designate the area drained by a river and its tributaries.

Rock bolting Process of sustaining rock volumes through the installation, from a free surface, of risers (metallic or not) that are fixed in depth and installed tensioned.

Rockfill Material composed of large rocks or stones loosely placed.

488 Glossary

Rockfill dam an embankment type dam in which more than 50% of the total volume comprises compacted or dumped pervious natural or crushed stone.

Rock weathering Generic name given to rock with its mineral constituents modified and transformed by the action of external agents. According to the degree and intensity of this modification, we have: healthy or little altered rock; moderately altered rock; very altered rock; extremely altered rock; decomposed rock.

Rolled fill dam See Embankment dam.

Roller In hydraulics, large-scale turbulent eddy: e.g., the roller of a hydraulic jump.

Roller compacted concrete (RCC) It is defined as a no-slump consistency concrete which is placed in horizontal lifts and compacted by vibratory rollers. RCC has been commonly used as construction material of gravity dams since the 1970s.

Roughness coefficient A factor in velocity and discharge formulas representing the effect of channel roughness on energy losses in flowing water.

Rubble dam A masonry dam in which the stones are unshaped.

Runaway speed The maximum rotational speed attained by a turbine with the generator load. It is attained in load rejection.

Run-of-river plant An HPP that operates based only on available streamflow in some short term (hourly, daily, or weekly). A development with little or no pondage regulation such that the power output varies with the fluctuations in the stream flow. See Storage reservoir.

Runner The rotating element which converts hydraulic energy into mechanical energy. For reversible pump-turbines, the element is called an impeller and converts mechanical energy into hydraulic energy for the pump mode.

Runner blades The contoured components of a propeller runner that radiate from the hub, deflect the flowing water and transfer energy to the runner hub. The blades may be angularly adjustable or rigidly fixed in the hub.

Runner buckets (impeller vanes) The contoured components of Francis and impulse runners that deflect the flowing water and transfer energy to the runner crown or disk when operating as a turbine.

Runner cone The extension of the runner crown, or runner hub, that guides the water as it leaves the runner.

Runner crown The upper axisymmetric portion (inner shroud) of the runner which provides a mechanical attachment to the main shaft and to which the top or inner ends of the runner buckets attach.

Runner crown seal The close running clearance between rotating runner crown and the stationary head cover. The close clearance restricts the flow of water into the chamber between the top of the runner and the bottom of the head cover.

Runner hub The axisymmetric portion of a propeller runner which provides the attachment to the main shaft and to which the inner ends of the runner blades attach.

Runoff (1) That part of precipitation, snow melt, or irrigation water that appears in uncontrolled surface streams, rivers, drains or sewers. It is the same of streamflow unaffected by artificial diversions, imports, storage, or other works of man in or on the stream channels. Runoff may be classified according to speed of appearance after rainfall or melting snow as direct runoff or base runoff, and according to source as surface runoff, storm interflow, or ground-water runoff; (2) The total discharge described in (1), above, during a specified period of time; (3) Also defined as the depth to which a drainage area would be covered if all of the runoff for a given period of time were uniformly distributed over it.

For more details, see Water Words Dictionary, Nevada Division of Water Planning (2000).

Runoff cycle That portion of the hydrologic cycle between incident precipitation over land areas and the subsequent discharge through stream channels or Evapotranspiration.

Saltation (1) Action of leaving or jumping; (2) in sediment transport, particle motion by jumping and bouncing along the bed.

Sand Composed predominantly of coarse-grained mineral sediments with diameters larger than 0.074 mm and smaller than 2 mm in diameter.

Sand trap A device, often a simple enlargement in a ditch or conduit, for arresting the heavier particles of sand and silt carried by the water.

Sanding up See Accretion, Sediment; Silting act or effect; obstruct; fill up with sand or earth.

Scale effect Discrepancy between model and prototype resulting when one or more dimensionless parameters have different values in the model and prototype.

S-curve The mass curve corresponding to a Unit Hydrograph or a distribution graph.

Scour The erosive action of running water in stream, which excavates and carries away material from the bed and banks.

Scour pools A pool formed by the action of the flow over the river bed.

Scroll case (spiral case) A spiral-shaped steel inlet guiding the flow into the wicket gates of the reaction turbine.

Sediment Any material carried in suspension by the flow or as bed-load which would settle to the bottom in absence of fluid motion.

Sediment load Material transported by a fluid motion.

Sediment transport Transport of material by a fluid motion.

Sediment transport capacity Ability of a stream to carry a given volume of sediment material per unit time for a given flow conditions. It is the sediment transport potential of the river.

Sediment yield Total sediment outflow rate from a catchment, including bed-load and suspension.

Sedimentation Strictly, the act or process of depositing sediment from suspension in water. Broadly, all the processes whereby particles of rock material are accumulated to form sedimentary deposits.

Seepage The interstitial movement of water that may take place through a dam, its foundation, or abutments.

Semi-scroll case (spiral case) A concrete inlet directing flow to the upstream portion of the turbine with a spiral case surrounding the downstream portion of the turbine to provide uniform water distribution.

Separation In a boundary layer, a deceleration of fluid particles leading to a reversed flow within the boundary layer is called separation. The decelerated fluid

490 Glossary

particles are forced outwards and the boundary layer is separated from the wall. At the point of separation, the velocity gradient normal to the wall is zero.

Separation point In a boundary layer, intersection of the solid boundary with the streamline dividing the separation zone and the deflected outer flow. The separation point is a stagnation point.

Sequent depth In open channel flow, the solution of the momentum equation at a transition between supercritical and subcritical flow gives two flow depths (upstream and downstream flow depths). They are called sequent depths.

Setting The vertical distance between the tailwater level and the center of the turbine.

Sewage Refused liquid or waste matter carried off by sewers. It may be a combination of water-carried wastes from residences and industries together with ground water, surface waters and storm water.

Sewer An artificial subterranean conduit to carry off water and waste matter.

Sharp-crested weir A device for measuring water, featuring a notch cut in a relatively thin plate and having a sharp edge on the upstream of the crest.

Sheet piling Stake of concrete or steel pieces placed vertically in the ground to contain erosion or the lateral movement of groundwater.

Shoulder The upstream and downstream parts of the cross section of an embankment dam on each side of the core or core wall. Hence the expression upstream shoulder or downstream shoulder.

Side-channel spillway A side-channel spillway consists of an open spillway (along the side of a channel) discharging into a channel running along the foot of the spillway and carrying the flow away in a direction parallel to the spillway crest (e.g., the Hoover dam in the Colorado river (USA) has two side-channel spillways).

Sill (1) a submerged structure across a river to control the water level upstream; (2) the crest of a spillway; (3) the horizontal gate seating, made of wood, stone, concrete, or metal at the invert of any opening or gap in structure. Hence, the expressions: gate sill, stoplog sill.

Siltation See Sanding up; Sediment; Accretion;

Similitude Correspondence between model and prototype behavior, with or without geometric similarity. The correspondence is usually limited by scale effects.

Siphon Pipe system discharging waters between two reservoirs or above a dam in which the water pressure becomes sub-atmospheric. The shape of a simple siphon is close to an omega (i.e., Ω-shape). Inverted-siphons carry waters between two reservoirs with pressure larger than atmospheric. Their design follows approximately a U-shape. Inverted-siphons were commonly used by the Romans along their aqueducts to cross valleys.

Siphon-spillway Device for discharging excess water in a pipe over the dam.

Skimming flow Flow regime above a stepped chute for which the water flows as a coherent stream in a direction parallel to the pseudo-bottom formed by the edges of the steps.

Slickensides Striated and polished surface produced on rock by movement along a fault.

Slope (1) A side of a hill or a mountain; (2) The inclined face of a cut, canal, or embankment; (3) Inclination from the horizontal. Measured as to the ratio of the number of units of the vertical distance to the number of corresponding units of the horizontal distance.

Slope protection   The protection of embankment slope against wave action or erosion.

Sluice   An artificial channel for conducting water, with a gate or a valve to regulate the flow. A sluiceway.

Sluice gate   Underflow gate with a vertical sharp edge for stopping or regulating flow.

Slump   The sliding or gravitational movement of an overlying layer of soil, typically from becoming saturated, and lying on a rock layer or other relatively impermeable layer.

Slurry trench   A narrow excavation whose sides are supported by a mud slurry filling the excavation. Sometimes used incorrectly to describe the cutoff itself.

Soffit or obvert   Roof of the barrel of a culvert.

Solid head buttress dam   a buttress dam in which the upstream end of each buttress is enlarged to span the gap between buttresses.

Specific discharge   The rate of discharge per unit of length.

Specific energy   Quantity proportional to the energy per unit of mass, measured with the channel bottom as the elevation datum, and expressed in meters of water.

Specific speed   A universal number that indicates the machine design.

Speed increaser   The geared drive unit which increases turbine shaft speed to drive the generator at the economic speed for power generation.

Spherical valve (plug valve)   A cylindrical closing body, encased in a spherical housing, and rotating around a shaft perpendicular to the penstock axis.

Spilling surge tank   Different types of surge tanks whose riser shaft or upper chamber (if any) has an overflow berm discharging the excess water into a wasteway in order to limit upsurges.

Spillway   The channel or passageway around or over a dam through which excess water is diverted; A structure built into a dam or the side of a reservoir over or through which flood flows are discharged; there are several types of spillways (see Chapter 8); for example, Auxiliary spillway: a secondary spillway designed to operate only during exceptionally large flood; For more details see also Nevada Division of Water Planning, Water Words Dictionary.

Spiral case   See Scroll case; Semi-scroll case.

Splitter   Obstacle (e.g., concrete block, fin) installed on a chute to split the flow and to increase the energy dissipation.

Spray   (1) A cloud of mist of fine liquid particles, as of water from breaking waves; (2) A jet of fine liquid particles, or mist, as from an atomizer or spray gun.

Spring runoff   Snow melting in the spring causes water bodies to rise. This in streams and rivers is called "spring runoff".

Staff gage   A graduated scale used to indicate the height of the water surface in a stream channel, reservoir, lake or other water body.

Stage   The height of the water surface above some established point or Datum (not the bottom) at a given location. Also referred as Gage Height.

Stage capacity curve   A graph showing the relation of the surface elevation of the water in a reservoir, usually plotter as at ordinate, to the volume below that elevation, plotted as the abscissa.

492 Glossary

Stage-discharge curve (rating curve) A graph showing the relation between the gage height, usually plotted at ordinate, and the amount of water flowing in a channel, expressed as volume per unit of time and plotted as the abscissa.

Stagnation point It is defined as the point where the velocity is zero. See Separation point.

Staircase Another adjective for 'stepped': e.g., a staircase cascade; stepped cascade.

Stall Aerodynamic phenomenon causing a disruption (i.e., separation) of the flow past a wing associated with a loss of lift (sustention).

Stay ring The structural member surrounding the wicket gates having two annular rings connected by a number of fixed stay vanes in the water passages. Its function is to provide support and structural continuity between the upper and lower portions of the turbine distributor, while guiding the water as it enters or leaves the spiral case.

Stay vane One of the streamlined steel or cast-iron supports built at the cylindrical discharging surface of the scroll case. The stay vane serves mainly structural purposes, though it may have a hydraulic function as well.

Stay vane ring The stay vanes and the lower and upper speed ring holding them.

Steady flow Occurs when conditions at any point of the fluid do not change with the time.

Still water A flat or level section of a stream where no flow or motion of the current is discernible and the water is still.

Stilling basin Structure for dissipating the energy of the flow downstream of a spillway, outlet work, chute or canal structure. In many cases, a hydraulic jump is used as the energy dissipator within the stilling basin.

Stoplogs Large logs or timber or steel beams placed on top of each other with their ends held in guides on each side of a channel or conduit so as to provide a cheaper or more-easily handled means of temporary closure than a bulkhead.

Storage reservoir A reservoir that has volume for retaining water from springtime snowmelt or other hydrologic events. Retained water is released as necessary for multiple uses: power production, fish passage, irrigation, navigation, municipal or industrial supply. For more details, see Water Words Dictionary, Nevada Division of Water Planning (2000).

Storm water Excess water running off the surface of a drainage area during and immediately following a period of rain.

Storm waterway Channel built for carrying storm waters.

Streamline It is the line drawn so that the velocity vector is always tangential to it (i.e., no flow across a streamline). When the streamlines converge the velocity increases.

Subcritical flow In open channel the flow is defined as subcritical if the flow depth is larger than the critical depth. In practice, subcritical flows are controlled by the downstream flow conditions.

Supercritical flow In open channel, when the flow depth is less than the critical flow depth, the flow is supercritical and the Froude number is larger than one. Supercritical flows are controlled from upstream.

Surface tension Property of a liquid surface displayed by its acting as if it were a stretched elastic membrane. SI Units: N/m.

Surge A surge in an open channel is a sudden change of flow depth (i.e., abrupt increase or decrease in depth). An abrupt increase in flow depth is called a positive surge while a sudden decrease in depth is termed a negative surge.

Surge tank; surge chamber A structure erected in the power conduit of high-head HPP between the pressures tunnel and the penstock (or pressure shaft) to protect the tunnel from water-hammer effects, to diminish overpressures due to water hammer in the penstock itself, and to store water for sudden load demand. A surge tank can also be located between the draft-tube por and the tailwater tunnel.

Surge wave Results from a sudden change in flow that increases (or decreases) the depth.

Suspended load Transported sediment material maintained into suspension.

Tailrace The channel that runs from the powerhouse to the riverbed downstream.

Tailwater depth Downstream of the powerhouse flow depth.

Tailwater level Downstream of the powerhouse free-surface elevation.

Tainter gate It is a radial gate.

Thrust block (anchor) A massive block of concrete built to withstand a thrust or pull.

Tidal power plant A power station that utilizes the potential hydraulic power originating from the tidal cycles of the sea.

Toe of dam The junction of the downstream face of a dam with the ground surface. Also referred to as downstream toe.

Top of a dam The elevation of the uppermost surface of a dam, usually a road, or walkway excluding any parapet wall, railing, etc.

Training wall Sidewall of chute spillway.

Transition zone or semipervious zone A substantial part of the cross section of an embankment dam comprising material whose grading is of intermediate size between that of an impervious zone and that of a permeable zone.

Trash rack A screen, metal or reinforced concrete bars, located in the waterway at the intake so as to prevent the ingress of floating or submerged debris.

Turbine A device which produces power by diverting water through blades of a rotating wheel which turns a shaft to drive generators.

Turbine discharge capacity The maximum flow that can be discharge by a single turbine at full gate age.

Turbine efficiency The entire efficiency of the turbine, i. e., the product of hydraulic mechanical and volumetric efficiencies.

Turbine valve See Inlet valve.

TWRC - TWRL Tailwater rating curve – Tailwater rating level.

Underground power st. A development where the machine hall is located in an excavated cavern.

Unsteady flow The flow properties change with the time.

Uplift (1) The upward interstitial pressure or on the base of a structure. When the pressure acts uniformly around outside, e.g., the water pressure on the outside of a tunnel lining, the term "external pressure" is used. The external pressure causing complete structural failure is termed "collapsing pressure"; (2) An upward force on a structure caused by frost heave or by wind force.

Upstream Toward the source or upper part of a stream; against the current.

Upstream blanket An impervious blanket placed on the reservoir floor upstream of a dam. In the case of an embankment dam, the blanket may be connected to the impermeable element.

Upstream slope (dam) The part of the dam that is contact with the reservoir water.

Upstream toe (dam) The junction of the upstream face of a dam with the ground surface.

U-Shaped valleys U shaped valleys are characteristic of glacial erosion.

V-Shaped valleys V-shaped valleys are characteristic of stream action erosion.

Validation Comparison between model results and prototype data to validate the model. The validation process must be conducted with prototype data that are different from that used to calibrate and to verify the model.

Velocity head Energy fluid function of velocity, usually expressed in meter of fluid.

Velocity of a stream Rate of motion of a stream measured in terms of distance the water travels in a unit of time. Usually expressed in m/s.

Vena contracta Minimum cross-sectional area of the flow (e.g., jet or nappe) discharging through an orifice, sluice gate or weir.

Venturi meter A short tube with a constricted throat followed by a smooth expansion used to determine fluid pressures and velocities by measurement of differential pressures generated at the throat as a fluid traverse the tube.

Viscosity Property which characterizes the fluid resistance to shear, i.e., resistance to a change in shape or movement of the surroundings. SI Units: $N.s/m^2$.

Volt (V) The unit of electromotive force or potential difference that will cause a current of 1 ampere to flow through a conductor with a resistance of 1 ohm.

Volumetric efficiency An efficiency component of the turbine, expressing exclusively the power losses due to leakage (clearance losses, etc.).

Water-bearing stratum Geological formation capable of storing and transmitting water in appreciable quantities. See Artesian aquifer.

Waterfall Abrupt drop of water characterized by a free-falling nappe of water.

Water power A general term used for characterizing both power (kW) and energy (kWh) or watercourses, lakes, reservoirs, and seas.

Watershed See Basin; Hydrographic region.

Water stop (water bar) A strip of metal, rubber or other material to prevent leakage, through joints between adjacent sections of concrete.

Watt (W) W is the SI power unit. It is equivalent to 1J/s: kg/m^2s^{-3}.

Wave wall A solid wall built along the upstream side at the top of a dam and designed to reflect waves.

Wearing rings Replaceable rotating rings fastened to the runner or adjacent stationary rings fastened to the head cover and the bottom ring (or discharge ring), thus forming removable seals with small clearances.

Weighting a slope Additional material placed on the slope of an embankment.

Weir A low river dam used to raise the upstream water level. Termed "fixed-crest weir" when uncontrolled. A structure built across a stream or channel for the purpose of measuring flow. Types of weirs include "broad-crested weir", "sharp-crested weir", "drowned weir" or "submerged weir".

WES standard spillway Spillway shape developed by the US Army Corps of Engineers at the Waterways Experiment Station.

Wetted perimeter The wetted perimeter is the length of wetted contact between the flowing stream and the solid boundaries. For example, in a circular pipe flowing full, the wetted perimeter equals the circle perimeter.

White waters Non-technical term used to design the free-surface aerated flows.

Wicket gates The angularly adjustable streamlined elements which control the flow of water to the turbine from the pump.

Wing wall Sidewall of an inlet or outlet.

Zoned earthfill An earthfill type dam the thickness of which is composed of zones of selected materials having different degrees of porosity, permeability, and density.

Appendix

Chapter 3 – Additional examples of layouts

A.1 BALBINA HPP

Balbina HPP (1989), with an installed capacity of 257.50 MW, is located in the Uatumã river, in the Amazonas (Figure A.1). The general layout is shown in Figure A.2.

The reservoir in the maximum operating level, at El. 50.00 m, floods an area of 2,360 km². Its filling began in October 1987. The first machine went in operation in February 1989.

The axis of the dam, which cuts the center of the island, is in a place without natural fall, in plain, with an extension of approximately 3.2 km. With these characteristics, the

Figure A.1 Balbina HPP. Localization map.

Figure A.2 Balbina HPP Site (1977). Observe the Middle Island, where the spillway was implanted.

general layout selected contemplates the concrete structures, with 334 m of length, as well as earth dams, with 2,930 m of length, to close the river. For example, if there is a lot of gravel on the site, the section can be mixed.

The layout includes the following structures:

- dams on both banks of the river and rockfill dam on the river bed, 2.93 km long and 31 m high; the leakage points of the reservoir were closed with earth dams;
- right and left transition walls;
- spillway, as gravity dam, 74 m long, with 4 gates of 13.5 m wide by 13.5 m high each and a maximum flow capacity of 5,840 m^3/s; stilling basin energy dissipator in hydraulic jump;
- intake/powerhouse at the foot of the dam with a total length of 124 m, with five Kaplan machines with a net drop of 23.2 m and a unit power of 51.50 MW.

Observe, in Figure A.3, the characteristic back currents between the spillway and the dam, impacting the downstream embankment of the dam and entering the dissipation basin. In addition, note that the right-side wall of the basin, which separates the flows from the spillway and the tailwater channel, is larger than the left wall of the basin. The phases of river diversion are shown in Figures A.4 and A.5.

Figure A.3 Balbina HPP. Eletronorte Magazine (1998). The machine 5 went in operation in July/1990.

A.2 TUCURUÍ HPP

Tucurui HPP, 8,400 MW, inaugurated in 1984, is located in the lower Tocantins/Araguaia river basin (see Figure A.6). The axis was located in a place 7.0 km long, without natural fall. As in the previous case, the earth dam was used to create the head. The hydraulic structures are of conventional concrete. It is worth mentioning that the river bank is at +10.00 m and the riverbed is variable. The region upstream of the spillway is at 0.00 m. The river channel depth is 10 m.

The general layout contemplates structures with extraordinary dimensions:

- an earth dam in the right bank, 3.6 km long and 95 m high (maximum);
- a gravity dam ski jump spillway, 92 m high, 602 m long, with 23 gates 20 m wide × 21 m high each, and a maximum flow capacity of 110,000 m^3/s;
- a gravity intake/powerhouse structure, 1,000 m long, with 22 Francis machines;
- a left bank dam 520 m long;

Figure A.4 First phase of deviation (1981). Eletronorte (1985).

- two navigation locks 210 m long × 33 m wide; the level difference is 37.8 m in the upstream chamber and 35 m in the downstream chamber; the intermediate channel is 5.5 km long and has a minimum width of 140 m at the base; its width allows maneuvers of the trains and the independent operation of the locks;
- one left bank dyke 780 m long (Figure A.7).

Figure 8.2 shows a typical section of the powerhouse structure. The operational levels of the reservoir are: WLmax, El. 75.30m; normal WL, El. 72.00m; WLmin, El. 58.00m; Minimum operating level, El. 51.60m. For the extreme levels, a maximum depletion of 23.70m is observed. It is recorded that in the WLmax the flooded area is 2,875 km^2 (Figure A.8).

The main volumes of the civil works are: soil excavation 27.5×10^6 m^3; rock excavation 23×10^6 m^3; earthfill 55.5×10^6 m^3; rockfill 20.5×10^6 m^3; filters and transitions 5×10^6 m^3; and concrete 6.3×10^6 m^3 (Technical Memory, 1989) (Figure A.9).

A.3 ITAIPU HPP

Itaipu (1984) is located on the Paraná river, upstream from the confluence with the Iguaçu river, between Brazil and Paraguay (Figure A.10A).

Appendix 501

Figure A.5 Second phase of deviation (1985). Eletronorte (1987).

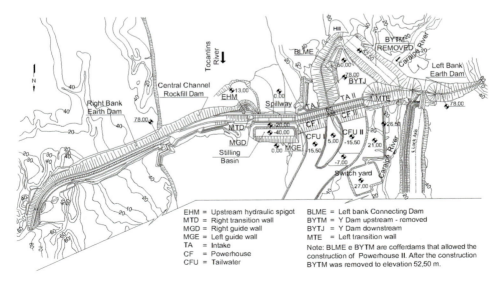

Figure A.6 Tucuruí HPP. Layout.

Figure A.7 Tucuruí. Spillway and right dam (July, 1984). Start of the Reservoir Filling. In the foreground, there is upstream hydraulic spigot. This structure was designed and tested in the reduced three-dimensional model in order to improve the accommodation of the approach flow to the spans at the right side of the spillway.

Figure A.8 Downstream view of the spillway.

The plant has 18 Francis turbines of 710 MW each, plus two Francis turbines of 700 MW each. The total installed capacity is 14,180 MW.

Appendix 503

Figure A.9 Tucuruí HPP. Eletronorte. Central dam, spillway, powerhouses 1 and 2, left dam/navigation lock.

Figure A.10A Itaipu HPP. Localization map.

Figure A.10B Itaipu HPP. CBDB (2006).

The channel of the Paraná river in the site is 90 m high (riverbed in the El. 35.00 m and banks in the El. 125.00 m). The layout has axis, 7.76 km long and the reservoir floods an area of 1,350 km^2 in WLmax normal, at El. 220.00 m. The minimum minimorum WL is 197.00 m.

The layout includes the following structures:

- a right bank earth dam, 872 m long and 90 m maximum height;
- a chute spillway in the right abutment 380 m wide, 483 m long and 44 m high; 14 spans 20 m wide by 21.34 m high; flow capacity of 62,200 m^3/s;
- right side concrete dam between the powerhouse and the spillway, 986 m long, 64.5 m high;
- main concrete dam (on river bed hollow gravity and left bank of buttress) with 1,064 m of total length and maximum height of 196 m; intake and powerhouse for 18 units;
- deviation structure with 170 m length and max. height 162 m; intake and powerhouse, units 16 to 18A;
- rockfill dam, left bank, 1,984 m long and maximum height 70 m;
- earthfill dam, left bank, 2,294 m long and 30 m high (Figure A.10B).

The main volumes of the civil works are: soil excavation 24×10^6 m^3; rock excavation 32×10^6 m^3; earthfill 6.3×10^6 m^3; rockfill 15×10^6 m^3; concrete 12.3×10^6 m^3 (Figure A.11).

A.4 SOBRADINHO HPP

Sobradinho HPP (1979), 1,050 MW of installed capacity, is located in the São Francisco river, in Bahia. The layout includes the following structures:

- dams and dykes with a length of 8,532 m and a maximum height of 41 m;
- surface spillway, 4 gates and a 12 gates bottom outlet; total flow capacity of 22,080 m^3/s;

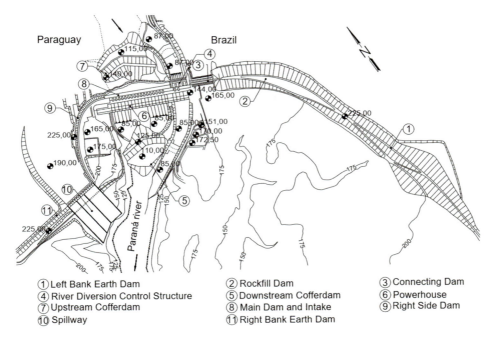

Figure A.11 Itaipu HPP. Layout. Moraes et al. (1979).

- intake/powerhouse with 6 blocks of 33 m wide, housing 6 Kaplan turbines of 175 MW of unit power, nominal drop of 27.2 m, flow 710 m^3/s; diameter of 9.5 m.
- water intake for irrigation on the left bank, at Dique B, for 25 m^3/s;
- lock on the right bank to overcome a 32.5 m the level difference; camera: 120 m × 17 m (Figure A.12).

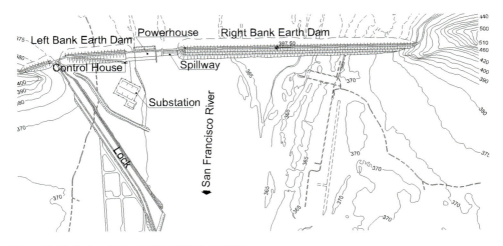

Figure A.12 Sobradinho HPP. CBDB (1982).

The reservoir has an area of 4,214 km^2; in the WLmax normal, El. 392.50 m, regulates a minimum flow of 2,060 m^3/s with a depletion of 12 m, when it reaches its minimum WL at El. 380.50 m. The river bank is at El. 365.00 m and the riverbed at El. 354.00 m. The depth of the river is 11 m. The dam has a zoned section, with clay core and gravel backrests, and function of the materials available in the borrow areas. The slopes are 1V:3H upstream and 1V:2.5H downstream. The main volumes of the civil works are: soil excavation 6.2 × 10^6 m^3; rock excavation 3.3 × 10^6 m^3; earthfill/rockfill 19 × 10^6 m^3; and concrete 1.72 × 10^6 m^3 (Figure A.13).

Figure A.13 (a–b) Sobradinho HPP. CBDB (2006). Bottom outlet, surface spillway, powerhouse and the right bank dam.

A.5 BARRA GRANDE HPP

Barra Grande HPP (2005) is located on the Pelotas river, upstream from the confluence with the Canoas river, between the states of Rio Grande do Sul and Santa Catarina (Figure A.14).

The reservoir floods an area of 92 km^2 in the WLmax, El. 647.00 m. The minimum WL is at El. 617.00 m. The layout includes the following structures:

- a concrete face rockfill dam with, El. 651.00 m, 665 m long and 185 m maximum high;
- a chute spillway on the left abutment, 119.4 m wide and 20 m high, with 6 gates 15 m and a discharge capacity of 23,840 m^3/s; the concrete chute is 274 m long with two aerator steps, to prevent cavitation; part of the energy dissipation is in ski jump;
- the intake also in the left abutment is 52.3 m high and 24.3 m wide; the structure has six wagon type gates of 6.54 m high by 6.20 m wide;
- 3 pressure tunnels, with diameter of 6.9 m and length of 39.2 m, with reinforced stretch of 117.1 m;
- the indoor powerhouse, 45 m high and 91.8 m long, houses 3 Francis turbines 236 MW each one; turbine flow is 165 m^3/s and the liquid head is 154 m; the generators have a capacity of 245 MVA and a power factor of 0.95.

It is a concrete face rockfill dam, which was more attractive than the alternative with clayey or asphalt concrete core. The total length of the structures is 900 m. It should be noted that the area of the reservoir is minimal when compared to the flooded areas of the dams located in plains (e.g., Balbina HPP – Uatumã river; Tucuruí HPP – Tocantins river, and Sobradinho HPP – São Francisco river).

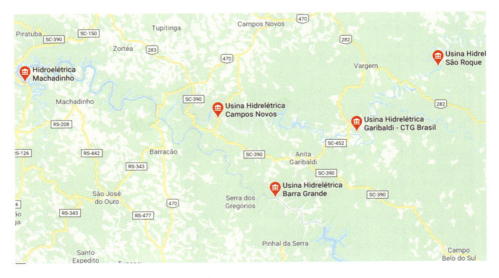

Figure A.14 Barra Grande HPP. Localization map.

508   Appendix

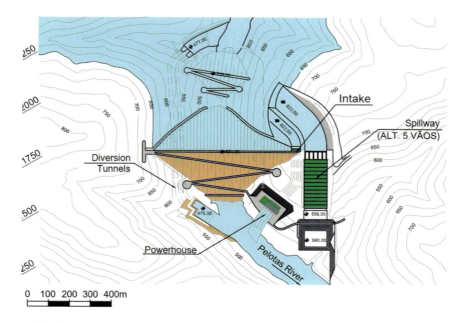

Figure A.15 Barra Grande HPP (RS/SC). Layout. MBD (2009).

The main volumes of the civil works are: soil excavation $2 \times 10^6 \, m^3$; rock excavation $10 \times 10^6 \, m^3$; tunnel excavations $5 \times 10^6 \, m^3$; rockfill/transitions $12 \times 10^6 \, m^3$; concrete $3.3 \times 10^6 \, m^3$ (Figure A.15).

A.6   CAMPOS NOVOS HPP

The layout of Campos Novos HPP (2006), on the Canoas river (Santa Catarina) 21 km upstream from its confluence with the Pelotas river, has similar characteristics to the layout of Barra Grande HPP (see Figure A.16a). The reservoir floods an area of 32.9 km² in the normal WL, El. 660.00 m. The WLmax of flood is at El. 666.00 m, and the minimum normal level is at El. 655.00 m. The general layout includes the following structures:

- concrete face rockfill dam with, with crest at El. 660.00 m, 590 m long and 202 m high;
- the skijump spillway at the right abutment, 83.10 m wide, has four gates 17.4 m wide and 20 m high; the concrete chute is 94 m long; the maximum flow capacity is 18,300 m³/s;
- intake 51.5 m high and 32 m wide, with three wagon type gates of 6.50 m high and 6.20 m wide;
- three pressure tunnels with diameter of 6.2 m, with stretches vertical of 173 m and a horizontal of 211 m;
- indoor powerhouse, 47.8 m high and 113 m long, including the assembly area, with 3 Francis turbines, vertical axis, 300 MW each, discharge of 175.6 m³/s, and 186 m net head; the operating speed is 200 rpm; the generators have a capacity of 311 MVA, an operating speed of 200 rpm and a power factor of 0.96 (Figure A.17).

Figure A.16 (a) Barra Grande HPP. (MBD, 2009). (b) Barra Grande. Spillway working (MBD, 2009).

Figure A.17 Campos Novos HPP. MBD (2009).

The diversion of the river was made by two tunnels, as exposed in CBDB (2009). As in Barra Grande, it is observed that the area of the reservoir is minimal when compared to the flooded areas of the dams located in plains (e.g., Balbina, Uatumã river; Tucuruí, Tocantins river; and Sobradinho, on the São Francisco river). The main volumes of the civil works: soil excavation $4 \times 10^6 m^3$; rock excavation $12 \times 10^6 m^3$; tunnel excavation $6 \times 10^6 m^3$; rockfill/transitions $13 \times 10^6 m^3$; concrete $3.5 \times 10^6 m^3$ (Figure A.18).

A.7 ITÁ HPP

Ita HPP, 1,450 MW, is located on the Uruguay river between the states of Santa Catarina and Rio Grande do Sul (Figure A.19).

The plant's layout is showed in the following figures. The reservoir floods an area of 141 km² at the normal WL, El. 370.00 m. The WLmax is at El. 375.70 m, and the minimum WL is at El. 364.00 m.

The layout includes the following structures:

- concrete face rockfill dam with, crest at El. 375.50 m, 880 m long and 125 m high;
- the plant has two spillways, one in each bank, with the maximum flow capacity of 49,940 m³/s: (1) the main surface spillway, 39 m long, 130.50 m wide, with a chute 193 m long; the auxiliary spillway, 54 m long, 85.50 m wide, with a chute 120 m long in concrete, plus 155 m unlined; the energy dissipation is in ski jump;
- the hydraulic power circuit in the left bank, intake, tunnels and powerhouse, is parallel to the auxiliary spillway;

Appendix 511

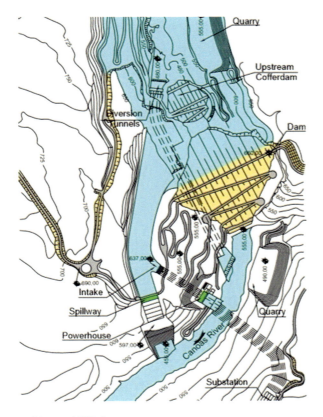

Figure A.18 Campos Novos HPP. Layout.

Figure A.19 Itá HPP. Localization map.

- the intake is 35.5 m high and 88.8 m wide, with 5 wagon-type gates of 9.30 m high by 7.20 m wide; penstocks in number of five, 18.35 m in length, with internal diameter of 6.8 m;
- the indoor powerhouse, 56 m high and 120.5 m long, including the assembly area, with five Francis turbines, vertical axis, 290 MW each; turbine flow of 313 m^3/s and head of 102 m; the generators have a capacity of 305 MVA

The main volumes of the civil works are: soil excavation 10×10^6 m^3; rock excavation 8×10^6 m^3; tunnel excavations 6.5×10^6 m^3; rockfill/transitions 11×10^6 m^3; concrete 5×10^6 m^3 (Figure A.20).

A.8 CORUMBÁ HPP

Corumbá HPP – 375 MW (1997), on the Corumbá river (Goiás), has a layout similar to that of Barra Grande and Campos Novos.

The dam, 540 m long and 90 m high, is made of sandstone with a waterproof clay core. In normal operation WL, El. 595.50 m, the reservoir is 65 km^2. The minimum operating WL is at El. 570.00 m, which characterizes depletion of 25.5 m. Total stored volume is 1.5×10^9 m^3.

The intake is 42 m wide. The penstocks are apparent and have an internal diameter of 6.8 m and lengths varying between 131 and 144 m (source: MDB, 2000).

The powerhouse, with dimensions of 22×100 m, houses 3 Francis turbines, 125 MW each one, rotation of 150 rpm, flow of 196 m^3/s; rotor has 4.6 m of diameter.

The spillway, with four sluices of 13.0 m by 17 m, has a flow capacity of 6,800 m^3/s (Figure A.21).

Figure A.20 (a–c) Itá HPP. Layout. MBD (2009).

(Continued)

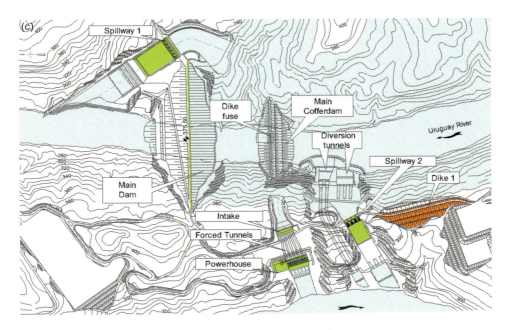

Figure A.20 (Continued) (a–c) Itá HPP. Layout. MBD (2009).

Figure A.21 Corumbá HPP. MBD (2000).

The main volumes of the civil works are: soil excavation $5 \times 10^6 \, m^3$; earth and rockfill/transitions $4.5 \times 10^6 \, m^3$; and concrete $3 \times 10^6 \, m^3$.

A.9 FUNIL HPP

Funil HPP – 223 MW in the Paraíba do Sul river (Rio de Janeiro), near the city of Resende, has been operating since 1969. The axis site is 400 m wide. The alternative chosen was as an arc dam, double curvature, 385 m long and 85 m high (Figure A.22).

The reservoir floods an area of 40 km² in the WLmax, at El. 466.50 m. In the layout there are two spillways in tunnel: one on the right abutment to $1,700 \, m^3/s$, taking advantage of the bypass tunnel, and another on the left abutment to $2,700 \, m^3/s$, as shown in the following photo (Figure A.23).

A.10 DONA FRANCISCA HPP

Dona Francisca HPP (2001), 125 MW, is located on the Jacuí River, in the Rio Grande do Sul. The reservoir at the WLmax, at El. 94.50 m, floods an estimated area of 22.30 km². The minimum WL is at El. 91.00 m.

The initial project envisaged a layout with a rockfill dam with a clay core and a spillway with gates. For economic reasons, this layout was changed to another with dam in CCR incorporating a free spillway.

The layout includes the following structures:

- CCR dam 610 m long and 50.5 m high;
- gravity dam on the right bank, between the spillway and the powerhouse, containing the diversion culverts;
- free spillway 335 m long and a flow capacity of $1,700 \, m^3/s$;

Appendix 515

Figure A.22 Funil HPP. Light S. A.

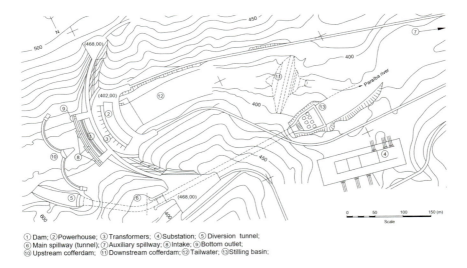

①Dam; ②Powerhouse; ③Transformers; ④Substation; ⑤Diversion tunnel;
⑥Main spillway (tunnel); ⑦Auxiliary spillway; ⑧Intake; ⑨Bottom outlet;
⑩Upstream cofferdam; ⑪Downstream cofferdam; ⑫Tailwater; ⑬Stilling basin;

Figure A.23 Funil HPP. Layout (CBDB, 1978).

- intake with two gates of 6.30 m × 6.54 m;
- penstocks, two of 6.30 m in diameter and 85 m in length;
- indoor powerhouse on the right bank, with two Francis turbines of 62.5 MW each (Figure A.24).

The main volumes of the civil works are: soil excavation $2 \times 10^6 m^3$; earth and rockfill/transitions $7.5 \times 10^6 m^3$; concrete $6.5 \times 10^6 m^3$.

Figure A.24 Dona Francisca HPP (MBD, 2009).

A.11 MONJOLINHO HPP

Monjolinho HPP, 74 MW, Figures A.25 and A.26, located in the Passo Fundo river, in Rio Grande do Sul, floods an area of 5.46 km^2 with the reservoir in the normal WL. The layout of the project contemplates:

- a concrete face rockfill dam 420 m long and 80 m maximum height;
- an intake on the left bank, two force tunnels diameter 3.85 m and 110 m long;

Figure A.25 Monjolinho HPP.

Figure A.26 Monjolinho spillway.

- an indoor powerhouse, with two Francis turbines of 37 MW each;
- a side free spillway on the right bank, with a length of 210 m and a capacity of 6,755 m³/s.

A.12 SERRA DA MESA HPP

Serra da Mesa HPP (1997), on the Tocantins river, Goiás, is owned by Furnas. The layout is shown in following figures. The reservoir WL is at El. 460.00 m and the WLmin at El. 417.30 m. The WLmax was set at El. 461.50 m. The total area is 1,784 km², and the total volume is 54.4×10^9 m³. The upstream WL is at El. 334.50 and the WLmin at El. 328.00 m. The gross head is 125.5 m (Figure A.27).

According to CBDB (2000), the Serra da Mesa reservoir, due to its capacity and location, was planned to operate as a strategic reserve for FURNAS, integrated with SIN. The stored volume would be used during critical periods. During normal periods, the reservoir was designed to operate at a normal WL at El. of 460.00. This reservoir, if emptied, would take 3 years to be refilled, according to the simulations carried out by Furnas (CBDB, 2000) (Figure A.28).

The dam is rockfill with clay core, 154 m high. The total length at the crest is 1,510 m and the width of about 11 m. The upstream slope is 1.6 H: 1.0 V and the downstream slope is 1.4 H: 1.0 V (Figure A.29).

The surface spillway at the right abutment, with five gates of 15 × 19 m, has a length of 96 m and a maximum capacity of 14,750 m³/s. The chute has a length of only 20 m unloading the flow in granite talvegue of the abutment, an excellent solution that provides significant savings of concrete.

Figure A.27 Serra da Mesa HPP (2008). Main dam. Underground powerhouse on the left bank. Spillway at the right abutment: outflow is restored over the rocky massif, providing significant concrete savings.

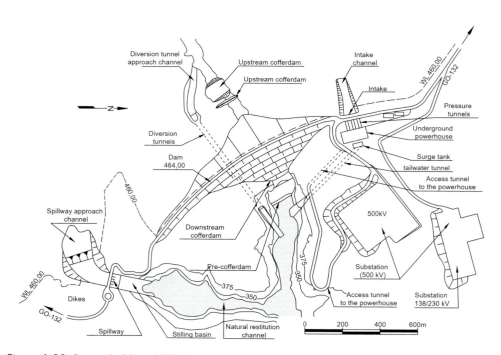

Figure A.28 Serra da Mesa HPP. Layout.

Appendix 519

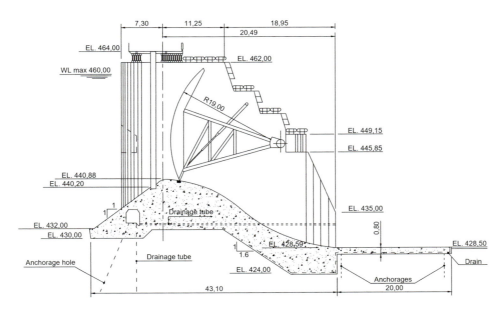

Figure A.29 Serra da Mesa spillway (MBD, 2000).

The adduction system consists of three tunnels of 9.0 m in diameter. The underground powerhouse, shown in Chapter 8 (Figures 8.14, 8.31 and 8.32), was excavated in hard granite. It is 149 m long, 29 m wide and 70 m high. It houses 3 Francis machines of 431 MW each. The total turbine flow is 1,116 m³/s.

The main volumes of civil works are as follows:

- soil excavation 3,226,000 m³;
- rock excavation 9,974,000 m³;
- dams and dykes 12,439,000 m³;
- concrete 218,000 m³.

A.13 NOVA PONTE HPP

Nova Ponte HPP (510 MW) is located on the upper Araguari river, upstream of Miranda HPP. The general layout is presented below. The work began in 1987 and finished in 1995.

The WL of the reservoir is at El. 815.00 m and the WLmin at El. 775.50 m. The plant gross head is 118.60 m. The reservoir has been dimensioned with a depletion of approximately 40 m. The regularized flow rate is 304 m³/s (51% of total maximum turbine flow). The evolution of reservoir WLs in the first 20 years of operation can be seen in Figure A.30. It is observed that, after the closure, the WL took almost 2 years to reach the El. 810.00 m. The WLmax, El. 815.00 m, was only reached in the years from 2005 to 2009. The area of reservoir is 450 km² and the total volume is 17.8×10^6 m³. The dam is of rockfill with clay core is 1,600 m long, 142 m high above the foundations and slopes of 1.65:1.0 upstream and 1.3:1.0 downstream.

Figure A.30 Nova Ponte HPP. CBDB (2006).

The chute spillway on the right abutment 61.6 m long has four gates of 11 × 17.35 m. The maximum flow capacity is 6,140 m³/s.

The chute of the spillway is 699.50 m long, and the initial stretch of 164.50 m is concrete lined. The remainder of the chute, 535 m, is unlined. This solution, like that of Serra da Mesa, deserves to be highlighted, since the flow flows for a long time over the rocky massif. It was not lined in concrete, expensive, which brought savings to the enterprise.

The adduction system on the left abutment consists of a classic three-block tower intake, 39 m long and 63 m high, three gates, followed by three penstocks in tunnels of 6.8 m in diameter.

The indoor powerhouse, 60 m long, 37 m wide, 53 m high, houses three Francis machines of 170 MW each. The total turbine flow is 597 m³/s.

The main volumes of civil works are as follows:

- soil excavation 14,308,000 m³;
- rock excavation 3,744,000 m³;
- dams and dykes 13,370,000 m³;
- concrete 208,000 m³.

A.14 PAULO AFONSO'S COMPLEX

Paulo Afonso (PA) Hydroelectric Complex on the São Francisco river, in the states of Bahia, Alagoas and Pernambuco, is composed of the PA I, PA II, PA III and PA IV plants, plus the Moxotó HPP, owned by CHESF (MBD, 1982).

The regional geology is composed of amphibolite and gneissic amphibolite, as well as of vein of aplite (granites of very fine texture) and migmatites usually parallel to the layers of the gneiss.

The construction of PA I (180 MW) started in 1949 and ended in 1954. The plant is composed of a gravity dam, 4,215 m long, a spillway with a capacity of 22,000 m^3/s, and an underground powerhouse with three 60 MW machines. It is noteworthy that PA I was planned to allow for expansions. The PA II plant (420 MW), built between 1955 and 1961, has three Francis units of 75 MW and three Francis units of 85 MW. The PA III plant (794.2 MW), built between 1967 and 1971, has four Francis machines of 198.55 MW.

In 1971, construction of the Moxotó HPP 3 km upstream of PA I was started. The Moxotó plant (440 MW) consists of a powerhouse with 4 Kaplan machines of 110 MW and an earth dam. The reservoir has 1.2 billion cubic meters to regulate the São Francisco river. In the right margin of the Moxotó reservoir the water is diverted by a canal with 5.6 km of extension and capacity of flow of 10,000 m^3/s to supply PA IV and with the possibility to supply PA V in future.

The construction of PA IV (2,460 MW) began in 1974 and ended in May 1983. The gross head is 112.5 m and the plant have 6 Francis units of 410 MW each. PA IV was planned to operate with the continuous flow of Sobradinho (470 km upstream).

All the plants of the Paulo Afonso Complex have underground powerhouses. The cave of PA IV is 222.6 m in length, by 54 m in height and maximum width of 24 m. Its lowest point is 110 m from the surface of the reservoir. The general layout of these plants is shown in Figures A.31–A.34.

A.15 PINALITO HPP

Pinalito HPP, 46.5 MW of installed capacity, was installed in the Tireo River, central region of the Dominican Republic, La Veja Province, Constanza, 150 km from the capital Santo Domingo. The project was prepared by PCE (RJ) and the construction

Figure A.31 Paulo Afonso complex. Google Earth (2014).

Figure A.32 Paulo Afonso I, II e III HPPs. Reservoir Paulo Afonso IV in the background. (Chesf.)

Figure A.33 Paulo Afonso IV HPP. CBDB (1982).

Figure A.34 Paulo Afonso IV HPP. CBDB (1982).

was done by CNO, Construction Company Norberto Odebrecht. The layout includes the following structures:

- a CCR dam, 198 m long and 57 m of height; the Pinalito reservoir is situated at a maximum El. 1,180 m;
- a free-spillway, incorporated on the dam, 44.5 m, for a flow capacity of 950 m^3/s;
- an intake, sand trap and conduct of the waters of Sonador stream and Blanco river and lead to the exit portal to the penstock; a surge tank of the cylindrical choke type: diameters are 2.4/8.0 m and the heights are 58/45 m; a penstock, 1.24 km length, diameter 1.95/0.85 m, until the indoor powerhouse.

The main tunnel, 8.15 km long and 3.6 m in diameter, is part of the Pinalito dam. For this tunnel the waters of the Sonador and Blanco streams are diverted. The intake and the hydraulic works to conduct the waters of the Sonador stream are made up of sand-trap, bottom intake and tunnel with a length of 3.2 km and dimensions of 3.5 m (b × H). The tunnels were excavated on two fronts: one using the tunnel boring machine (TBM) method, with a shield of 3.6 m in diameter, and another using the drill and blast method (Figure A.35).

The gross head is of 543 m and the net head is of 528 m, which means a loss of load of 3.0%. The design flow of the plant is 9.8 m^3/s. The average inflow is 3.2 m^3/s. The outdoor powerhouse has two vertical-axis Pelton turbines (Figures A.36 and A.37).

A.16 YELLOWTAIL HPP

Yellowtail HPP on the Bighorn river (Montana, USA) began commercial operation in 1966 with an installed capacity of 250 MW. The powerhouse at the foot of the dam has four Francis machines of 62.5 MW each.

Figure A.35 Pinalito HPP. Dam and spillway.

Figure A.36 Pinalito HPP. Penstock.

The plant has multiple uses: generation, irrigation, supply, flood control, sediment retention, fish and wildlife enhancements, and recreational/tourism activities (Figure A.38).

The arched concrete dam, 450 m long and 160 m high, $ec = 6.7$ m; and $ef = 45$ m, is based on sedimentary rocks (sandstones, limestones, shales and siltstones - USCOLD, 1988). Volume of concrete is on the order of $1.2 \times 10^6 \text{m}^3$ (Figure A.39).

Appendix 525

Figure A.37 Penstock and powerhouse.

The spillway, with a discharge capacity of 2,605 m³/s, is in a concrete-lined tunnel, 9.75 m in diameter, on the left abutment, taking advantage of the diversion tunnel (Figure A.40).

A.17 SHASTA HPP

Shasta HPP (539 MW) is located on the Sacramento river, Redding, California. The owner is the USBR – United States Bureau of Reclamation. The purposes are flood control, irrigation, municipal and industrial water, navigation, power and prevention of intrusion of ocean water.

The general layout is presented in the following figure. The work of the dam began in 1938 and finished in 1945. The WL of the reservoir is at El. 324.60 m. The area of the reservoir is 120 km² and the total volume is 5,615 Hm³ (Figure A.41).

Figure A.38 Yellowtail HPP. Localization map.

    The dam is concrete curved gravity dam with an embankment wing 1,055 m long, 183.52 m high above the foundations; has 18 outlet works 2.6 m diameter conduits through dam in three tiers with 1,840 m^3/s of maximum discharge capacity. The main volumes are: 4.8 million m^3 of concrete and 1.7 million m^3 of earth and rockfill in wing dam. The spillway has an overflow section near center of the dam with three drum gates. The crest length is 100.6 m and the maximum discharge capacity is 7,080 m^3/s.

Appendix 527

Figure A.39 Yellowtail HPP. USBR (2010).

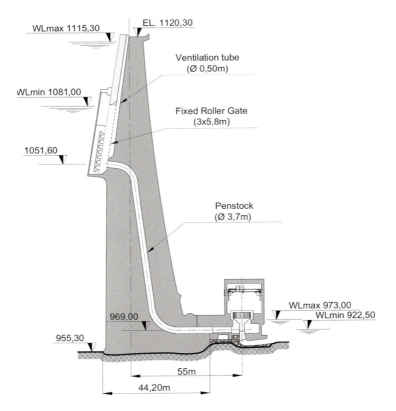

Figure A.40 Yellowtail HPP. Cross section.

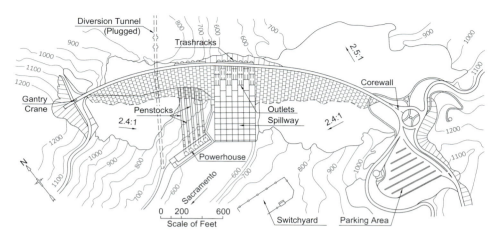

Figure A.41 Shasta dam. Plant.

Figure A.42 Shasta dam. USBR (1988).

The plant has an indoor powerhouse with a total installed capacity of 539 MW in seven Francis turbines with a rated head of 148.4 m. The five penstocks are 4.6 m diameter (Figure A.42).

A.18 DWORSHAK HPP

Dworshak dam (600 MW) is located on the North Fork Clearwater river, Orofino, Idaho. The owner is the United States Army Corps of Engineers, Walla Walla District. The purpose is flood control, power generation, water supply and recreation.

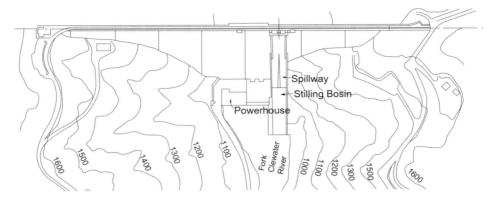

Figure A.43 Dworshak dam. Plant.

The general layout is presented in the following figure. The work of the dam began in 1965 and finished in 1973. The WL of the reservoir is at El. 488.90 m. The area of the reservoir is 69.2 km^2 and the total volume is 4.28 Hm^3 (Figure A.43).

The dam is a straight concrete gravity dam with a structural height of 218.54 m and a crest length of 1,002 m and contains 5.03 million m^3. The spillway has an overflow, tainter-gate control with stilling basin. The crest length is 30.5 m and the maximum discharge capacity is 4,000 m^3/s. The plant has an indoor powerhouse with a total installed capacity of 400 MW in three Francis turbines (two of 90 MW and one of 220 MW) with a rated head of 170 m. The five penstocks are 4.6 m diameter (Figure A.44).

Figure A.44 Dworshak dam. USBR (1988).

A.19 COOLIDGE HPP

Coolidge dam (10 MW) is located on the Gila river, Peridot, Arizona. The owner is the United States Department of the Interior Bureau of Indian Affairs. The purpose is power generation, regulatory storage, flood control, recreation. The dam was completed in 1928. The multiple-dome dam was designed in order to provide greater stability and massiveness and to overcome objections to the multiple-arch type dam for high dams (Figures A.45 and A.46).

Figure A.45 Coolidge dam.

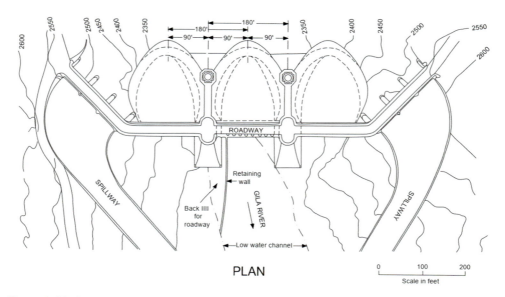

Figure A.46 Coolidge dam. Plant.

A.20 FORT PECK HPP

Fort Peck dam (185.3 MW) is located on the Missouri river, Glasgow, Montana. The owner is the United States Army Corps of Engineers. The purpose is navigation, flood control, power generation, irrigation and recreation. The general layout is presented in the following figure. The dam was completed in 1940. The first powerplant in 1943 and the second in 1961 (Figures A.47 and A.48).

The hydraulic fill dam is 76 m high and has a crest length of 6.40 km. The spillway, controlled by 16 gates, has 1.6 km long concrete chute. The maximum capacity is 7,075 m^3/s. The chute width varies from 244 to 37 m (Figure A.49).

The power facilities are composed of two concrete tunnels, 7.3 m diameter, 1.7 km long. In the powerhouse, there are five generators with an installed capacity of 185.3 MW (Figure A.50).

A.21 WOLF CREEK HPP

Wolf Creek HPP (270 MW) is located Russel County in southeastern Kentucky, on the Cumberland river, 460.9 miles above its confluence with the Ohio river. The owner is the United States Army Corps of Engineers. The purpose is flood control and power generation. The general layout is presented in the following figure. The dam was completed in 1950 (Figures A.51 and A.52).

The dam is combined rolled earth embankment and concrete gravity at 79 m high. The spillway, a concrete gravity ogee with 10 tainter gates, has a crest length of 180 m and a maximum discharge capacity of 15,310 m^3/s. The plant has six Francis turbines and six penstocks are of 6.1 m diameter. The rated head is 49 m.

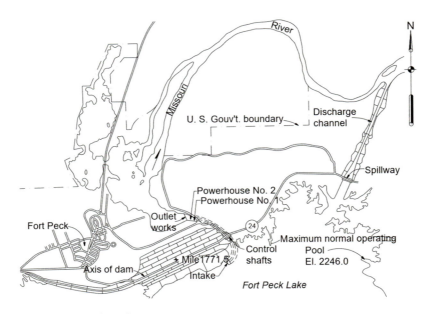

Figure A.47 Fort peck dam. Layout.

532 Appendix

Figure A.48 Fort peck dam. Google.

Figure A.49 Fort peck dam.

A.22 CETHANA HPP

Cethana HPP (100 MW) was constructed in northern Tasmania, on the Forth river, between 1967 and 1971. The purpose was power generation. The general layout is presented in Figures A.53–A.55.

Appendix 533

Figure A.50 Fort peck spillway.

Figure A.51 Wolf creek dam. Localization map.

Figure A.52 Wolf creek dam.

Figure A.53 Cethana. Localization map.

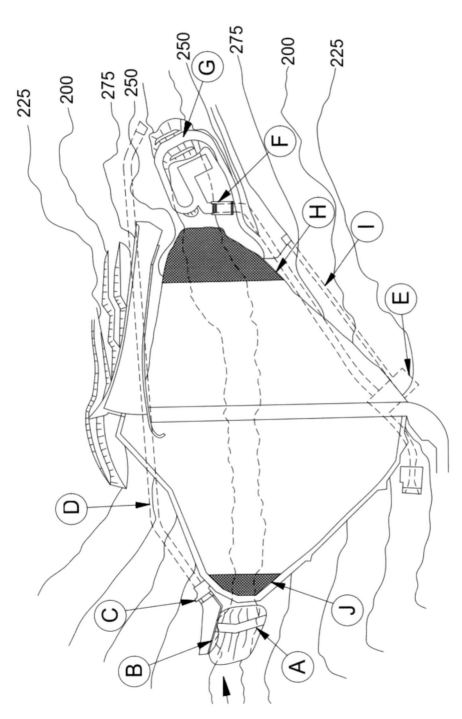

Figure A.54 Cethana dam. Layout. (a) Upstream first stage cofferdam. (b) Diversion wall; (c) Diversion tunnel portal. (d) Diversion tunnel. (e) Power-station. (f) Tailrace tunnel portal and gate structure. (g) Downstream cofferdam. (h) Downstream mesh protection. (i) Access. (j) Thick bituminous treatment (second-stage cofferdam).

Figure A.55 Cethana dam. Downstream view.

The dam is a concrete-faced rockfill structure 110 m high. The powerhouse is underground, housing a single 100 MW machine immediately downstream of the dam and a tailrace tunnel which joins the original river just below the toe of the dam.

Difficult upstream tunneling conditions, with several floods and repairs to the concrete lining, and the need to maintain design thickness, caused serious delays in the completion of the diversion tunnel and portal works. Since construction of the power-station was critical to meet the "Power ON" date. It was decided to excavate the power-station from the assembly bay level via permanent access tunnel, both of which were above the design flood level. During the 2.5 year of its use, this cofferdam was overtopped many times and required only minor repair in the latter stages of dam construction.

Dam construction proceeded with the downstream slope protected by mesh, higher than the upstream rockfill protection.

A flood of 460 m^3/s overtopped the dam in November 1968, when the downstream face was 16 meters high destroying part of the protection, with loss of 15,000 m^3 of rock. In the reinstatement and subsequent construction, a more reliable (and expensive) system of gabions was used. No further overtopping occurred.

By the start of the winter of 1969 the dam rockfill was above the level of 176.00 m and safe against overtopping by the 1,000 years of flood (200 m^3/s approx.).

Concrete face construction commenced in the summer of 1969–1970 with sufficient protection behind the first stage cofferdam for the lower part of the face. By the winter of 1970, the lowest part of the concrete face was above the elevation of 140.00 m, to the level of 15 years recurrence interval flood. Only one further interruption occurred in August 1970 when the 100 years recurrence interval flood occurred, raising the upstream level to 55 meters above the riverbed and inundating the tailrace works downstream.

Neither the dam itself nor the face suffered any damage, but a considerable quantity of debris required removal.

A.23 HÖLJES HPP

Höljes HPP is situated on the river Klarälven in Sweden, some 13 km downstream of the Norwegian border. In this project, it was necessary to provide a flume for timber logging past the site during construction (Figure A.56 and A.57).

The plant has an underground powerhouse on the left bank that houses three units. The total installed capacity is 190 MW (Figure A.58).

The dam is 70 m high, creating a reservoir of $270 \times 10^6 m^3$ in volume under a regularization drop of 34 m. The chute spillway, positioned on the right abutment, has a flow capacity of 1,160 m^3/s in the normal operating level 5 m below the dam crest. The intake to the timber flume next to the spillway has a capacity of 120 m^3/s (Figure A.59).

A.24 IVAILOVGRAD HPP

Ivailovgrad HPP, 108 MW, was completed in 1964 and is situated in the last stage of the Arda river cascade in Bulgaria. It regulated almost the entire run-off of the river (Figure A.60).

The geological and topographic conditions at the site of the dam permitted the construction of any type of concrete or rockfill dam. However, the exceptional flood sizes limited to a large extent the scope of their application. The concrete dams proved most appropriate for carrying the floodwaters, due to which a concrete gravity dam with expansion joints was chosen. The maximum height of the dam is 71 m, the gross head 44 m and the turbine flow is 280 m^3/s (Figures A.61 and A.62).

The spillway, 7,350 m^3/s, is located in the central section of the dam, and the chute over the roof of powerplant's machine building (Figures A.63 and A.64).

Figure A.56 Höljes dam. Localization map.

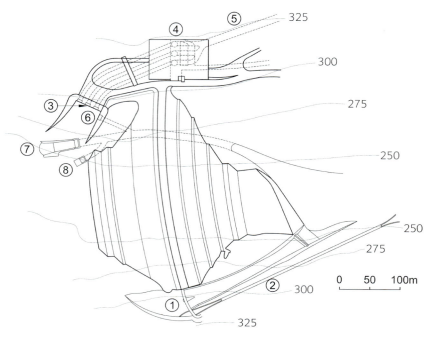

Figure A.57 Höljes HPP. Layout. (1) Spillway; (2) Timber flume; (3) Intake; (4) Powerhouse; (5) Tailrace tunnel; (6) Low-level outlet; (7) Diversion tunnel intake; (8) Bottom outlet.

Figure A.58 Höljes HPP. Downstream view.

Appendix 539

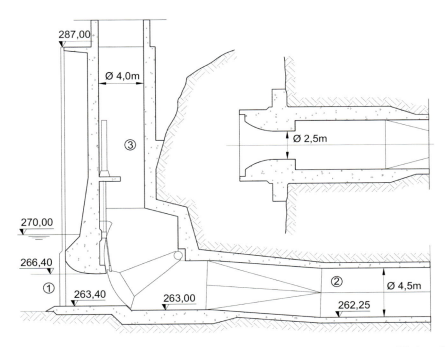

Figure A.59 Höljes HPP. Low-level outlet. (1) Low-level outlet; (2) Tunnel; (3) Air shaft.

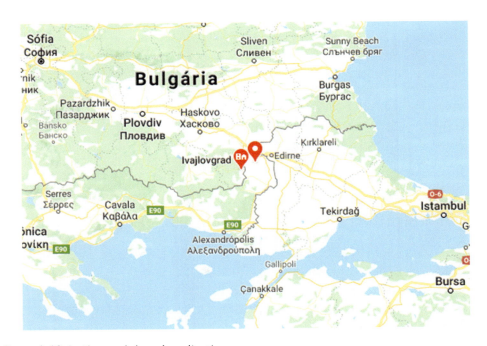

Figure A.60 Ivailovgrad dam. Localization map.

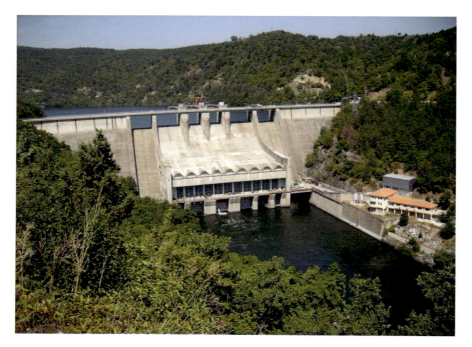

Figure A.61 Ivailovgrad dam.

Figure A.62 Ivailovgrad spillway.

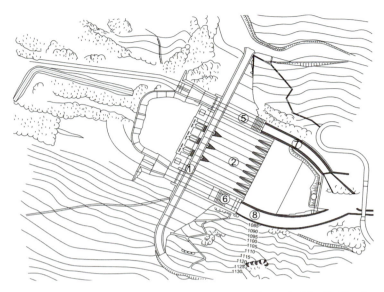

Figure A.63 Ivailovgrad plant. (1) Dam; (2) Chute; (3) and (4) Cofferdams; (5) and (6) Diversion galleries; (7) and (8) Diversion channels.

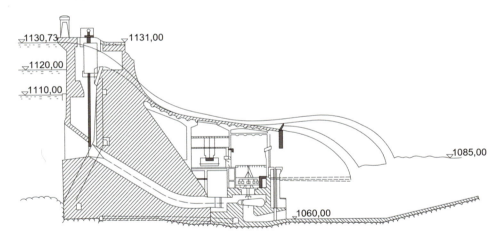

Figure A.64 Ivailovgrad section.

A.25 L'AIGLE HPP

L'Aigle dam, 349 MW (4×54 MW, 1×133 MW), was constructed from 1935 to 1945 on the Dore river, a tributary of the Garonne river, in France (Figures A.65 and A.66).

The gravity-type semi-circular concrete dam is 95 m high and 290 m long. The powerhouse housed 4 vertical-axis Francis turbines of 54 MW each. In 1982, the fifth 133 MW turbine was installed. The plant's gross head is 84 m. The ski jump spillway, with four spans without floodgate control, has a maximum capacity of 3,800 m^3/s (Figure A.67).

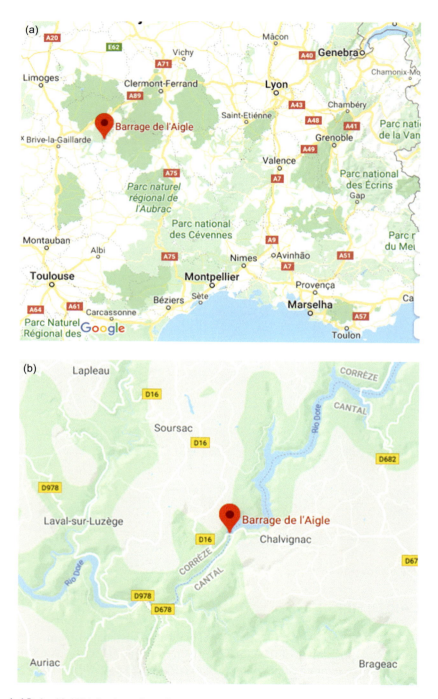

Figure A.65 (a–b) L'Aigle dam. Localization map.

Figure A.66 L'Aigle dam. Google view. Plant.

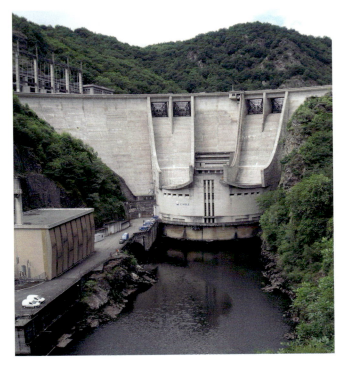

Figure A.67 L'Aigle dam. Downstream view.

A.26 SAINTE-CROIX HPP

Sainte-Croix dam, 136 MW, was constructed from 1971 to 1974 on the Verdon river, on a tributary of the left bank of the Rhône river, in France (Figure A.68).

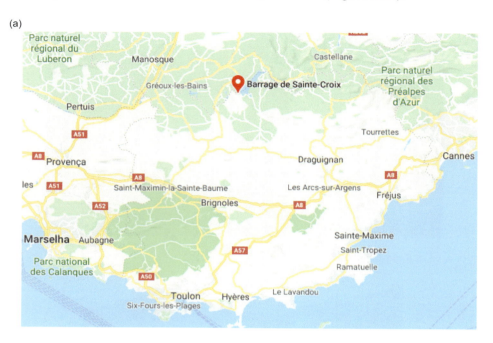

Figure A.68 (a–b) Sainte-Croix dam. Localization map.

The double-curved arch dam is 95 m high and has a crest length of 138 m. The powerhouse houses 4 Francis vertical-axis turbines of 54 MW each. The turbine flow is 123 m^3/s and the plant's gross head is 132 m. The orifice spillway has a maximum flow capacity of 1,100 m^3/s. It is recorded that the 10,000 years of recurrence time flood is of the order of 1,880 m^3/s (Figures A.69 and A.70).

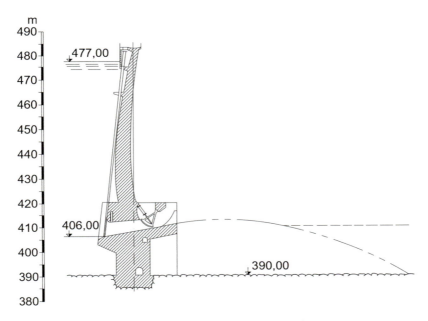

Figure A.69 Sainte-Croix dam. Orifice spillway (two orifices).

Figure A.70 Sainte-Croix dam. Downstream view.

A.27 SERRE DE PONÇON HPP

Serre de Ponçon HPP with a 380 MW generator is underground and was installed in 1955 on the left bank of the Durance river, in the departments of Hautes-Alpes and Alpes-de-Haute-Provence, Provence-Alpes-Côte d'Azur region, in southern France, one of the largest in Western Europe (Figure A.71).

The barrage is a 123 m high earth core dam. The lake gathers the waters of Durance and Ubaye rivers, flowing down through the Hautes-Alpes and the Alpes du Sud to the Rhône river. In addition to the power generation, the dam provides the reservoir and overall water management to facilitate 15 HPPs along the Durance and Verdon rivers in south-eastern France with total capacity of 2,000 MW (Figures A.72–A.75).

A.28 PIEVE DI CADORE HPP

Pieve di Cadore dam, 35 MW, was constructed from 1946 to 1949 on the Piave river, on Belluno Province, Italy. Note that the town is next to Monte Castelo (Figures A.76 and A.77).

The layout was designed with a classic dam on a curved axis upstream and a powerhouse downstream, served by a headrace tunnel. The dam is of concrete gravity type, 174 m high. The two spillway tunnels, 10 m in diameter and variable length, has a flow capacity of $1,500\,m^3/s$ (Figures A.78 and A.79).

Figure A.71 Localization map.

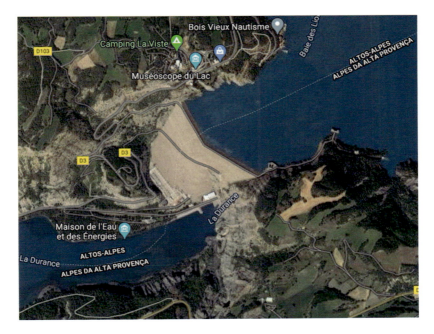

Figure A.72 Serre de Ponçon. Layout.

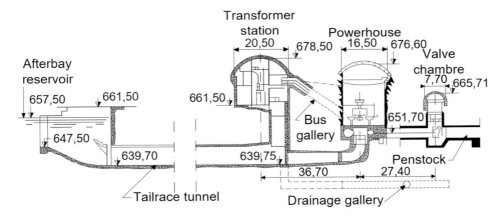

Figure A.73 Serre de Ponçon. Cross section.

A.29 EDLING HPP

Edling dam, 87 MW – 2 generating Kaplan, was installed between 1959 and 1962 on the Drau river, next to the city of Edling, Carinthia, Austria. The layout is shown in Figures A.80 and A.81. The average rate of flow in the site is $270 \, \text{m}^3/\text{s}$. It should be noted that the river diversion was designed for a flow of $1,600 \, \text{m}^3/\text{s}$, corresponding to a recurrence period of 14 anos.

Figure A.74 Downstream view.

Figure A.75 (a–c) Serre de Ponçon. Construction view.

(Continued)

Figure A.75 (Continued) (a–c) Serre de Ponçon. Construction view.

Figure A.76 Pieve de Cadore dam. Localization map.

Figure A.77 HPP Pieve di Cadore. Downstream view.

Appendix 551

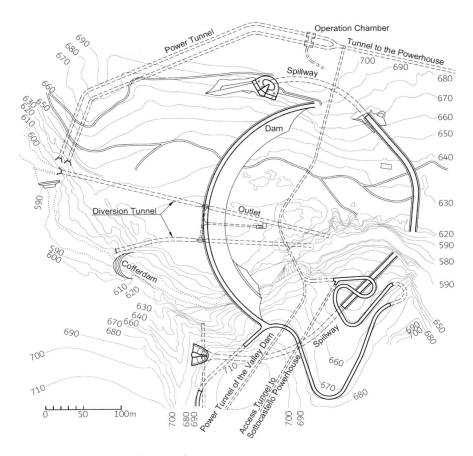

Figure A.78 HPP Pieve di Cadore. Layout.

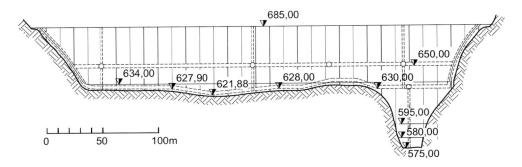

Figure A.79 HPP Pieve di Cadore. Longitudinal section.

552 Appendix

Figure A.80 Edling plant.

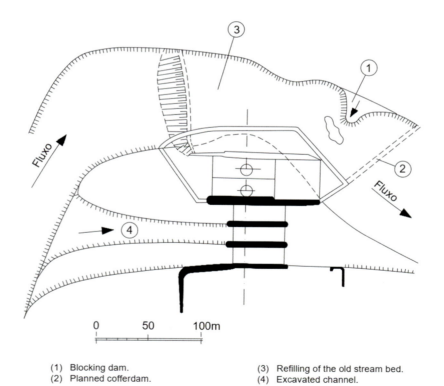

(1) Blocking dam.
(2) Planned cofferdam.
(3) Refilling of the old stream bed.
(4) Excavated channel.

Figure A.81 Edling dam. Plant.

A.30 YBBS-PERSENBEUG HPP

Ybbs-Persenbeug HPP, 32 MW – two generating Kaplan, was installed between 1952 and 1956 on the Danube river, next to the city of St. Martin am Ybbsfelde, Austria (Figures A.82 and A.83).

Figure A.82 Ybbs-Persenbeug HPP. Localization map.

Figure A.83 Ybbs-Persenbeug HPP.

The layout is shown in Figures A.84–A.86. The bulb turbines were designed for the discharge of 360 m³/s and for the head of 10.6 m.

A.31 KRASNOYARSK HPP

Krasnoyarsk HPP, 6,000 MW – ten units, was implanted in between 1956 and 1972 in the Yenissei river downstream of Sayano Shushensk SSH HP, Khakassia, Russia, as shown in the previous figure (Figure A.87).

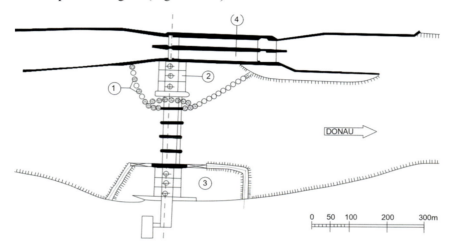

Figure A.84 Ybbs-Persenbeug HPP. Plant. (1) Cell cofferdam; (2) Northern powerplant; (3) Southern powerplant; (4) South lock.

Figure A.85 Ybbs-Persenbeug HPP. Construction view. 1956.

Figure A.86 Lowering the cell cofferdam.

Figure A.87 Krasnoyarsk HPP. Downstream view.

The dam, founded on hard and cracked granitic rock, is 1,073 m long at the crest.is 128 m high and has a volume, approximate, of $5.0 \times 10^6 \, m^3$ (Grishin, M.M.). The powerhouse and spillway cross sections are showed in Figures A.88 and A.89. The spillway has seven spans, with a hydraulic head of 10 m.

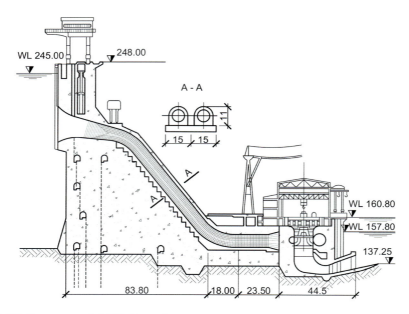

Figure A.88 Krasnoyarsk HPP. Powerhouse cross section.

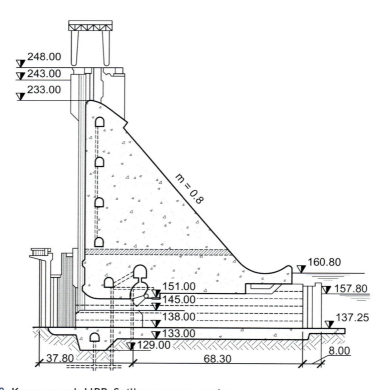

Figure A.89 Krasnoyarsk HPP. Spillway cross section.

A.32 IRKUTSK, BRATSK AND UST-ILIMSK HPPS

Hydroelectric plants Irkutsk, Bratsk and Ust-Ilimsk were installed on the high stretch of the Angara river. As shown in the figure, the river rises in Lake Baikal and is a tributary on the right bank of the Yenissei river, where it flows into the region of the city of Lesosibirsk (Figure A.90).

Irkutsk HPP, 662.4 MW (8 × 82.8 MW), is located adjacent to Irkutsk, Irkutsk Oblat. Construction on the dam began in 1950 and the plant was opened in 1956. It has a long rockfill embankment, 2,740 m in length and 56 m high. The complete filling of the reservoir took 7 years. It was the first large HPP constructed in Eastern Siberia. The power station is contained in a 240 long, 77 m wide and 56 m high reinforced building. Power is supplied to the Irkutsk Aluminium Factory in Shelekhov and local areas for residential use. A complete rehabilitation of the plant was done in 2010 (Figure A.91).

Bratsk HPP, 4,515 MW (15 × 250 MW; 3 × 255 MW), construction began in 1954 and the plant was opened in 1967. The hydraulic head is 108 m. The structure is a concrete gravity dam 123 m high and 924 m in length. The spillway has 10 radial gates and a flow capacity is 6,500 m^3/s (Figures A.92–A.95).

Ust-Ilimsk HPP, 3,840 MW (16 × 240 MW), is a concrete gravity dam 1,475 m in length and 105 m high with a spillway of 242 m in length. It is flanked by two earth-fill auxiliary dams: the one on the left bank that is 1,710 m long and 28 m high; on the right bank, the auxiliary dam is 538 m long and 47 m high (Figure A.96).

The construction began in 1963. Its reservoir began filling in 1974 and the powerplant was commissioned in 1980 (Figure A.97).

A.33 PLANTS ON DNIEPER RIVER

The following figure shows the plants on Dnieper river, in the Republic of Ukraine, which flows into the Black Sea (Figure A.98).

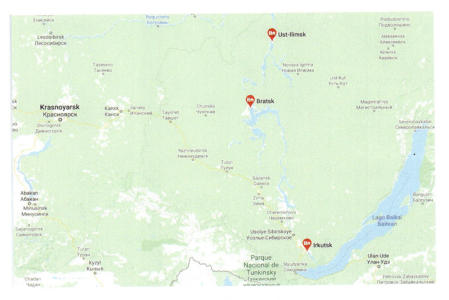

Figure A.90 Irkutsk, Bratsk and Ust-Ilimsk HPPs. Localization map.

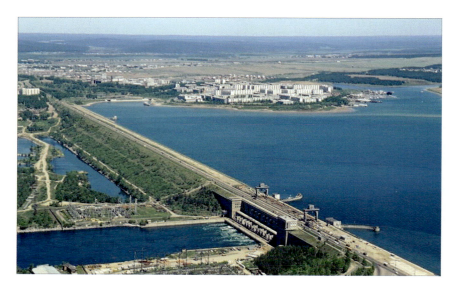

Figure A.91 Irkutsk HPP view.

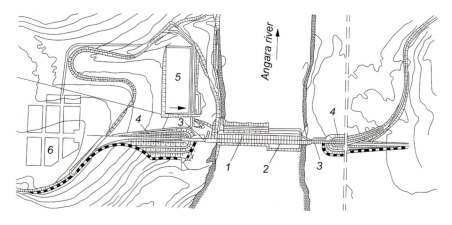

Figure A.92 Bratsk HPP. Plant (Google).

Kiev HPP, 418.8 MW (4 units of 22.7 MW; 16 units Bulb of 20.5 MW), was opened in 1954.

Kanivs'ke HPP, 444 MW (24 units of 18.5 MW) has a height of 39.5 m and a length of 343 m. The spillway has a capacity of 13,200 m^3/s.

Svitlovodsk or Kremenchuk HPP 624 MW was constructed between 1954 and 1959.

Dnipropetrovsk HPP, or Dnieper HPP, was constructed between 1927 and 1932, but was destroyed during the second world war. The plant was reconstructed between 1944 and 1950. It has an installed capacity of 1,548 MW (Figures A.99–A.102).

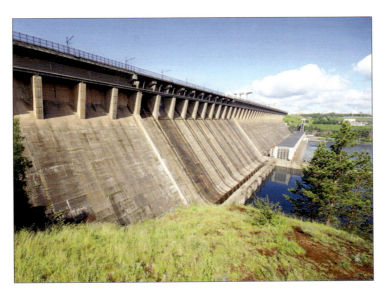

Figure A.93 Bratsk HPP. Downstream view.

A.34 CHIRKEYSK HPP

Chirkeysk HPP, 1,000 MW (four units of 250 MW), was implanted, between 1964 and 1976, on Sulak river, which flows into the Caspian Sea, in the Republic of Dagestan, Russia (Figure A.103).

In view of the characteristics of the site, the dam was designed with structural concrete in double curvature arch, 233 m high and 338 m long. The reservoir surface is of the order of 42 km², and the reservoir volume is of about $3.0 \times 10^9 \, m^3$. The hydraulic head is of 205 m (Figures A.104 and A.105).

The spillway consists of a 509 m long non-pressure bottom outlet in the left bank with a discharge capacity of 2,900 m³/s (Figure A.106).

A.35 HIRAKUD HPP

Hirakud HPP, 347.5 MW, was implanted, between 1947 and 1957, on the Mahanadi river, about 15 km from Sambalpur in the state of Orissa. It is the longest dam in India. The plant has two powerhouses: PH I – Burla: 2 × 49.5 MW, 3 × 37.5 MW, 2 × 32 MW; and PH II – Chiplina: 3 × 24 MW (Figures A.107 and A.108).

The main earth dam is 61 m high, 4.8 km long and the entire earth dam is 25.8 km long. The spillway, with a discharge capacity of 42,450 m³/s, has 34 crest gates and 64 sluices-gates. In the right abutment is positioned the PH1-Burla (Figure A.109).

It should be noted that for the young engineers, the excellent condition of the structures for a dam is over 60 years, which reveal the high quality of maintenance services (Figure A.110).

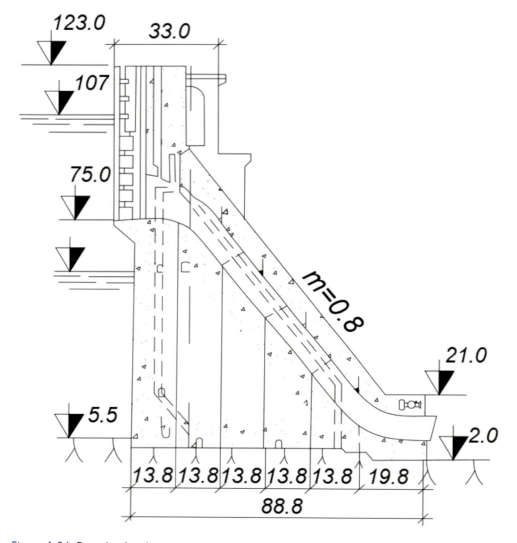

Figure A.94 Bratsk - Intake cross section.

A.36 BAKUN HPP AND MURUM HPP

Bakun HPP, 2,400 MW, was implanted on the Balui river, Sarawak, east Malaysia, between 1996 and 2011. The plant has 8 Francis turbines of 300 MW each (Figure A.111).

The dam is a rockfill type with a concrete face, is 205 m high and 750 m long at the crest. The chute spillway, on the left abutment, has two chutes and four spans controlled by gates and the discharge capacity is of order of 15,000 m^3/s (Figures A.112 and A.113).

Appendix 561

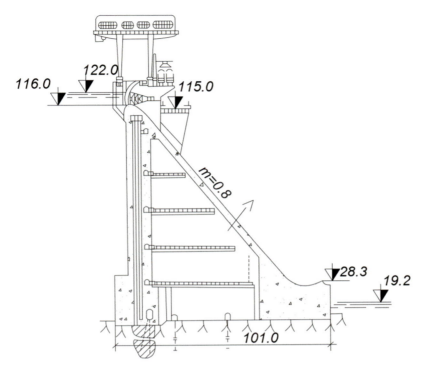

Figure A.95 Bratsk. Spillway cross section.

Figure A.96 Ust-Ilimsk HPP. Google view. Plant.

Figure A.97 Ust-Ilimsk HPP. Side view.

Figure A.98 Plants on the Dnieper river. Localization map.

Appendix 563

Figure A.99 Kiev HPP.

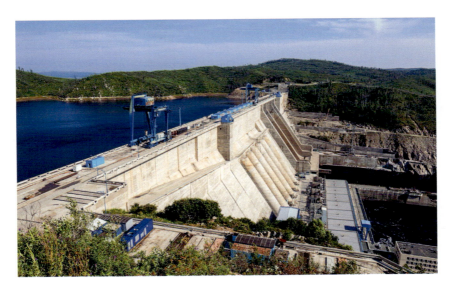

Figure A.100 Kanivs'ke HPP.

Figure A.101 Kremenchuk HPP.

Figure A.102 Dnieper HPP.

Appendix 565

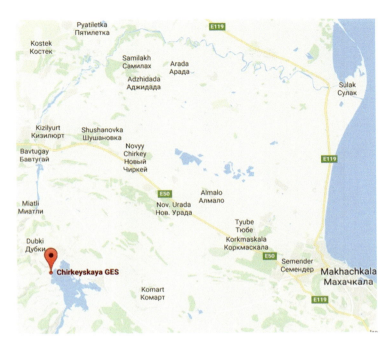

Figure A.103 Localization map.

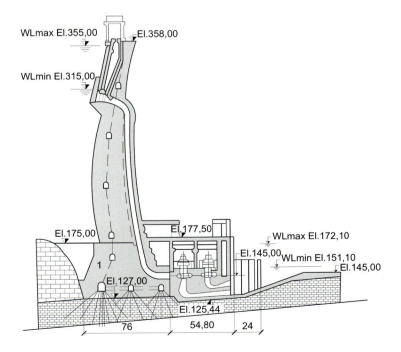

Figure A.104 Chirkeysk powerhouse.

Figure A.105 Chirkeysk HPP. Google plant.

Figure A.106 Chirkeysk HPP. Downstream view. (Note the bottom outlet, spillway, in the left bank.)

Appendix 567

Figure A.107 Hirakud HPP. Localization Map 7.

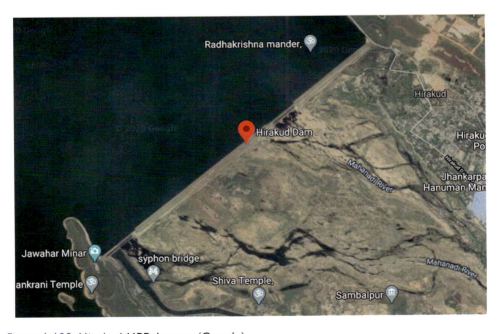

Figure A.108 Hirakud HPP. Layout (Google).

Figure A.109 Hirakud HPP. Downstream view.

Figure A.110 Sluice gates in service.

Figure A.111 (a–b) Localization map.

Figure A.112 Bakun HPP. Studied powerhouse alternative on the right bank.

(a)

(b)

Figure A.113 Bakun HPP. View.

A.37 RICOBAYO HPP

Ricobayo HPP, 291 MW – eight generating, on the Esla river, before it joins the Douro river, near Ricobayo de Alba, Spain. Commissioning work was done between 1929 and 1935 (Figure A.114).

The gravity dam is 99.4m high and 270m long. The local average discharge is 140 m^3/s. The plant has a powerhouse at the foot of the dam, 133 MW, and another underground on the right bank, 150 MW (Figure A.115).

Figure A.114 Ricobayo HPP. Localization map.

Figure A.115 Ricobayo HPP. Downstream view.

A.38 HPPS IN DOURO RIVER BETWEEN SPAIN AND PORTUGAL

In this item, we will present a summary of the following plants: Miranda (369 MW), Picote (426 MW), Bemposta (431 MW) and Aldeadávila (46.3 MW) (Figure A.116).

Miranda HPP, 369 MW, on the Douro river, in the municipality Miranda do Douro, in Bragança District, Portugal. Commissioning work was done between 1957 and 1964. Consists of a dam with a spillway at the crest. The arch concrete dam is 87 m

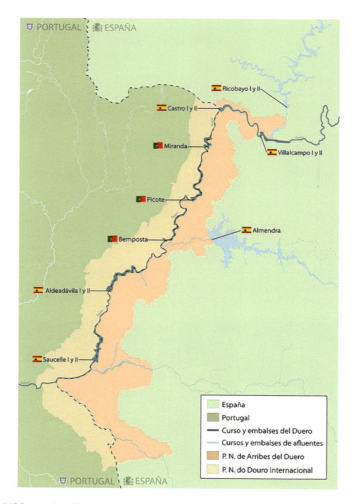

Figure A.116 HPPs in the Douro river. Localization map.

high and 297 m long. The spillway over the dam and has four gates and a capacity of 11,500 m³/s. Plant 1, 240 MW, went into operation in 1958. Plant 2, 191 MW, went into operation in 2008 (Figure A.117).

Picote HPP, 426 MW, downstream of Miranda, the work was done between 1953 and 1957. Consists of a dam with a spillway at the crest and two underground plants. The arch concrete dam is 100 m high and 139 m long. The spillway over the dam and has four gates and a capacity of 10,400 m³/s. The plant went into operation in 1958 (Figure A.118).

Bemposta HPP, 431 MW, downstream of Picote, the work was done between 1957 and 1964. Consists of a dam with a spillway at the crest and two underground plants at the right shoulder of the dam. The arch concrete dam is 87 m high and 297 m long. The spillway over the dam and has a capacity of 11,500 m³/s. Plant 1, 210 MW went into operation in 1964. Plant 2, 191 MW went into operation in 2011 (Figure A.119).

Figure A.117 Miranda HPP. Downstream view.

Figure A.118 Picote HPP. Downstream view.

Figure A.119 Bemposta HPP. Downstream view.

Figure A.120 Aldeadávila HPP. Downstream view.

Figure A.121 Saucelle HPP. Downstream view.

Aldeadávila HPP, 1,150 MW, on the Douro river, near Aldeadávila de la Ribera, in the Salamanca province, Autonomous Community of Castile and Leon, Spain. The first phase of the plant was completed in 1962. The second in 1986. The dam is a mixed vault-to-gravy type, 140 m high and 250 m long. The spillway has a capacity of 11,700 m^3/s (Figure A.120).

Saucelle HPP, 520 MW, was implemented between 1950 and 1956 downstream of Aldeadávila. The dam is 83 m high and 289 m long. The spillway over the dam has a capacity of 12,940 m^3/s. Plant 1 (1956) has four units and an output of 251 MW. Plant 2 (1985) has two units and an output of 191 MW. The spillway over the dam and has a capacity of 12,940 m^3/s (Figure A.121).

A.39 CRESTUMA HPP

Crestuma dam, 105 MW – three generating bulge, on the Douro river near Oporto, Portugal, mainly hydroelectric power and improvement of navigation. The movable dam, is an eight double-fixed roller gates (hook type) 28 m clear span, 13.7 m in height, supported by 6 m thick, 49 m large concrete piers, founded by concrete-wall technique on the schist bed-rock 40 m below the alluvium river bed. The stilling basin is a plain concrete poured on alluvium. In view of the geological and geotechnical conditions of the site, the spillway, for 26,000 m^3/s maximum flood flow, was designed with a stilling basin (Figures A.122–A.125).

Figure A.122 Crestuma HPP.

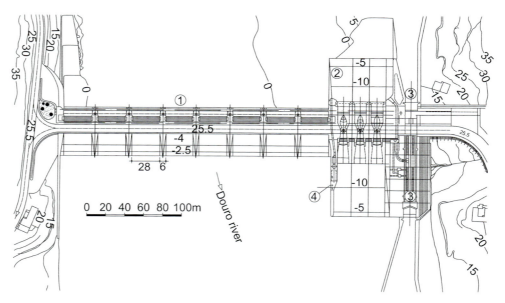

Figure A.123 Crestuma Plant. (1) Dam/spillway; (2) Intake/powerhouse; (3) Navigation lock; (4) Fish ladder. Ribeiro (1973).

Figure A.124 Hydraulic model. Scale 1:80. Ribeiro (1973).

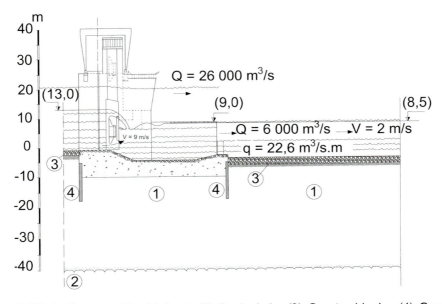

Figure A.125 Spillway profile. (1) Sand; (2) Rock-shale; (3) Granite blocks; (4) Concrete cutoff. Ribeiro (1973).

A.40 MEKONG DAMS AND HPPS

This item presents a brief commentary on the hydroelectric plants of the Mekong river and its tributary Nam Ou on the left bank in Vietnam, which are shown in Figure A.126. The estimated potential is about 60,000 MW (Mekong River Commission).

Xiaowan HPP, 4,200 MW – six units, was implanted in Lancang river, name given by Chinese to the Mekong river in their territory, in Nanjian County, Yunnan Province, southeast China, between 2002 and 2010. The double-curvature arch dam 292 m high and 902 m long is the third largest in the world. The spillway has a discharge capacity of 10,014 m^3/s (Figure A.127).

Figure A.126 Dams on the Mekong river (The Thirdpole Net).

Figure A.127 Xiaowan plant (researchgate.net).

Nuozhadu HPP, 5,850 MW, nine units, was implanted also in Lancang river, in Yunnan province, between 2004 and 2014. The rockfill dam, earth core, is 261.5 m high and 608 m long. The crest width is 18 m. The spillway has a discharge capacity of 31,320 m^3/s (Figure A.128).

Xayaburi run-of-river HPP, 1,285 MW, 8 units, was implanted on medium Mekong river, approximately 30 km east of Xayaburi in northern Laos, between 2012 and 2019. The concrete dam, 32.6 m high and 820 m long, has a spillway with ten radial gates, has a discharge capacity of 3,980 m^3/s (Figure A.129).

A.41 JINPING-I

One of the tallest arch dams in the world, Jinping-I HPP, 3,600 MW, on the Yalong river in Liangshan, South Sichuan, China, is 305 m high and 570 m long. Construction began in 2005 and was completed in 2014. The dam has three spillways with a total discharge capacity of 12,150 m^3/s: the crest spillway has a capacity of 3,000 m^3/s; the bottom outlets 5,500 m^3/s; and the tunnel spillway 3,650 m^3/s (Figures A.130–A.133).

A.42 LONGYANGXIA

The last one, the Longyangxia HPP, 1,280 MW – four units, was implanted, between 1976 and 1992, on the Yellow river in Gonghe County, Qinghai Province, China. The arch gravity dam is 178 m high and 396 m long. The dam has two service spillways on the right abutment (gates 12 m wide), and another single-chute spillway on the left abutment (Figures A.134 and A.135).

Figure A.128 Nuozhadu plant (youtube.com).

Figure A.129 Xayaburi plant (The Nation Thailand).

Appendix 581

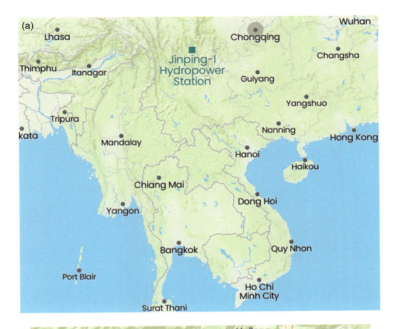

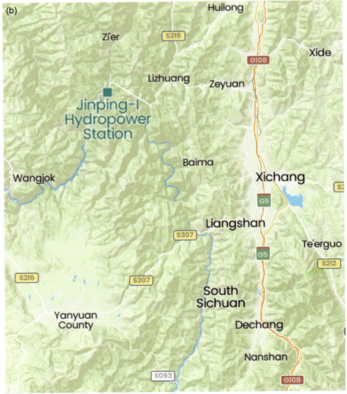

Figure A.130 (a–b) Jinping-I HPP Localization map.

Figure A.131 Jinping-1 plant. Downstream view (researchgate.net).

Appendix 583

Figure A.132 Jinping-I plant. Bottom outlet working (researchgate.net).

Figure A.133 Jinping-I. Bottom outlet working (researchgate.net).

Figure A.134 Longyangxia plant. Localization map.

Figure A.135 Longyangxia plant (eng.spic.com.cn).

Index

Note: **Bold** page numbers refer to tables and *italic* page numbers refer to figures.

aeration 223–234
    aeration devices 224
    aerator geometry 225, *225*
    erosion tests 232
    incipient cavitation index 224, **225**
    water vaporization level 224, *224*
aerator geometry 225, *225*
"affinity laws" 339
Aimores dam 98, *100*, 117
Aldeadávila HPP 571, *574*, 575
alliance, construction planning 428
Allievi Equation 261
alternating current system (AC) 390
America's power plant 7, *7*
ANA hydrometric network 32
anchor block 252–253, *253*,
    254–257, *258*
Anderson, D. 237
Aneel/Eletrobras standards 29
Araguari river 519
arch dam 160–170
    first trial load method 169
    narrow valley/wide valley 165–166
    profiles 161
    slender arch dams 161, 164
    β-form factor 161
Arndt, R. E. A. 6, 279, 330
asphalt concrete face rockfill dams 146–149,
    147, 148
asphalt core rockfill dams 149, *150, 151*
assembly/erection planning 422
auxiliary electrical systems
    alternating current system 390
    direct current system 390–391
average flow 53

backwater studies 69
Baecher, G. B. 92
Bakhmeteff, B. A. 247
Bakun HPP 560, *569–570*
Balbina HPP 497
    eletronorte magazine *499*
    general layout 498
    localization map *497*
    powerhouse *293*
    spillway *498*
Ball, J. W. 212
Barra Grande HPP 297, *299, 300*
    Brazilian Aluminum Company 297, *299, 300*
    layout and structures 507
    localization map *507–508*
    Pelotas river 507
Barton, N. R. 271, 281
"base power plants" 404
Basic Dam Safety Guide 455
basin characterization
    hydrological and energetic studies
        basin shape 54–55
        concentration time 55, **56**
        drainage area 54
        mean bed slope 55
battery chargers 390–391
bearing lubrication 373–375, *375*
bear-trap gate *313,* 316
Bemposta HPP 571, 572, *574*
Bieniawski, Z. T. 271
Bieudron HPP *452, 453*
Bighorn river 523
"black box" effect 459
"bonneted and unbonneted" gates 318,
    321, 322

586 Index

bonneted slide gate *321*
borehole/drillhole log 98, *101*
Bratsk and Ust-ilimsk HPPS 557
Brazilian Aluminum Company (CBA) 297, *299, 300*
breaker failure protection 395
Brekke, T. L., Rippley, B. D. 269, 272, 275
Brito, R. J. R. 189, 208, 229
Broch, E. 41, 269, 272
budget
    planning hydropower generation
        after privatization 37–38
        standard budget 37
        technical and economic evaluation 38
bulb turbines 352, *353*
    application range 352–353
    basic principle 353
    dimensions 353
    performance data 353–354
butterfly valves 323, 325, *326*

Campos Novos HPP
    Canoas river 508
    layout and structures *508*
    spillway working *509*
canal drop layouts 42–47, *44–46*
Canambra criteria, surge tanks 284
Canoas river 507
Carter, Jimmy 439
cartographical surveys 27, 53, 68
cartographic plans 32
Carvalho, N. O. 75, 77, 79
Castro Alves HPP *304*
catastrophic dam breaks 431–432
    El Guapo dam 440
    Grande Dixence dam 445–448
    lower San Fernando dam 440–443
    Malpasset dam *433–436*, 433–437
    Sayano – Shushensk HPP 443–445, *448*
    Teton earth dam 438–440
    Vajont dam *437*, 437–438, *438*
cavitation 341–343
    cases 218–219, *218–223*, 221–222
    conceptualization and characteristic
        parameters 211–212
    erosion 342–343
    irregularities 212–213, *212–214*
    limits 343
    protective measures specifications *213*,
        213–216, *215, 216*, **217**
Cedergren, H. R. 117, 130
CERAN Complex 299, **302, 303**
Cethana HPP 532, 536
Chadwick, W. L. 41
Chavarri, G. 218
Chirkeysk HPP 559, *566*

Chow, V. T 59
Christian, J. T. 92
Churchill curve 77
circuit breaker failure scheme 395
clay-core rockfill dams 140–143, *141, 142*
Coelho 79
Colebrook and White 246, 247
concrete blocks
    anchor block 252–253, *253*, **254–257**, *258*
    support block 250–252, **251**, *251*
    types of 250
concrete face rockfill dams 143–146, *144*
Conduit Tainter gates 317–318
construction phases
    first phase diversion 407
    second phase diversion 407–409
construction planning 418–422
    assembly/erection planning 422
    construction phases
        first phase diversion 407
        second phase diversion 407–409
    infrastructure designs 423
    procedures
        alliance 428
        guaranteed maximum price 428–429
        traditional (classical) modality 423–425
        turn-key 425–428
    river diversion design
        closed-valley plants 415
        discharges and risks 409–415
        execution 416–418
        hydraulic dimensioning 416
        hydraulic models 418
        open valleys plants designs 415
        river diversion phases 415–416
    site logistics 423
Conti-Varlet method 68
control and protection system 402
control gates 258, *319*
control system failure 340
Cooke, J. B. 143
Coolidge dam 529–530
cooling system 375–376
Corumba HPP 575–577
    Corumba river 512
    layout *512–514*
Costa, W. D. 102, 110
crestuma plant *576*
critical period 81–82
Cruz, P. T. 130, 131, 134, 135, **137,** 139,
    141, 143
Cumberland river 531
current protection criteria 391–392

dams 310
    alternative geomechanical model *111*

Index   587

arch dams 160–170
dam accidents 92
earth dams
    design 127–128, *129*
    percolation analysis 130–133
    tension and strain analysis 134–138
foot protection 190
gravity dams
    conventional concrete 150–155,
        153–155, **155**
    roller compacted concrete 155–160,
        157–159, **158, 159**
layouts 42, *42, 43*
non-technical factors 92
project criteria manual 152
rockfill dams 138–150
    asphalt concrete 146–149, *147, 148*
    asphalt core 149, *150, 151*
    concrete face 143–146, *144*
    impermeable membrane (clay core)
        140–143, *141, 142*
safety system 91, 432
standardized norms 128
technical terms 125, *126*
types 125
Dam Safety Plan 455
Dams and Public Safety 4
"Dams Breaks" theme 431
Darcy's law/Dupuit's theory 120, 130, 132
Darcy-Weisbach equation 280
Darcy-Weisbach formula 246
Datei, C. 248
Davis, C. V. 310
DC auxiliary service system 390
decision-making power 424
Deppo, L. 248
design conditions 331–335
design head, operation out of 339–340
design principles, hydropower generators
    370–376
    bearing lubrication 373–375
    cooling system 375–376
    poles and pole windings 373
    stator core 372–373
    stator winding 373
differential surge tank 283
Digital Supervisory and Control System 392
dimensioning factors 367–370
direct current system (DC) 390–391
discontinuity systems **115**
Dnieper river 557–559
Dona Francisca HPP
    Jacui River 514
    layout and structures 514–516, *515–516*
double bar
    double breaker 402–403

one breaker and a half 401–402
single breaker 398–400
single circuit breaker
    with bypass disconnecting switches
        400–401
    and transfer bar 401
Douro river 571–575
downstream back pressure 331
drainage systems
    internal drainage system 117–120
        percolation analysis 119, 120, *121*
    main drainage system 122
DRI (Drilling Rate Index) 421
drum gate *311*
Dworshak dam 218, 528–529

earth dams
    design
        compatibility principle 127–128, *129*
        flow control principle 127
        stability principle 127
    empirical methods 125
    percolation analysis 130–133
        filter and transition materials 132–133
        foundation waterproofing devices 133
        internal drainage system 132
    slopes protection 138
    stability analyses 133
    tension and strain analysis 134–138
        displacement analyses 135–138, *136,*
            136–137
        suffer deformations 135–138, *136,*
            136–137
ecological balance 40
economic diameter, penstocks 248–249
    annex support and anchor blocks 250–253,
        251, *251, 253,* **254–257,** *258*
economic feasibility studies 69
Edison, T. 7
Edling dam 547, *552*
Edvardsson, S. 41, 269
electrical equipment
    auxiliary electrical systems
        alternating current system 390
        direct current system 390–391
    generating unit layout 379–388
    main transformers 388–389
    operation and maintenance 405–406
    protection systems
        breaker failure protection 395
        current protection criteria 391–392
        elevator transformers 393–394
        generating units 392–393
        protective relays 391
        substation bar protection 395–396
        transmission line protection 394–395

588 Index

electrical equipment (*cont.*)
   substation interconnection
      components and installations 397
      equipment arrangements 397
      maneuvering schemes 397–403
      maneuvering scheme selection criteria 403–404
      plant switchyard/substation 396–397
      powerplant connection to electrical system 404–405
      switchyard types 397
   synchronous generator
      design principles 370–376
      dimensioning factors 367–370
      electromagnetic characteristics 365
      energy conversion 362–364
      generator main elements 365–366
      generator rated capacity 366–367
      monitoring and instrumentation 377
      synchronous machines 361–362
      tests 379
      turbine-generator and assembly 377–378
electrical faults 392–393
The Electrical Sector 68
electromagnetic characteristics 365
Eletrobras/ANEEL Hydroelectric Power Plant Design Guidelines (1999) 36, 81
Eletrobras Manual (2000) 54
Eletrobras Manual (2007) 28
Elevatorski, E. A. 198, 200, 202
elevator transformers 393–394, 396, 397
El Guapo dam 440, *443*
Emborcação HPP powerhouse *295, 296*
energy conversion 362–364
energy cost 38
energy-economical design 85, **85–86**
Energy Reallocation Mechanism (MRE) 87
environmental impact studies (EIA) 29, 35, 36
equipment arrangements 397
Erbisti, P. C. E. 310, 314, 318
erosion tests 232
Euler equation 337–338
Excavation Support Ratio (ESR) 271, **272**
expansion surge tank 283
extreme flows
   hydrological and energetic studies
      diversion flows 67
      exponential distribution 64–67, **65–66**
      powerhouse design flow 67
      river management 67

failure cause statistics **432**
Falcao thesis 146, 150
Falvey, H. T. 229
Faraday, M. 6
feasibility studies 35, 81

Ferros station 78
Finger of God 95
Finite Element Method 120, 273
Fionnay Power Plant 43
firm system energy 41, 81
Flaming Gorge HPP *307*
flood routing studies 69
fluviometric stations 32
   hydrological and energetic studies *56,* **57,** *57–59*
      discharge measurement procedures 57, *58*
      important characteristics 59
Fort Peck dam 531
Francis turbines 83, 345, *347–348,* 370
   application range 346
   basic principle 346
   dimensions 346–348
   performance data 348
free board studies **70,** *70–74,* **71,** *72–74*
French hydraulic park (EDF) 465
Fuller's formula 64
Funil HPP 324, 514
fuse gate *314, 315*

Garrison tunnel model 317
gate-controlled spillways 181
Gate Operation Plan (GOP) 234
gates
   classification 312, **315–316**
   construction of 309
   discharge coefficients 318–322
   gate size and heads 314
   outlet discharge coefficients 314–318
   selection of 312–313
   spillways segment gates 318–322
   types of 311–312
   usage limits 313–314
generalities
   action turbines 331
   reaction turbines 331
generating unit layout 379–388, 392–393
   electrical faults 392–393
   mechanical faults 393
   power output *85,* 85–86, *86,* **88**
generation voltage 393–394
generator main elements 365–366
generator rated capacity 366–367
geo-hydrological model *111*
geological and geotechnical investigations
   construction materials 123
   drainage systems
      internal drainage system 117–120, *121*
      main drainage system 122
   feasibility studies 94
   foundation instrumentation 122–123
   foundation sealing device 117

foundation treatment project 112–113,
116–117, *118–119*
inventory studies
geological structures *96–102*, 96–107, *109*
geological units 95–96
geomechanical classification systems 100,
102
maximum possible precision 94
measure physical properties 94
subsurface exploration 94
material parameters 109–110, **112–114**, *115*,
115–116
percolation analysis 119
planning 93, *93*
geological units 95–96
geomechanical classification systems 100
Ginocchio, G. 446, 458
Gireli, 79, 80
Gomide, F. L. S. 68
Grande Dixence dam 445–448
gravity dam
conventional concrete 150–155, *153–155*,
155
roller compacted concrete 155–160,
157–159, **158, 159**
Grishin, M. M. 146, 164, 173
guaranteed maximum price 428–429
guard valves **326, 329**
Gulliver, J. S. 6, 279, 330

Hartung, F. 188, 237
Hausler, E. 188
head losses
intake 244–245, **245**
penstocks 246–248, *247, 249, 250*
HEC-RAS model 69
HEC-RAS program 416
high-head power plants 362, 370
high-speed turbines (Kaplan) 338
high-voltage equipment 397
Hirakud HPP 559–560, *567–568*
Hoeg, K. 98, 271
Höljes HPP 537
downstream view *538*
localization map *537*
low-level outlet *539*
Hoover dam 322
Howell-Bunger valves 323
hydraulic conveyance design
intake
geometry 241, *242*, 243
head losses 244–245, **245**
minimum submergence 243, *244*
ventilation duct 243
vibration in trashracks 244
penstocks 246

economic diameter 248–253, **251,** *251,*
253, **254–257,** *258*
head losses 246–248, *247, 249, 250*
waterhammer 258–269
power canal 239, 241, **241**
powerhouse
elevations of 292
outdoor powerhouses 292, *296–300,*
297–299
standard design types 289, *290–291*
underground powerhouses 299, *301,* **302,**
303
structures of 239, *240*
surge tanks 282
Canambra criteria 284
criteria used in Inventory Studies
(Canambra) 283
interconnected system operation
286–288, *287*
minimum dimensions of *288,* 288–289
rotating masses inertia 284–286, *285, 286*
surge tank need 288
types of 282–283, *283*
tailrace 302, *304–307*
tunnel
assumptions for lining dimensioning
281–282
criteria for hydraulic tunnel dimensioning
274–278
design application 278–281
general design criteria 269–274
hydraulic design 338
criteria specification 175–178, 181–182,
181–183
physical model studies 185, *186–187*
Tucurui HPP spillway 182–184, **184**
hydraulic energy 330, 331
hydraulic jacking 272
hydraulic jump dissipators 197–203, *198–207,*
205–206
hydraulic model *577*
hydraulic similarity 338–339
hydraulic thrust 341
hydraulic turbine 331
hydrodynamic forces downstream of
dissipators 208
hydroelectric patrimony management
asset management issues 458
context evolution 457–458
risk hierarchy 459–466
define unwanted events 460–461
evaluate occurrences 461–462
impact and occurrence **463**
operations prioritization process 460
risk management 458–459
actions, coordination of 458–459

590 Index

hydroelectric patrimony management (*cont.*)
  decision approaches development 459
  decision support 459
  technical questions 458
hydroelectric plants risks
  dam breaks risks 448–454
  flood risks 456
  geological and geotechnical risks 456
  operation and maintenance 457
  penstocks, risks related to 457
  regulatory and legal aspects 455
  risk criteria **456**
  risk prevention 455
  risks, constructive aspects 457
  turbine start-up, risks related to 457
hydroelectric power plants (HPPs)
  base and peak use 40–41
  in brazil 8–18, *9–16*, **16–17**, *18*
  component structures 20
  energy transformation **19**, 19–20
  events 6–8, *7–8*
  function of the head 41
  history **2**, 2–6, *3, 5, 6*
  largest hydroelectrics in the world 21, **21**
  largest hydro powerplants in the world 21,
    21–23
  layouts
    canal drop layouts 42–47, *44–46*
    dam layouts 42, *42, 43*
    spillway positon 47–52, *47–52*
  operation type *39*, 39–40
  regularization reservoir 39, *40*
hydrograms 86, *86*
hydrological and energetic studies
  backwater studies 69
  basin characterization
    basin shape 54–55
    concentration time 55, **56**
    drainage area 54
    mean bed slope 55
  curves quota x area x reservoir volume *68,*
    68–69, **69**
  extreme flows
    diversion flows 67
    exponential distribution 64–67, **65–66**
    powerhouse design flow 67
    river management 67
  flood routing 69
  flow-duration curves 60–63, *61,* **62, 63,** *64*
  fluviometric stations 56, **57,** 57–59
    discharge measurement procedures 57, *58*
    important characteristics 59
  free board studies **70,** 70–74, **71,** *72–74*
  gumbel and exponential parameters 64
  hydrometeorology
    climate classification 57

    climatological stations 56, **56**
    precipitation 57, **57**
    relative humidity **56,** 57
    temperature values **56,** 57
  minimum flows 67
  power transmission system 68
  regularization of discharges 67–68
  reservoir filling studies 74
  reservoir useful life studies 75–80, **76,** *76,*
    77, *78, 79*
  sanitary/ecological flow 68
  tailwater elevation curve 59–60, *60*
hydrometeorology
  climate classification 57
  climatological stations 56, **56**
  precipitation 57, **57**
  relative humidity **56,** 57
  temperature values **56,** 57
hydropower inventory studies 25

ICOLD Bulletin 61 125
ICOLD Dam Failures Statistical Analysis
  Bulletin No. 99 449
The Iguacu falls 25, *27*
incipient cavitation index 224, **225**
infiltration tests 94
Installation License (LI) 29
Instrumentation and Behavior of Foundations
  of Concrete Dams (Silveira) 122
intake, hydraulic conveyance design
  geometry 241, *242,* 243
  head losses 244–245, **245**
  minimum submergence 243, *244*
  ventilation duct 243
  vibration in trashracks 244
integrated environmental assessment 34
interconnected system operation
  surge tanks 286–288, *287*
inventory hydroelectric studies 32–33, **33,** *33*
  geological structures *96–102,* 96–107, *109*
  geological units 95–96
  geomechanical classification systems
    100, 102
  maximum possible precision 94
  measure physical properties 94
  subsurface exploration 94
Inventory Studies (EIH) 29
Irkutsk HPP 557
Isbash formula 416
Itaipu HPP
  generating unit *388*
  layout and structures *501,* 504, 510–512, *511*
  localization map *503*
  Parana river 500
  spillway *317*
  Uruguay river 510

Itumbiara HPP powerhouse *294*
Ivailovgrad HPP 537, 539–541
  localization map *539*

Jaeger, C. 265
Jansen, R. B. 4, 436
Jinping-I HPP 579, *581–583*
Jirau HPP Brazilian machines 352
Joukowsky Equation 261
July 14 HPP *303, 304*

Kaplan turbines 341, 343, *387*
  application range 349–350
  basic principle 350
  dimensions 350–351
  performance data 351
Karakaya HPP, Euphrates river (Turkey) 297, *297–298*
kinetic energy 331
Kjaernsli, B. 71, 73
Kolesnikov, E. 457
Kollgaard, E. C. 41, 169
Koppen 57
Kramer, K. 229
Krasnoyarsk HPP 554–556

L'Aigle dam 541–543, *542–543*
Lanna 68
layouts
  canal drop layouts 42–47, *44–46*
  dam layouts 42, *42, 43*
  spillway positon 47–52, *47–52*
Leps curves 110
"level of impact" 461
Libby dam spillway 218
light electricity services 10
light power plants *11*
Linsley, R. K. 59
local firm energy 82
Longyangxia HPP 579, *584*
Lowe III, J. 221
lower San Fernando dam 440–443
low-head power plants 362
low-speed turbines (Pelton) 338

Magela 189, 208
magnetic flux 364
Mahmood, K. 79
main transfer bar, single breaker 398
Malpasset dam *433–436*, 433–437
maneuvering schemes 397–403, *398*
  double bar
    double breaker 402–403
    one breaker and a half 401–402
    single breaker 398–400
    and transfer bar 401

main transfer bar, single breaker 398
  selection criteria 403–404
  simple bar 398
  single circuit breaker with bypass
    disconnecting switches 400–401
Manning coefficient **241**, 275, **276**, 278, **280**
Marsal, R. J. 139
Martins, R. 188
Mason, P. J. 209
masonry gravity dams 150
MAUT (Multi-Attribute Utility Theory)
  approach 465
Maxwell, J. 6
Mays, L. W. 4, 75
mechanical auxiliary equipment 360
mechanical auxiliary systems 359
mechanical energy 363
mechanical faults 393
mechanical power 337
medium flows 53
Mekong dams and HPPS 578–579
Micro Power Plant (MCH) 29
mini hydroelectric plant 34
minimum coverage criteria 272, *273*
minimum flows 67
Miranda HPP *573*
Miranda, J. C. M. 79
mobile bottom tests 185
modeling aerator devices 226
Mohr-Coulomb equation 102
momentum principle 337
monitoring and instrumentation 377
Monjolinho HPP
  Passo Fundo river 516
  spillway *517*
Moody, L. F. 247, *247, 248*
multicriteria decision system 464–465

narrow valley/wide valley arc dams 165
National Dam Safety Information System 455
National Water Resources Council 455
Natural Resources Conservation Service
  (NRCS) 55
Naturno Plant 379
negative waterhammer gradient 259, *260*
Neidert, S. H. 202
Newton's second law 200
Nielsen, B. 273, 421, 422
Nikuradse, J. 247
nominal data 331–335
nonpermanent flow 246
North Fork Clearwater river 528
Norwegian Institute of Technology (NIT)
  41, 273
Nova Ponte HPP
  Araguari river 519

592 Index

Nova Ponte HPP (*cont.*)
 civil works 520
 indoor powerhouse 520
Nunez, D. R. 139
Nuozhadu plant *580*
Nurek dam 91

oblique elements 372
Ohio river 531
Oliveira, A. R. 222
ONS Network Procedures 392
open flume turbine 356–357
orifice surge tank 283
outdoor powerhouses 292, *296–300*, 297–299

packages 366, 425
Pampulha dam 432
Paraiba do Sul river 514
Parana river 500, 504
Passo Fundo river 419, 516
Paulo Afonso (PA) Hydroelectric Complex *305, 306*, 520–521
Pelotas river 507
Pelton turbines 341, 345, 370, 379, *381*
 application range 344
 basic principle 345
 dimensions 345
 performance data 345–346
penstocks 246, 259
 dimensions of 309
 economic diameter 248–253, **251**, *251, 253*, **254–257**, *258*
 head losses 246–248, *247, 249, 250*
 throttling gates 314–315
 waterhammer 258–269
percolation analysis
 earth dams 130–133
  filter and transition materials 132–133
  foundation waterproofing devices 133
  internal drainage system 132
performance tests
 field test 358
 model tests 358
 performance guarantees 357–358
permanent flow 246
Peterka, A. J. 198
Picote HPP *573*
Pieve di Cadore dam 546–547, *546–547*
 downstream view *550*
 layout and section *551*
 localization map *550*
Pigott, R. J. S. 247
Pinalito HPP 521–523
Pinto, N. L. 59, 68, 222, 223, 225
planning hydropower generation
 budget

after privatization 37–38
 standard budget 37
 technical and economic evaluation 38
catchments areas 26–28, *27*
expansion planning *28*, 28–29
geological and geotechnical investigations 93, *93*
surveys and basic studies 29–36, *30, 31, 33, 33*
 basic engineering project 35–36
 detailed project 36
 environmental basic project 36
 environmental impact studies 35
 feasibility studies 35
 integrated environmental assessment 34
 inventory hydroelectric studies 32–33, **33**, *33*
 mini hydroelectric plant 34
 small hydroelectrics plants 34
plant efficiency 335–336
plant switchyard/substation 396–397
Poço Fundo SHP
 energy 86, *87*
 energy benefits 85, **85**
 hydrograms 86, *86*
 incremental analysis–benefit/cost 85, **86**
 investment cost 85, **85**
Poco Fundo SHP project 60, *60*
Poisson's formula 367–368
poles and pole windings 373
positive waterhammer gradient 259, *259*
power canal 239, 241, **241**
power dissipators 185–209
 erosion pit dimensions assessment 208–209, *210*
 hydraulic jump dissipators 197–203, *198–207*, 205–206
 hydrodynamic forces downstream of dissipators 208
 ski jump dissipators 187–192, *188–197*, 196
Power, G. 79
powerhouse
 downstream of dam 297, *299*
 elevations of 292
 at foot of dam 292, *296*
 outdoor powerhouses 292, *296–300*, 297–299
 at part of dam 297, *297–298*
 standard design types 289, *290–291*
 underground powerhouses 299, *301*, **302**, *303*
power output
 critical period 81–82
 energy criterion or deterministic criterion 81
 energy-economical design 85, **85–86**
 energy simulation 83–84, **84**, *84*

Index 593

feasibility studies 81
firm system energy 81
generating units *85,* 85–86, *86,* **88**
physical guarantee 87, 89
turbine type 83
powerplant connection to electrical system
404–405
receiving substation 405
transmission line 405
power plants, operation and maintenance
405–406
power transmission system 402
Prior License (LP) 29
protection systems
breaker failure protection 395
current protection criteria 391–392
elevator transformers 393–394
generating units 392–393
electrical faults 392–393
mechanical faults 393
protective relays 391
substation bar protection 395–396
transmission line protection 394–395
protective relays 391

Q System 100
Quintela, A. C. 212, 216

Ramos, C. M. 212, 216
receiving substation 405
recurrence time (TR) 411, **414**
Reinius, E. 208
Reliability and Statistics in Geotechnical
Engineering (Baecher
and Christian) 92
reservoir filling studies 74
daily or weekly regulating reservoirs 40
interannual reservoirs 40
seasonal reservoirs 40
reservoir useful life studies 75–80, **76,** *76,* **77,**
78, 79
Reynolds number 247, 248, 276
Reyran river 432
Ricobayo HPP 570, *571*
downstream view *571*
localization map *571*
rim/magnetic ring 366
Ripley, B. D. 269, 272, 275
river diversion design 59
closed-valley plants 415
concrete structures **419**
discharges and risks 409–415
execution 416–418
hydraulic dimensioning 416
hydraulic models 418
open valleys plants designs 415

risk criteria **414**
river diversion phases 415–416
tunnel excavation *422*
rockfill dams 138–150
asphalt concrete 146–149, *147, 148*
asphalt core 149, *150, 151*
concrete face 143–146, *144*
impermeable membrane (clay core)
140–143, *141, 142*
Rockfill Dams 92
rock mass rating (RMR) 100
Rogun dam 91
Roig 79
roller compacted concrete (RCC) 155–160,
157–159, **158, 159**
rotating masses inertia
surge tanks 284–286, *285, 286*
rotor pole *374*
Rouse, H. 247
Rudavsky, A. B. 198, 205
runaway speed 340, **340,** 340–341

safety factors 133
Sainte-Croix dam 544–544, *544–545*
San Carlos Plant 379, *381*
sanitary/ecological flow
hydrological and energetic studies 68
Sao Francisco river 504, 520
Sarkaria, G. S. 248, 275
Saucelle HPP *575*
Saville Jr. T. 70, 72
Sayano – Shushensk HPP 443–445, 457
Schleiss, A. 237
Scientific Research Institute of Hydrotechnics
(SRIH) 416
sediment production 75
Seepage, Drainage and Flow Nets
(Cedergren) 117
Serra da Mesa HPP
civil works 519
layout *518*
spillway *519*
Tocantins river (Goiás State) 299, *301,* 517
Serre de Poncon HPP 546
Sharma 272
Shasta HPP
localization map *526*
Sacramento river 525–528
Sherard, J. L. 141, 143
Shihmen Dam 155
short-circuit defects 364
side-channel spillways 182
Signer 139
simple bar 398
siphon-type spillways 173
site logistics 423

594 Index

ski jump dissipators 187–192, *188–197,* 196
sliding gate Fixed-wheel gate *312*
slopes protection 138
small hydroelectrics plants 34
smooth pole machine 366
Sobradinho HPP
  layout and structures 504–505, *505*
  powerhouse *506*
  Sao Francisco river 504
  surface spillway *506*
soils and rock masses
  classification 98, 105–108, **109**
  common defects 103–104
specific numbers 339
speed controller 359
speed numbers 338, **338,** 338–339
spillway 309
  aeration 223–234
    aeration devices 224
    aerator geometry 225, *225*
    erosion tests 232
    incipient cavitation index 224, **225**
    water vaporization level 224, *224*
  cavitation
    cases 218–219, *218–223,* 221–222
    conceptualization and characteristic
      parameters 211–212
    irregularities 212–213, *212–214*
    protective measures specifications *213,*
      213–216, *215, 216,* **217**
  hydraulic design
    criteria specification 175–178, 181–182,
      181–183
    physical model studies 185, *186–187*
    Tucurui HPP spillway 182–184, **184**
  operating aspects 234–238, *236–238*
  positon 47–52, *47–52*
  power dissipators 185–209
    erosion pit dimensions assessment
      208–209, *210*
    hydraulic jump dissipators 197–203,
      198–207, 205–206
    hydrodynamic forces downstream of
      dissipators 208
    ski jump dissipators 187–192,
      188–197, 196
  selection criteria 171–174, *172–180*
    siphon-type spillways 173, *174–176*
    typical sections 173, *177–179*
    underwater inspections 174, *179–180*
  surge tank 283
Spurr, K. J. W. 188
standard budget 37
Standard Step Method (SSM) 416
standby 69
stator core 372–373

stator frame 366
stator winding 373
Straflo turbines 354–356, *356*
Stucky, A. 263, 264, 265
substation bar protection 395–396
substation interconnection
  components and installations 397
  equipment arrangements 397
  maneuvering schemes 397–403
    double bar and transfer bar 401
    double bar, double breaker 402–403
    double bar, one breaker and a half
      401–402
    double bar, single breaker 398–400
    main transfer bar, single breaker 398
    selection criteria 403–404
    simple bar 398
    single circuit breaker with bypass
      disconnecting switches 400–401
  plant switchyard/substation 396–397
  powerplant connection to electrical system
    404–405
    receiving substation 405
    transmission line 405
  switchyard types 397
SUCS-USBR plasticity chart 98, *102*
suction height 341–343
support block 250–252, **251,** *251*
surge tanks 282, 288
  Canambra criteria 284
  criteria used in Inventory Studies
    (Canambra) 283
  interconnected system operation 286–288,
    287
  minimum dimensions of *288,* 288–289
  rotating masses inertia 284–286, *285, 286*
  surge tank need 288
  types of 282–283, *283*
surveys and basic studies
  planning hydropower generation 29–36, *30,*
    31, **33,** *33*
    basic engineering project 35–36
    detailed project 36
    environmental basic project 36
    environmental impact studies 35
    feasibility studies 35
    integrated environmental assessment 34
    inventory hydroelectric studies 32–33,
      33, *33*
    mini hydroelectric plant 34
    small hydroelectrics plants 34
switchyard types 397
synchronous generator
  design principles 370–376
    bearing lubrication 373–375
    cooling system 375–376

poles and pole windings 373
stator core 372–373
stator winding 373
dimensioning factors 367–370
electromagnetic characteristics 365
energy conversion 362–364
generator main elements 365–366
generator rated capacity 366–367
monitoring and instrumentation 377
synchronous machines 361–362
tests 379
turbine-generator and assembly 377–378
synchronous machines 361–362

tailrace 302, *304–307*
tailwater elevation curve 60, *60*
tailwater key-curve characterizes 53
tainter/radial gate *311, 320, 324*
Taquaruçu HPP *386*
TBM (Tunnel Boring Machine) method
419, 421
"technical" urgency 461–462
teleprotection schemes 395
tension and strain analysis
earth dams 134–138
displacement analyses 135–138, *136,*
136–137
suffer deformations 135–138, *136,*
136–137
Tesla coil 8
Tesla, N. 8
Teton earth dam 91, 432, 438–440, *439*
The Grande Dixence dam 44
Thidemann, A. 273, 421, 422
three-phase alternating current (AC) 361
three-phase electrical energy 363
throttling gates **327–328**
Tocantins river 407, 499, 517
traditional (classical) modality 423–425
transformers 363, 388–389, *389*
three-phase transformer 389
transmission system 388
transmission line protection 394–395, 405
transmission voltage 405
trashracks flow velocity
limits for 243
vibration in 244
triaxial tests 94
Tubular turbines 354, *355, 356*
Tucuruí HPP *383–386, 408–410, 412*
downstream view *502*
general layout 499
powerhouse structure *296, 306, 500*
spillway and right dam *502*
Tucuruí spillway *318*
tunnel

assumptions for lining dimensioning
281–282
criteria for hydraulic tunnel dimensioning
274–278
design application 278–280, *281*
excavation 422
general design criteria
alignment 269–270, *270*
covering criteria *271,* 271–274, **272,**
273, 274
tunneling boring machine (TBM)
279, 280
"turbine center" 341
turbine control 359–360
turbine design 331
turbine efficiency 335–336, 345
turbine equation 336–338
turbine-generator and assembly 377–378
turbines
bulb turbines
application range 352–353
basic principle 353
dimensions 353
performance data 353–354
cavitation limits 341–343
design conditions 331–335
design head, operation out of 339–340
efficiency 335, *336*
francis turbines
application range 346
basic principle 346
dimensions 346–348
performance data 348
generalities
action turbines 331
reaction turbines 331
hydraulic similarity 338–339
hydraulic thrust 341
kaplan turbines
application range 349–350
basic principle 350
dimensions 350–351
performance data 351
mechanical auxiliary equipment 360
nominal data 331–335
open flume turbine 356–357
pelton turbines
application range 344
basic principle 345
dimensions 345
performance data 345–346
performance tests
field test 358
model tests 358
performance guarantees
357–358

596 Index

turbines (*cont.*)
    plant efficiency 335–336
    rotor configuration *333,* 334
    runaway speed 340–341
    specific numbers 339
    speed calculation 334
    speed number 338–339
    straflo turbines 354–356
    suction height 341–343
    tubular turbines 354
    turbine control 359–360
    turbine efficiency 335–336
    turbine equation 336–338
turbine shaft 330
turbine torque 340
turn-key 425–428
two-pole machine *362*

Uatuma river 497
underground powerhouses 299, *301,* **302,** *303*
unified soil classification system 105
uniform flow 246
United States Geological Survey 57
unit system 388
"unit turbine" 339
unlined tunnels
    design chart 274, *274*
    hydraulic equivalence between 280, *280*
    methods for reducing head losses 279, *279*
    resistance coefficient 276, *277*
unwanted event, definition of 461
US Army Corps of Engineers (USACE) 60
US dam safety 439–440

Vajont dam *437,* 437–438, *438*
valves 322–330
    butterfly valves 323, 325, *326*
    fixed-cone valves 323
    guard valves **326**
vapor bubbles 342
Veiga, A. P. 436, 437
ventilation duct, intake 243
Veronese equation 209
Vigario dam 117
Vinogg, L. 330

Viollet, P. 446, 458
von Kármán and Prandtl 247
von Kármán-Prandtl equation 276, 278

water, chronology 2, **2**
waterhammer 258–260
    negative waterhammer gradient 259, *260*
    overpressure calculation due to gradual
        closure without surge tank 264–265, **265,**
        266, **267,** 267–269, *268, 269*
    overpressure calculation due to instant
        closing 261, *262,* 263–264, **264**
    positive waterhammer gradient 259, *259*
"waterhammer regulation" 286
watermark engineering 93, *93, 96, 99*
waterproofing devices 133
watertightness, classification **116**
"water to wire" generation process
    335–336
Westgaard, A. K. 361, 375
White, W. R. 79
Whittaker, J. G. 237
Wilfred Airy's Law 416
Wilson, D. 146
wind-generated waves 70
WLmax max elevation 132
Wolf Creek HPP 531, 534
World Meteorological Organization
    (WMO) 184
World Water (1988) 75

Xavier 190
Xayaburi plant *580*
Xiaowan plant *579*

Ybbs-Persenbeug HPP 553–554
Yellowtail HPP 523
    Bighorn river 523
    cross section 527
    dam and spillway 524
    penstock 524
    penstock and powerhouse *525*
Yuditskii, G. A. 209